Akansha Singh • Krishna Kant Singh
Editors

Multimodal Generative AI

Springer

Editors
Akansha Singh
School of Computer Science Engineering and Technology
Bennett University
Greater Noida, Uttar Pradesh, India

Krishna Kant Singh
Delhi Technical Campus
Greater Noida, Uttar Pradesh, India

ISBN 978-981-96-2354-9 ISBN 978-981-96-2355-6 (eBook)
https://doi.org/10.1007/978-981-96-2355-6

This Springer imprint is published by the registered company Springer Nature Singapore Pte Ltd.
The registered company address is: 152 Beach Road, #21-01/04 Gateway East, Singapore 189721, Singapore

Preface I

Multimodal Generative AI is meticulously designed for a readership well-versed in the intricacies of machine learning and artificial intelligence. This text delineates itself by delving into the fusion of two traditionally distinct AI disciplines: generative models for visual data and natural language processing.

Multimodal Generative AI is driven by the growing need for AI systems that not only process but also synthesize novel content that spans both visual and linguistic elements. In an era where digital information is overwhelmingly multimodal, the development of AI that can interpret and generate such content is not only revolutionary but also essential. This book fills a gap in the current literature, providing a holistic exploration of how disparate generative technologies can be interlinked to produce more sophisticated and versatile AI systems.

The core contents of *Multimodal Generative AI* include:

- An examination of the evolution of generative models, from early neural networks to contemporary architectures like GANs and VAEs and their application in creating realistic images and videos.
- A thorough analysis of language models, particularly transformer-based designs, and their unprecedented success in understanding and generating human-like text.
- A detailed discourse on the integration of visual and textual models, presenting state-of-the-art techniques for creating cohesive multimodal systems.
- Case studies showcasing the application of multimodal generative AI across various sectors, highlighting breakthroughs in areas such as autonomous systems, content creation, and human–computer interaction.
- An insightful discussion on the ethical and societal ramifications of generative AI, promoting a dialogue on responsible innovation.

What sets *Multimodal Generative AI* apart is its focus on the intersectionality of generative visual and language models. It offers readers a unique vantage point on how these models can be harmoniously combined to engender AI with a more profound understanding and creative capability. This book is designed to catalyze further research and development in the field, serving as a springboard for innovation.

The intended readership stands to benefit from a comprehensive understanding of how multimodal systems can be employed to solve complex problems that require an amalgamation of visual understanding and language proficiency. It is a text that will not only inform but also inspire its audience to push the boundaries of what is possible in AI.

Prerequisites for fully grasping the contents of *Multimodal Generative AI* include a foundation in machine learning concepts, familiarity with neural network architectures, and an understanding of the basics of computer vision and natural language processing. This ensures that readers are equipped to appreciate the advanced methodologies and novel insights presented in this book.

Greater Noida, Uttar Pradesh, India Akansha Singh
Greater Noida, Uttar Pradesh, India Krishna Kant Singh

Preface II

The landscape of artificial intelligence has undergone an extraordinary evolution in recent years, with breakthroughs in machine learning and deep learning pushing the boundaries of what we once thought was impossible. Among the most exciting developments is the rise of generative AI—systems that do not just analyze data but create new content, be it text, images, or even entire virtual environments. What was once the realm of science fiction is now a reality shaping industries, scientific research, and human–computer interactions.

Multimodal Generative AI represents a timely and crucial contribution to this fast-growing field. As we move further into the era of digital innovation, the need for AI systems that can understand and generate content across multiple modalities—visual, linguistic, auditory, and more—is becoming increasingly apparent. The ability of machines to interpret the world in a way that mirrors human perception, by processing images, sounds, and language concurrently, is transformative.

This book fills a critical void in current AI literature. While there has been significant work done on both visual generative models (such as generative adversarial networks or GANs) and language models (like GPT), the integration of these systems remains in its infancy. The fusion of these modalities, explored in-depth in this text, opens new doors for AI applications that require both sophisticated visual understanding and natural language proficiency. The potential impact spans diverse sectors—autonomous vehicles, virtual assistants, healthcare, entertainment, and beyond.

For researchers, practitioners, and advanced students, *Multimodal Generative AI* offers not only a deep dive into the theoretical underpinnings of generative systems but also practical insights on how these technologies can be deployed in real-world scenarios. It bridges the gap between cutting-edge research and its tangible applications, making it an indispensable resource for anyone looking to explore the future of multimodal AI.

What stands out in this text is its balance of rigor and accessibility. While the subject matter is advanced, this book is written with a clarity that ensures readers will not only grasp complex ideas but also feel inspired to contribute to the field.

The inclusion of case studies and discussions around the ethical implications of generative AI further underscores the book's relevance in today's world.

As someone deeply engaged in the world of AI research and its applications, I find the potential for multimodal generative AI to be not just fascinating but essential. The ability to synthesize data across multiple forms and create cohesive, intelligent systems represents the next major leap in AI's evolution. It is with great enthusiasm that I recommend this book to those looking to be at the forefront of this transformation.

Greater Noida, Uttar Pradesh, India Akansha Singh
Greater Noida, Uttar Pradesh, India Krishna Kant Singh

Contents

About the Editors

Akansha Singh Professor at the School of Computer Science and Engineering, Bennett University, Greater Noida, boasts a comprehensive academic background, with a B.Tech, M.Tech, and Ph.D. in Computer Science. Her doctoral studies, conducted at the prestigious IIT Roorkee, focused on the cutting-edge fields of image processing and machine learning. Dr. Singh has an impressive research portfolio with over 100 published research papers and more than 30 authored or edited books in advanced computer science domains. Her expertise spans image processing, deep learning, machine learning, remote sensing, and the Internet of Things (IoT). She is currently the Principal Investigator of a funded project titled Autimate: an assistive tool for autistic children. A recognized leader in her field, Dr. Singh holds several prestigious editorial positions, including Associate Editor of IEEE Access, Discover Applied Sciences, and Academic Editor for PLOS ONE, Computational Intelligence and Neuroscience. She has also served as Guest Editor for numerous special issues in reputed journals. Her research interests are diverse and influential, spanning image processing, remote sensing, the Internet of Things (IoT), blockchain, and machine learning. Prof. Singh's work in these areas not only advances the field of computer science but also significantly contributes to the broader scientific and technological community. Her recent work focuses on hybrid deep learning models, federated learning, blockchain integration for smart cities, and generative AI. She actively contributes to research excellence through international collaborations and leadership roles in organizing international conferences.

Krishna Kant Singh Director of Delhi Technical Campus in Greater Noida, India, is a highly experienced educator and researcher in the field of engineering and technology. Dr. Singh holds a B.Tech and M.Tech degree, a Postgraduate Diploma in Machine Learning and Artificial Intelligence from IIIT Bangalore, a Master of Science in Machine Learning and Artificial Intelligence from Liverpool John Moores University, United Kingdom, and a Ph.D. from IIT Roorkee. Dr. Singh has made significant contributions to the academic and research community. With over 18 years of teaching experience, he has played a vital role in educating and mentoring future professionals. Dr. Singh has published over 150 research papers, with more than 3,700 citations, and serves on the editorial board of Applied Computing and Geosciences and as Senior Editor of IEEE Access. He has also served as the guest editor of several reputed journals like Complex and Intelligent systems, Journal of Real Time Imaging, Big data analytics, Open computer science etc. His commitment to knowledge dissemination is further evidenced by his authorship of more than 25 technical books, which serve as significant resources in the fields of engineering and technology. Dr. Singh's dedication to his field is evident through his involvement in various capacities, from research and writing to editorial responsibilities. His work not only advances the field of engineering and technology but also inspires future generations of engineers and researchers.

Chapter 1
Introduction to Multimodal Generative AI

R. Brindha, R. K. Pongiannan, A. Bharath, and V. K. S. M. Sanjeevi

Abstract A state-of-the-art method in machine learning is called Multimodal Generative Artificial Intelligence (AI), which aims to produce a variety of outputs in many modalities, including text, audio, and images. This survey explores multimodal generative AI's developments, techniques, and uses. First, we clarify the main concepts behind such models, such as adaptive autoencoders (VAEs), generative adversarial networks (GANs), and their various extensions. The methods for fusing and aligning different modalities—such as conditional generation, cross-modal embeddings, and attention mechanisms—are then covered. A cutting-edge machine learning method called multi-modal generative artificial intelligence (AI) generates multiple methods of outputs such as text, audio, and visuals. This survey examines the progress, methods, and applications of generative AI. The article also examines various applications of multimodal generative AI, including image captioning, music composition, and text-to-picture synthesis. We also discussed challenges and ethical considerations related to multimodal content, such as privacy issues and prejudice reduction. Finally, we focus on a research roadmap that emphasizes the need for a multimodal, socially responsible, controlled, and comprehensible AI system.

Keywords Multimodal generative artificial intelligence · Machine learning · Generative intelligence · AI chatbots · Language models

R. Brindha (✉) · R. K. Pongiannan · A. Bharath · V. K. S. M. Sanjeevi
Department of Computing Technologies, SRM Institute of Science and Technology, Kattankulathur, India
e-mail: brindhar@srmist.edu.in; pongiank@srmist.edu.in; ba6311@srmist.edu.in; Vs4382@srmist.edu.in

A. Singh, K. K. Singh (eds.), *Multimodal Generative AI*,
https://doi.org/10.1007/978-981-96-2355-6_1

1.1 Introduction

A generative AI model is a type of machine learning architecture that uses AI algorithms to create new data examples based on patterns and relationships observed in training data. A generative AI model is of a central but critical nature, but it is incomplete, as it requires further adjustments to specific tasks through systems and applications. Starting with the basics, we first describe the general design methods: variable-autoencoding (VAE), generative antagonistic networks (GANs), and their different improvements (Huzaifah et al., 2021). We also deal with ethical considerations and difficulties associated with the creation of multimodal material, including issues of reducing prejudices and privacy. Finally, we conclude with a research road map that highlights the need for socially responsible, controlled, and comprehensible multimodal generative AI systems. The survey is a wide-ranging resource for researchers, practitioners, and policy makers interested in the future of multimodal genetic artificial intelligence. The term generative AI refers to computational techniques capable of generating new and important content from training data, such as text, images, and audio.

Generational and prediction AI are two different types of artificial intelligence technologies, each with different functions. Generative AI is an artificial intelligence that creates new content, such as images, music, and text.

Use complex algorithms and deep learning technologies to generate new content, such as training data supplied. Predictive AI analyses data with statistical algorithms and machine learning and predicts future events and behaviour. It learns from historical data to identify patterns and predict future results. The main difference between predictive AI and generative AI is that predictive AI uses historical data to make predictions and generative AI to create new content and data based on existing patterns and trends. Prediction models provide information on different data points to help them make decisions. Is this a dog or a cat image? Is this tumour benign or malignant? A human monitors the training of the model and tells him whether his output is correct. Based on training data, models learn to respond in different ways to different scenarios (Kola, 2019).

Generative models generate new data points based on the learnings they get from training data. These models usually train without supervision, analyse data without human input, and draw their own conclusions. For years, the generation model has faced more difficult tasks, such as learning to produce photos and accurate answers to questions in text, and progress has been slow. Further, the increase in the availability of computer power enabled the Machine Learning (ML) team to build a basic model: a large unsupervised model of large amounts of data training (sometimes all available data on the Internet). Over the past few years, ML engineers have calibrated these generative basic models, providing annotated data subsets to target outputs to specific objectives, and used them in practical applications. ChatGPT-3/4 is a good example. It is a version of Chat GPT, a basic model developed on a large amount of unlabelled data. To create ChatGPT, OpenAI hired thousands of annotations to mark appropriate data subsets, and its ML engineers then refined the model to teach it to generate specific information. With such fine-tuning methods, generative models are beginning to produce outputs that were not previously possible, resulting in a rapid expansion of functional generative models. Progress in the field of large-scale language models and advanced AI in health care is rapid, and it is difficult to stay updated with the changing news. For example, Insilco has developed an AI platform that uses deep-learning models, reinforcement learning, transformers, and other modern machine learning techniques for discovering new targets and producing desired molecular structures. There are many examples of this type in the whole spectrum of health care and clinical research.

1.2 AI Evolution

The history of artificial intelligence is a story of human genius and technological progress. It began with a primitive mechanical calculator created in 1642 by French mathematician Blaise Pascal and has evolved into a series of important milestones that shape the trajectory of AI research. This chapter will explore key dates and events that opened the way to the development of artificial intelligence, including Ada Lovelace's first programmed machine, the introduction of neural networks, Alan Turing's pioneering work, and the development of modern artificial intelligence technologies. Pascaline is not an AI system by current standards, but marks the first time that mathematical operations have been automated and provides the way for more complex tasks to be automated in the coming centuries. B. Ada Lovelace and the analytical engine 1837 brought another important development in the history of artificial intelligence, thanks to Ada Lovelace. Mathematician and writer Lovelace collaborated with Charles Babbage to design analytical engines.

We can say that the widespread spread of this technology, such as the Dall-E 2, GPT-4, and Copilot, is currently revolutionizing the way we work and communicate with each other. Generative AI systems can not only be used for artistic purposes to create new texts that mimic authors and new illustrations that mimic illustrators, but also help humans as intelligent question-answer systems. The application includes an Information Technology (IT) help desk that supports the transitional work tasks of knowledge and mundane needs such as cooking recipes and medical advice.

Artificial crashes indicate that generative artificial intelligence can boost global GDP by 7% and replace 300 million jobs for knowledge employees. This is really having a major jolt not only on the Business and Information System Engineering (BIS) community, where we face revolutionary openings, but also on the expostulations and pitfalls we've to face and take to guide technology shown in Fig. 1.1.

Deep Learning is an improved portion of engine literacy algorithms grounded on artificial neural networks. This paper discusses colourful operations of game proposition. The games are played between single or multitudinous players, and the games are managed by colourful manners of players. The players are codified into two manners, collaborative and non-cooperative, depending on players' position of

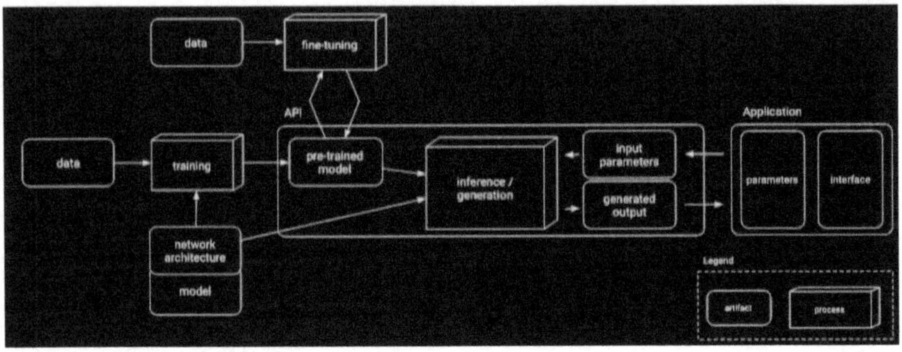

Fig. 1.1 Generative AI elements

knowledge. Game proposition is a conception of the game proposition, and it can be applied to a wide variety of fields. In this paper, we concentrate on generative inimical neural networks (GAN), a type of deep literacy, and a class of neural network infrastructures. The vantages of the system are enhancement in celerity, stability, and interpretation of the artificial neural network (Singh et al., 2024).

1.3 Inspiration for Game Theory

The generation AI process begins with a variety of data sets. AI training processes the dataset iteratively to learn and internalize existing patterns. Fine tuning is used to further refine the pre-trained model. Once properly trained and finely adjusted, the model is ready to infer. Artificial intelligence can generate new data by providing input parameters or specific guidelines.

Game theory, is where the concept of opponent training originates, with two participants (discriminators and generators) participating in the game. While the generators try to generate real data, the discriminators try to distinguish between real data and generated data.

The classification corresponds to the steps shown in Fig. 1.2, which show the process of generating artificial intelligence from data input to new content creation and application creation in real-world scenarios.

World Scenario Overview of the Generative Artificial Intelligence Process: Every Generative Artificial Intelligence Process begins with data, including a variety of data sets, including text, images, sounds, or other data sets. This set is a basic input for the recognition and understanding of patterns by AI.

AI Training After data collection, the next crucial step is AI training. During this stage, AI processes the dataset in iteration to learn and internalize existing patterns.

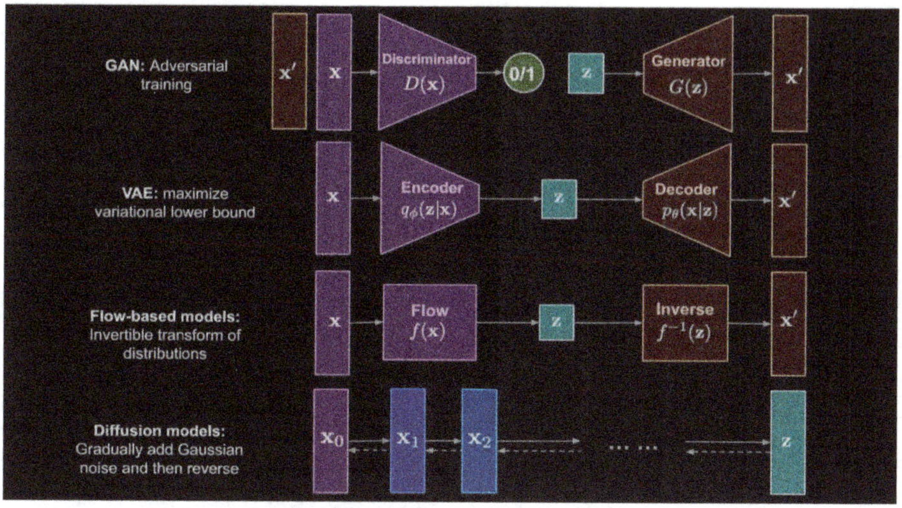

Fig. 1.2 Overview of Generative AI Models

This learning process culminates in the creation of a "model" which acts as a digital representation of the acquired knowledge.

Fine-Tuning In some scenarios, AI needs to focus more deeply on specific nuances or attributes. Here, fine tuning occurs when pre-trained models are further refined using additional datasets. This process increases the model's capability in the desired direction.

The Use of Models Once the model has been sufficiently trained and, optionally, finely adjusted, the model is ready for inference. Inference refers to the use of knowledge acquired to analyse and process new data inputs to generate relevant outputs. This inference process can take place locally on the machine or remotely via an API, depending on various factors such as computational resources and application needs (Ali, 2024).

New Data Generation With the model prepared and ready, AI can begin generating new data. By providing a specific input parameter or rule, AI generates "generated output", which represents newly created content.

Application The AI results are integrated into various applications, including web sites, mobile applications, and other digital platforms. The front-end aspects of these applications facilitate user interactions, enabling them to use and benefit from AI capabilities.

In essence, Generative AI involves furnishing an AI system with vast amounts of data, training it to discern fundamental patterns, and subsequently harnessing that knowledge to engender new data. The implications and potential applications of this technology are extensive, evolving continuously as the field progresses. This classification aligns with the depicted steps in Fig. 1.2 elucidating the journey of Generative AI from data input to the creation of novel content and its real-world application.

1.4 Early AI: Chatbots Beginnings

Despite the recent announcement of generative AI models, this technology is in its twenties and has undergone its first research since the 1960s. We are now in an age of AI with human cognitive abilities, such as OpenAI ChatGPT and Bing Chat, which use GPT models for web searches and Bing conversations. Back up a little, the first AI prototype was a type-written chatbot based on a knowledge base extracted from a group of experts and expressed on a computer. The knowledge base answers were generated by the keywords appearing in the input text. However, it quickly became clear that this method using type-written chatbots is not suitable for large-scale usage. A statistical approach to artificial intelligence: Machine learning in the 1990s, statistical approaches were applied to text analysis. This led to the development of a new algorithm called machine learning that can learn patterns from data without being explicitly programmed. This method allows machines to simulate human language comprehension: The statistical model is trained on a combination of text labels, allowing the model to classify unknown input texts, and a pre-defined label represents the message's intention. Recent advances in hardware technology that can handle large amounts of data and complex computations have led to the development of advanced artificial intelligence algorithms called artificial intelligence networks or deep learning algorithms (Souza, 2020).

Neuronal networks (especially regular neuronal networks) significantly improve the processing of natural languages, allowing text to be expressed more clearly, and improving the context of words in sentences. This technology, born in the early days of the twenty-first century, is capable of understanding human language, identifying needs, and performing actions to meet those needs, such as responding to a predefined script, or using third-party services. Today, Generative AI So that is the way to Generative AI today, which can be seen as a subset of deep learning. AI, ML, DL, and generation AI after decades of research in the field of AI, a new model architecture called a transformer has overcome the limits of RNN and can obtain text sequences longer as inputs. Transformers are based on attention mechanisms that allow models to give different weights to inputs they receive, giving "more attention" to the most relevant information centered in the text sequence, regardless of the order in which it is placed. Most of the most recent generative AI models are also known as large-scale language models (LLMs) because they work with text input and output, and are actually based on this architecture. The interesting part of these models is that they can be trained with a large amount of unlabelled data from different sources such as books, articles, websites and can be adapted to many tasks and generate grammatically correct texts with some creativity. Therefore, they not only dramatically improved the machine's ability to "understand" input text, but also enabled it to generate original responses in human language. How does a large-scale language model work? In the next chapter, we will explore the different types of Generative AI models, but we will now focus on the OpenAI GPT (Generative Pre-trained Transformer) model and the process of working with large-scale language models.

Tokenizer, text to numbers: Large-scale language models receive texts as inputs and generate texts as outputs. However, as a statistical model, they are better at numbers than with text sequences. Therefore, every input of the model is processed by the tokenizer before it is used in the core model. Tokens are text pieces, consisting of variable characters, and the main task of the tokenizer is to divide input into token arrays. Afterwards, each token is mapped with the token index, which is the full-length encoding of the original text block. Example of tokenization Predicting Output Tokens: If you give n tones as input (the max n varies from model to model), the model can predict one tones as output.

This token is incorporated into the next iteration input, and in the window expander model, the user has a better experience when he receives one (or multiple) sentences as an answer. If you've played with ChatGPT before, you might notice that it sometimes looks like it's stopped in the middle of a sentence. Selection process, probability distribution: The output token is selected by the model according to its probability to occur after the current text sequence, the highest probability token is always not chosen from the result. By adding a certain degree of randomness, the model acts non-deterministically—we do not get the exact same output from the same input. This degree of randomness can be adjusted using a model parameter called temperature, which simulates the creative thought process.

How can we use our startup's large-scale language model? Now that we have better understood the inner functioning of a large language model, let us see some practical examples of the most common tasks they can perform fairly well, with regard to our business environment.

The main functionality of large-language models is to create a text from scratch, starting with a text input written in natural language. But what type of text input and output? Large language model inputs are called prompts, outputs are called completions, which refer to the model mechanism that generates the next token to complete the current input. We will go in depth into what prompts are and how to design them so that we can get the most from our models.

But for now, just say that prompts can be: In addition, generative AI models' outputs are not perfect, and models' creativity can sometimes affect them, creating outputs that can be interpretable as a combination of words by human users as a mysterious reality or as offensive. Generational AI is not intelligent, at least in a more comprehensive definition of intelligence, including critical and creative reasoning or emotional intelligence, because it is not determined, but because it is not trusted, because fabrications such as false references, content, statements can be combined with correct information and presented convincingly and confidently.

In the following lesson, we will address all these limitations and see what we can do to mitigate them.

1.5 Building Generative AI Models

Erecting a generative AI rendering device requires training AI models on voluminous quantities of law across programming languages via deep literacy. (Deep literacy is an expressway to train computers to reuse data like we do—by feting patterns, making connections, and drawing consequences with restricted guidance.) To emulate the expressway, humans get patterns, these AI models exercise vast networks of bumps, which process and weigh input data, and are aimed to serve like neurons. Once trained on voluminous quantities of data and suitable to produce useful law, they're erected into tools and operations. The models can also be plugged into rendering editors and IDEs where they respond to natural language prompts or law to suggest new law, places, and expressions.

Before we talk about how generative AI rendering tools are made, allow' s outline what they're first. It starts with LLMs, or voluminous language models, which are sets of algorithms trained on voluminous quantities of law and mortal language. Like we mentioned over, they can prognosticate rendering sequences and induce new content utilizing being law or natural language prompts. Moment's state-of-the-art LLMs are mills. That means they exercise commodity called an concentration medium to make adjustable connections between nonidentical commemoratives in a stoner's input and the affair that the model has formerly generated. This allows them to give responses that are more contextually applicable than former AI models because they're good at connecting the blotches and monumental-picture thinking. Then's an illustration of how a motor works. Let's enunciate you encounter the word log in your law. The motor knot at that position would exercise the concentration medium to contextually prognosticate what sort of log would come next in the conclusion.

Let's say-so, in the illustration below, you input the statement from calculation import log. A generative AI model would also infer you mean a logarithmic function. And if you append the prompt from registering import log, it would infer that you're utilizing a registering function. Though occasionally a log is precisely a log. LLMs can be erected utilizing fabrics besides mills. But LLMs utilizing fabrics, like an intermittent neural network or long short-tenure mind, struggle with processing long rulings and paragraphs. They also generally bear training on labelled data (making training a labour-ferocious process). This limits the complication and applicability of their labours, and the data they can get from.

Motor LLMs, on the other phase, can train themselves on unlabelled data. Once they're given away introductory literacy objects, LLMs take a portion of the new input data and exercise it to exercise their literacy pretensions. Once they've achieved these pretensions on that portion of the input, they apply what they've learned to understand the rest of the input. This tone-supervised literacy process is what allows motor LLMs to dissect massive quantities of unlabelled data and the larger the dataset an LLM is trained on, the further they gauge by recycling that data. Why should inventors watch about mills and LLMs? LLMs like OpenAI's

GPT-3, GPT-4, and Codex models are trained on an enormous quantum of natural language data and intimately accessible source law.

This is portion of the argument why tools like ChatGPT and GitHub Copilot, which are erected on these models, can produce contextually accurate labours. Then's how GitHub Copilot produces rendering suggestions All of the law you've penned consequently so far, or the law that comes before the cursor in an IDE, is fed to a series of algorithms that decide what corridor of the law will be reused by GitHub Copilot. Since it's powered by a motor-grounded LLM, GitHub Copilot will apply the patterns it's distracted from training data and apply those patterns to your input law. The result contextually applicable, initial coding suggestions. GitHub Copilot will indeed filter out given screen susceptibility, liable law patterns, and law that matches other systems.

Keep in mind creating new content similar as textbook, law, and images is at the heart of generative AI. LLMs are complete at abstracting patterns from their training data, applying those patterns to being language, and also producing language or a line of law that follows those patterns. Given away the sheer scale of LLMs, they might induce a language or law conclusion that doesn't indeed live yet. Precisely as you would reconsider a coworker's law, you should charge and support AI-generated law, too. Why environment matters for AI rendering tools Developing good advisement casting ways is important because input law passes through commodity called an environment window, which is present-day in all motor-grounded LLMs.

The environment window represents the capacity of data an LLM can reuse. Though it can't reuse an horizonless quantum of data, it can grow larger. Right now, the Codex model has an environment window that allows it to reuse a couple of hundred lines of law, which has formerly advanced and accelerated rendering tasks like law completion and law revise summarization. Inventors exercise details from draw queries, a brochure in a design, open effects and the list goes on—to contextualize their law. Consequently, when it comes to a rendering device with a restricted environment window, the challenge is to figure out what data, in extension to law, will conduct to the stylish suggestions lately, GitHub made updates to its brace programmer so that it considers not only the law incontinently before the cursor, but also some of the law after the cursor.

The paradigm which is called charge-in-the-middle ground (FIM) leaves a gap in the middle ground of the law for GitHub Copilot to charge, furnishing the device with further environment about the inventor's intended law and how it should align with the rest of the program. This helps produce advanced quality law suggestions without any appended quiescence. Illustrations can also contextualize law. Multimodal LLMs (MMLLMs) scale motor LLMs so they reuse images and vids, as well as textbook. These models are aimed to respond to natural language and images, like interspersing textbook and images, image-caption dyads, and textbook data (Sohn, 2020).

1.6 Language Modelling and Evolution

From the discussion, we come to an understanding that it begins with an explanation of generative artificial intelligence (AI) that creates new content based on existing data. We also include a variety of forms, including text creation and natural language processing, and examples such as Google Translate and Siri illustrate its widespread use. So we can say that the introduction of OpenAI's gp4 model highlights its abilities in overtaking human performance in tasks such as SAT tests and text generation. It is pointed out that the development of generative AI is based on language modelling and the prediction of sequences of next words, highlighting the evolution of a simple tool to a more sophisticated system such as gp4. The neural network language model introduced represents its structure and parameters, and transformers are a common type. The discussions deal with the training process of generative AI models, emphasizing the importance of scale and training data. However, there are concerns about the behaviour of the model, including cases of undesirable content generation and the impact of energy consumption on the environment. It is believed that AI regulation is inevitable, as are other regulated technologies such as nuclear energy, to mitigate the associated risks while maximizing its benefits.

1.7 Importance of GPU Innovation

Dealt with colourful aspects of the democratization and commercialization of voluminous-scale language models, as well as the pitfalls and ethical considerations associated with them. The deliberation stressed the practicality and cost-forcefulness of structure and training language models eased by advances in chip technology and effectiveness. The Palm Prompt2 model from Google has been mentioned as an illustration of this availability, and lower models are indeed adaptable to movable phones. The deliberation also concentrated on the significance of nonstop chip advancements and competitiveness in the GPU, requesting to stimulate invention in this area. Pitfalls and regulations relating to artificial intelligence are also being managed, pressing the want to balance invention and security and the significance of addressing being detriment similar as partisanship and dependence. Conversations were extended to the jolt of voluminous-scale language models across diligence, especially in tours of content coinage and overcritical use cases similar as transportation, and asked for indemnification for generators who exercise workshop. Moreover from the discussion we can say that the implicit jolt of AI on employment openings has also been explored, taking into account the evolving nature of technological places and the want for humans to acclimatize to the process of AI. Also it is said that the deliberation rounded with a reflection on the originality capabilities of ingenious AI and the gregarious adaptations that may be demanded as AI becomes more widespread.

1.8 Debates Held on Generative AI

Debate: "Generative AI Should be Stopped!", Chris Berg, the director of RMIT's Blockchain Innovation Hub sets the stage for a debate on the disruptive potential of generative AI and whether it should be stopped or not. He acknowledges the traditional owners of the land and introduces the participants, with Shirag Shah and his team arguing for the affirmative side that generative AI should be stopped or regulated, and Rachel and her team representing the negative side, advocating for letting it rip. Berg outlines the ground rules for the debate, which involves each team having 3 min to present their arguments before engaging in a conversation. Shah begins the affirmative team's presentation by emphasizing three reasons why generative AI must be stopped or regulated: The concentration of power is bad, self-regulation doesn't work, and the amplification of existing problems with information, including misinformation and disinformation.

The negative team argues against the proposition that generative AI, specifically ChatGPT, should be stopped. They emphasize that the evolution of language and digital search have become embedded in our daily lives, and generative AI represents the future of search. The negative team believes that to contend that it must be stopped is to contend that human progress must be stalled. They argue that ChatGPT is a gift for our times, improving productivity and yielding benefits in various areas such as cyber security, energy efficiency, and agricultural yield. The team also points out that humans have the ability to control and manage the AI, and it is not a new concept, as digital co-pilots have been around since Y2K. The economic benefits of generative AI are estimated to be 115 billion annually by 2030.

The speaker raises concerns about the increasing use of generative AI in spreading misinformation, disinformation, and deep fakes, which could potentially threaten democratic elections and societal harmony. By 2026, 90% of the information on the internet is forecasted to be generated or manipulated by generative AI, making it difficult to validate the authenticity and integrity of data. The speaker also warns about the potential for scams and financial losses, as well as the ease with which malicious actors can weaponize open-source generative AI models. The speaker argues for more transparency and accountability through better regulation and safeguards. However, the opposing side challenges that there is already regulation in place, and it is unlikely that unregulated harm from generative AI exists. The debate goes by each side giving their perspectives on the issue.

Jason Potts argues for the regulation of AI to protect individuals and businesses from potential threats, including corporate and nation-state actors. He uses the political science theory of the Stationary Bandit model to support his argument, which posits that individuals seek protection from a powerful entity, such as a regulatory body, to shield them from even worse agents. Potts acknowledges that there may be differing opinions on the reasons for regulation and its benefits, but ultimately asserts that the power and importance of AI necessitate its regulation.

Speaker Chris discusses the evolution of humans as a tool-making species and the impact of technology on our lives. He argues that humans have always made tools that save labour and brain power, leading to advancements like language and mathematics. The speaker also acknowledges the challenges of misinformation in the digital age but believes that individuals should be responsible for evaluating the truth of what they see online. He emphasizes the potential benefits of AI in enhancing human creativity and objectivity.

The debate continues with Chris attempting to address unanswered arguments from both sides. The speakers discuss the issue of misinformation and its regulation, specifically in the context of political campaigns and generative AI. While acknowledging the existence of laws against misleading advertising and authorizing content, one speaker argues that the focus should not be on stopping bad actors but rather addressing the widespread user-generated misinformation.

They argue that anyone can create realistic and truthful-sounding pieces of information using generative AI, making it a significant concern. Despite this, some argue that regulating or stopping generative AI entirely may not be the solution, as other countries may continue to use it for disinformation purposes. Instead, they suggest implementing content provenance labelling as a way to ensure transparency and accountability. The discussion shifts towards the regulatory approach to combating potential harms caused by generative AI. The speaker explains that the current focus of regulations, such as the misinformation and disinformation Bill, is on content moderation rather than stopping generative AI development altogether. They argue that a more effective regulatory approach is needed, one that involves identifying new harms and ensuring accountability when things go wrong. The European Union's AI regulation model is mentioned as an example, which includes identifying different kinds of risks and harms and assigning specific rules and obligations based on the level of risk.

The speaker also emphasizes the importance of involving marginalized and vulnerable groups in the regulatory process to identify harms and ensure societal needs are met. The speakers discuss the argument that there are no new harms caused by AI as it cannot think outside the box and generates new harms on its own. One speaker argues that the focus should be on enforceability and regulating harms at the front end, as it becomes increasingly difficult to pin down individuals responsible for generating harmful content with the proliferation of AI.

The opposing team's best argument is that there is a significant power imbalance in the face of this new technology, and they advocate for safety, security, and social welfare benefits through regulation. The negative team argues for the need of regulation in the AI domain, emphasizing the power imbalance that new technology brings and the need for an agency to act on behalf of individuals against potential threats. The affirmative team, however, asserts that AI should not be regulated as there are no novel harms brought about by algorithmic technology and that it is simply a human tool for the twenty-first century. The debate concludes without a clear winner, with both teams making compelling arguments.

1.9 Inovations

The foundation of multimodal generative AI lies at the intersection of many fundamental concepts and ideas in the field of artificial intelligence and machine learning. Below is a detailed study of the principles:

1.9.1 Generative Models

Figure 1.2 provides a visual representation of three key generative AI models: GANs, VAEs, and BMs, each with its unique approach to generating new content. The illustration helps in understanding the architecture and functionality of these models in the context of generative artificial intelligence.

Generative models form the cornerstone of multimodal generative AI: This model aims to capture and learn the consequences of the distribution of the data set to create new models similar to the data shown. Two well-known types of neural networks are:

Generative Adversarial Network (GAN): GAN was proposed by Ian Goodfellow in 2014. It consists of two neural networks (generator and discriminator) and is involved in minimax. Game. The purpose of the generator is to create the actual model, while the discriminator tries to separate the model from the model. Through feedback training, GANs learn to produce good synthetic information (Dwivedi, 2023).

Variable Autoencoder (VAE): VAE is a powerful design that learns to encode input data into latent space and then decodes it back to the original data record. VAEs are trained to be able to reproduce the training data while ensuring that the latent space follows a predefined distribution (usually a Gaussian distribution).

1.9.1.1 Multimodal Data Representation

Multimodal generative AI deals with data that encompasses multiple modalities, such as images, text, audio, and more. To effectively model such data, it's essential to devise methods for representing and fusing different modalities. Key techniques include:

Cross-Modal Embeddings: These techniques learn joint embeddings that capture semantic similarities across modalities. For instance, methods like Word2Vec and GloVe learn embeddings for words that encode semantic relationships, which can be extended to other modalities.

Attention Mechanism: Attention Mechanism permit models to center on significant parts of input information. Within the setting of multimodal generative AI, consideration instruments can specifically go to particular modalities or districts inside modalities to produce coherent yields.

1.9.2 Fusion and Alignment Strategies

- Coordination and integration of information from multiple disciplines is crucial to achieving consistent multimodal results. In this context, several strategies have been developed:
- Conditional Generation: Adapt a generation model to a form input to produce another form output. For example, based on text description, create image-generating conditions for generating this image.
- Cross-Modal Translation: Information to be transmitted from one form to another. Translations between text and images or different languages, for example.

1.9.3 Applications and Use Cases

Multimodal generative AI has many applications:

- Image Captioning: Generating natural language descriptions for images.
- Text-to-Image Synthesis: Generating realistic images from textual descriptions.
- Music Generation: Creating new music compositions based on textual or audio input.
- Cross-Modal Retrieval: Retrieving relevant data across different modalities, such as finding images based on textual queries or vice versa.

1.9.4 Challenges and Future Directions

While multimodal generative AI provides a great promise, it still has some challenges:

Interpretation and Control: Making sure that the output is interpretable and controllable, especially while processing complex multimodal information.

Ethical Considerations: These address the ethical issues related to the creation of multimodal content, such as neutrality, fairness, and privacy.

Scalability and Efficiency: Build scalable and efficient algorithms that can solve the complexity of multimodal data and produce data in real time.

Continuous Learning: This allows designs to frequently learn from new information and adapt to changing environments. It is very important to consider these challenges as research into multimodal generative AI continues, unlocking its full potential and creating change in many ways.

1.10 Evolution of Generative AI Models

1. **Early Models**: Early models such as Markov chains and random graph models were basic and not much suitable for modelling real-world data and they can produce more accurate models.
2. **Autoencoders**: In the past, early autoencoder models were more used for data compression and other removal tasks. Today's autoencoders, such as variable autoencoders (VAE), provide requirements and normalization techniques to create new data models in addition to restructuring input data (Singh et al., 2024).
3. **Generative Adversarial Networks (GAN)**: Goodfellow et al. in (2014), GANs made significant progress in design but were limited by security and crash issues. Over the years, GAN architecture and training methods have been greatly improved to solve security issues, leading to the development of powerful and versatile models such as StyleGAN and BigGAN.
4. **Transformer-based Models**: Transformers were first introduced to programming languages and models such as BERT and GPT were used. Transformer architecture has been modified and extended to solve problems of previous generation language design such as image synthesis and music generation. They demonstrate a great ability to maintain long-term prospects and produce consistent, valuable content.
5. **Autoregressive Models**: Autoregressive models such as PixelCNN and WaveNet are the oldest methods of generating connected data such as images and audio (Gao, 2020). Recent advances in autoregressive modeling have led to the development of more effective and efficient models that can produce high-resolution images and interconnect audio with improved quality and accuracy.
6. **Flow Models**: Flow-based models were introduced as an alternative design method focusing on flexible and accurate calculations. Streaming architecture has evolved to include powerful and flexible transformations that can model complex data distribution and achieve high performance on tasks such as image processing and density estimation.
7. **Hybrid Method**: The hybrid method, which combines elements of different models, is less common and often involves ad hoc combining of methods. Today's designs increasingly use hybrid architectures that offer the benefits of diversity, such as combining GANs with autoregressive models or streaming models with variables, resulting in good performance and diversity. Training a variety of adaptive networks (GANs) and variable autonomous encoders (VAEs) can be a tricky difficult road. GAN, for all its achievements in generating real data, tends to be stuck in ruts. Imagine the bank training a GAN to create synthetic customer profiles for fraud detection. GAN could be stuck in producing profiles similar to existing honest customers, lacking subtle patterns of fraud

behaviour. This is known as the mode collapse, and GAN prioritizes misleading discriminators with limited output. In order to counteract this, researchers are applying techniques such as spectral normalization and gradient penalty functions, essentially driving GAN to explore broader possibilities.

1.10.1 Enhancing GAN and VAE Quality

Fortunately, there are solutions and techniques that help improve GAN and VAE performance and quality. For example, GANs can use alternative loss functions such as Waterstein distance, hinge loss, perception loss, beta-VAE, InfoVAE, and VQ-VAE. In addition, techniques such as dropout, weight degradation or spectral normalization can be applied to networks.

We can optimize model performance and quality through architecture and hyper-parameter adjustment, such as layer number and size, activation function type and order, learning rate and optimizer, or GAN latent dimension and prior distribution of VAE (Saxena et al., 2021).

1.10.2 Improving GAN Learning Stability

GAN and VAE are both powerful and flexible models, but training them also poses some challenges and limitations. Mode collapse, disappearing gradients, and blurry images are all common problems that may arise. Mode failure can be caused by a generator finding a shortcut to deceive the discriminator, or if the discriminator is too strong or too weak. The disappearing gradients are caused by the saturation of activation functions, or by the inconsistent matching between generators and discriminators. The blurry image may be caused by a loss in pixel reconstruction used by VAE or by a regularization limitation of the latent space. All these problems can make learning processes unstable or slow, lead to a lack of clarity and detail, and generate images that make them unrealistic or noisy.

GAN faces problems such as learning instability and mode failure, and produces limited samples. Solutions include the insertion of conflict training for better image quality. VAE is more difficult to train and perform slower than traditional automatic encoders. The solutions include improved latent space modeling and improved reconstruction accuracy. The combination of VAE and GAN in models such as VAE-GAN addresses the limitations of both, thereby improving image quality and diversity. The use of conditional VAEs allows the generation of targeted images based on specific attributes or labels. GAN and VAE play a key role in the creation of synthetic data for training models, the improvement of security algorithms, and the creation of realistic multimedia content.

1.11 Architechtural Improvements

Researchers have suggested a number of architectural modifications to GANs over time in order to solve different issues and improve their functionality:

1.11.1 Deep Convolutional GANs (DCGANs)

DCGANs allowed the generator and discriminator architectures to learn hierarchical representations of the picture data by introducing convolutional layers into them. This design greatly enhanced the quality of output photos and it also stabilized the training (Singh et al., 2013).

1.11.2 Progressive GANs (PGANs)

Wasserstein distance was developed by Wasserstein GAN (WGAN) as a more reliable tool for comparing probabilities. Compared with traditional GANs, WGAN produces better models and more stable training by optimizing the Wasserstein distance. During training, the generator and discriminator start with negative images and gradually increase the resolution due to the gradual evolution process of PGAN. This approach allows PGAN to generate stable, high-resolution, and variable content during learning.

1.11.3 Ethical Points to Remember

- The ethical use and implications of using GANs become very important as they become very much powerful.
- Deepfakes: By masking or creating a person's face from other people's faces, GAN videos and scammers are used to create images and raise concerns about theft, misinformation and fraud, Blackmail, privacy, problems.
- Privacy Concerns: Using GANs to create artificial intelligence increases worries about the security and privacy of those who may use the data to inform the model.
- Misuse: The possibility of using GANs to produce propaganda, fake news, or illegal content, is definitely an issue to worry about.

In order to solve these types of ethical problems, the development of intellectual property rights has to be based on a strong base, be open to practice education standards, and follow practice and legal standards.

1.11.3.1 Frontiers of Research

Research boundaries: At present, GAN research is trying to overcome modelling boundaries and solve some problems:

- Types of collisions: Researchers are looking for many ways to reduce this type of phenomenon, which occurs when generators offer low standards, resulting in less or equal diversity.
- -Safety during training: Regulation, increased penalties and other training goals are being worked to improve the safety of GAN training.
- Checking the design: for checking the accuracy of the final product, researchers are now developing ways to check aspects of the design (such as appearance, style, or function).

By addressing current issues and expanding the capabilities of GANs, these research areas hope to create more reliable and adaptable generative models.

1.12 Introduction of the Open Source Generative AI Index (GenOS)

These fields tend to create more reliable designs by solving many existing problems and expanding GAN capabilities. Figure 1.3 shows the Generation AI Open Source Directory (GenOS), which tracks most of the generative AI open source projects. The importance of community participation, the diversity of the project teams, and the regular updating of the index are highlighted. The aim is based on promoting

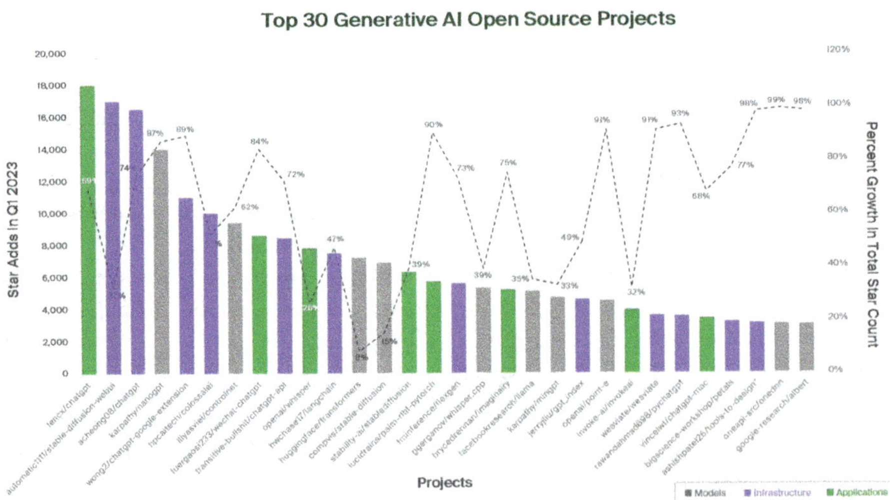

Fig. 1.3 Generative AI Models Growth Index

and collaboration, innovation, and transparency in the AI generation community. Background on Generative Artificial Intelligence:

Information about the generation of artificial intelligence: The paper explains about the rise of interest in artificial intelligence and it compares its development with significant changes in platforms such as the early Internet and mobile devices. This shows that generative artificial intelligence has a significant impact on all aspects of technology and innovation.

1.12.1 Community Contribution and Open-Source Projects

With the emergence of artificial intelligence, developers and users from all over the world have contributed to open projects in this field. This has remained same as the past development of large power plants and this also displays the potential for innovation and cooperation. Launch of Open-Source Generative AI Index (GenOS):

Launch of Open Source Generative AI Index (GenOS): The Open Source Generative AI Index (GenOS) is the index that tracks the most active open source projects in the field of generative AI depicts in Fig. 1.4.

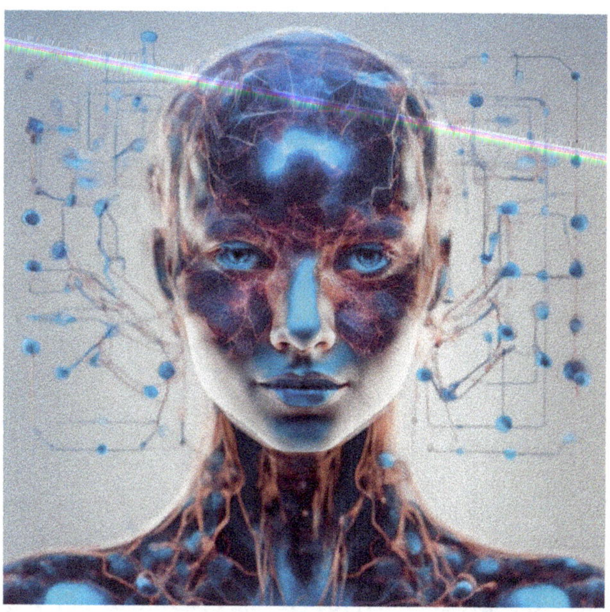

Fig. 1.4 Open AI

1.12.2 Index Criteria and Subcategories

GenOS Index identified the top 30 projects based on the growth of stars in GitHub over the past 90 days and included a minimum of 500 stars. The project is divided into three sub-categories: models, infrastructure/tools, and applications, reflecting the diversity of contributions and focus areas of generative AI. Monthly updates:

- The index is expected to be updated monthly, which enables continuous monitoring of many active projects in the field of generative AI. This frequent update cycle makes sure that the index will remain relevant and will reflect current trends and developments.

1.12.3 Obstacles and Prospective Paths

1.12.3.1 Interpretability and Manageability

Interpretability is said to be the ability to understand and explain the process of producing outputs by models, especially in complex multimodal situations. The ability to modify and control particular characteristics or properties of the generated outputs is referred to as controllability. Ensuring interpretability and controllability in multimodal generative AI presents a number of difficulties:

Complexity

Interpreting the inner workings of multimodal generative models can be difficult due to the complex interactions between many modalities.

Interpretation Methods

Gaining understanding of the behaviour of multimodal generative models requires developing methods for interpreting and visualizing the learnt representations and decision-making processes of these models (Harshvardhan, 2020).

Controllability Mechanisms

To enable real-world applications in fields like content production and modification, mechanisms that are designed to control some qualities or features of the created outputs, such as style, content, or sentiment, while maintaining other characteristics, are essential. To construct interpretable and controlled multi modal generative models, multidisciplinary research efforts combining methods from cognitive science, machine learning, and human-computer interaction are needed to address these issues.

New Trends in Generative AI

Generative models generate new data points based on the learnings they get from training data. These models usually train without supervision, analyse data without human input and draw their own conclusions. For years, the generation model has faced more difficult tasks, such as learning to produce photos and accurate answers to questions in text, and progress has been slow.

Further, the increase in the availability of computer power enabled the Machine Learning (ML) team to build a basic model: a large unsupervised model of large amounts of data training (sometimes all available data on the Internet). Over the past few years, ML engineers have calibrated these generative basic models, providing annotated data subsets to target outputs to specific objectives, and used them in practical applications. ChatGPT-3/4 is a good example illustrated in Fig. 1.5.

It is a version of Chat GPT, a basic model developed on a large amount of unlabelled data. To create ChatGPT, OpenAI hired thousands of annotations to mark appropriate data subsets, and its ML engineers then refined the model to teach it to generate specific information. With such fine-tuning methods, generative models are beginning to produce outputs that were not previously possible, resulting in a rapid expansion of functional generative models (Fui-Hoon Nah, 2023).

Progress in the field of large-scale language models and advanced AI in health care is rapid, and it is difficult to stay updated with the changing news. For example, Insilico has developed an AI platform that uses deep-learning models, reinforcement learning, transformers, and other modern machine learning techniques for discovering new targets and producing desired molecular structures. There are many examples of this type in the whole spectrum of health care and clinical research.

Fig. 1.5 Chat gpt

The intensive excrescency of generative AI has been extraordinary over the once time. It was launched by the public launch of ChatGPT (wasn't it a time ago?), now it's far and wide. Sweats to punch swells, all Office and eBay apps have appended generative capabilities, and further and further people are chancing usages in their everyday and professional lives. Given away its nature, it isn't astounding that content generators, especially happy generators, buy it's an important extension to the tools set. Marketing agencies, advertising creatives, news associations, and gregarious media influencers were among the most enthusiastic early adopters. It provides great openings to ameliorate effectiveness and automate repetitious homemade rudiments of innovational work, but it also poses important expostulations. Effects related to brand, spam content, visions, algorithm coinage formulas and impulses must be taken into account by professionals calculating to incorporate them into their workflows. With this in mind, I give an overview of the goods of generative AI formerly in this field and some ideas about what to anticipate as technology becomes more important and society adapts to the world of stoked AI.

Generative AI in practice: There are already many astonishing examples of generative AI used to create amazing things (sometimes terrible). For example, Coca-Cola's "Men's Show" is a collaborative effort between human artists and artificial intelligence, which brings to life the greatest works of art in history in a way never seen before. It also replaced the lyrics of John Lennon's partially recorded song with new material by Paul McCartney and was used to create a new Beatles song. Generative design is an emerging field where generative artificial intelligence is used to design and produce new product templates and production processes (Baidoo-Anu 2023).

For example, General Motors designed a new seatbelt bracket with 40% lighter and 20% stronger components than the existing component using Autodesk's generation tools. And it is also used to accelerate drug discovery, and a British company recently announced that it has created the world's first artificial intelligence-generated immunotherapy cancer treatment. Generative AI is also the latest technological phenomenon behind the deep fake, which blurs the lines between reality and fiction by appearing as if real people were doing or saying things fake.

Deepfake Tom Cruise is one of the first and most well-known examples. We can say that more insidiously, potential candidates, on both sides of the upcoming U.S. where we know that presidential elections in 2024 have played in deep deceptions aimed at discrediting them for political purposes. Even if propaganda is bad enough, there are other uses as well, including stealing money through cloned voices and fraudulently deceiving money by pretending to be the company's CEO.

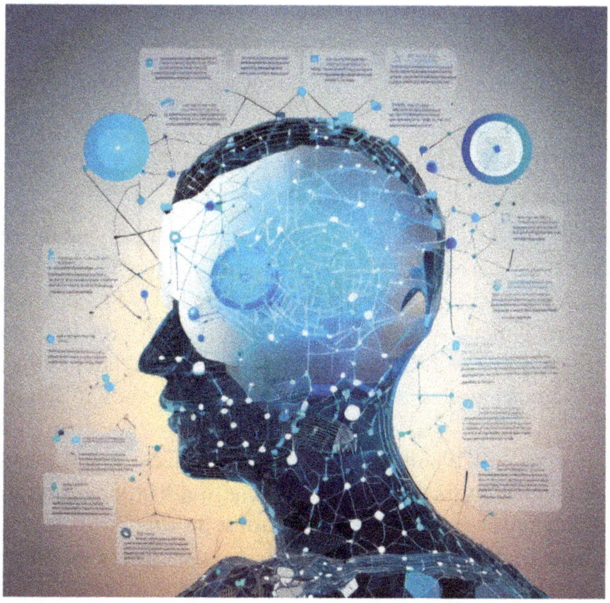

Fig. 1.6 AI in social media

1.13 Ethical Problems with Generative AI

Although generative AI is certainly capable of doing amazing things, it is clear that its existence forces us to confront some difficult problems and questions. Perhaps one of the most important is when we get to the point where it is impossible to distinguish between what is real and what is generated by artificial intelligence.

Hence we can say that given the incredible rapid pace of innovation in this field, it will probably happen sooner rather than later. So what should we do (if anything) to address this situation? Countries, including China, have already adopted laws that illegally fake people without their consent—should the world do the same? Then there is the question of how this will affect human jobs, and how creators' livelihoods will be threatened if the company that employs them can produce the image, sound, and video they need by simply telling the computer to do it? Another issue that needs to be addressed is copyright. If AI is used to create an art work, who owns it? Who creates art using AI? Who is the creator of the AI itself? Or all thousands of artists whose work has been used to train artificial intelligence (often without permission)? All these questions must be answered—and at the speed of the rapid development of this technology should be answered soon. The answer may play an important role in determining the future of generative AI in society and our lives shown in Fig. 1.6.

Drawback of generation AI among the content creation in addition to the hazards of boring and unproductive content, other problems must be taken into account. One of the highest priorities is copyright issues, which are two-sided. First, the jury remains to decide who owns the content generated by AI.

1.14 Copyright Challenges in AI Content

Are they developer of the aid that produced them? Or, what was the real data owner used to train AI? If you divert getting sued from a creator who declares that the AI you're using copies his work, you must decide whether you can apply copyright to your own creations. This thing may seem to be an issue for the line of work utilizing it to create their own assets and stuff. There also is the tendency for AI to erroneously mislead. This is called hallucination because they often seem to make things up.

Of course, no company wants to look stupid by releasing inaccurate information in fact. And there certainly is enough disinformation available online preliminarily without AI permitting it spread and even creating more!

Although it already has had a major influence, we clearly are just starting to use the generative AI. Eventually in future, we will even see more robust tools and, most importantly, even easier to use. It is very likely that most of the problems discussed here will be overcome, namely bland content, lack of emotional resonance, and false facts.

As linguistic models become stronger and more complex, we can see tools for creating content that inspire and stimulate us, which can be the same as humans. This makes issues such as the identification of deep fakes—very realistic artificial intelligence creations depicted to fool humans—and also the mitigation of the spread of AI-generated false information even a lot critical. However, the technology is also likely to become increasingly accessible, which means that its power can be used by a broader user base.

This creates a framework for generative tools and creations informed by the richest stories and experiences of humanity. Personally, I think that human beings are always needed in the process creating the content.

Basically, we are natural authors and creators. However, those who can learn to use creative tools to uplift our creative standards will have a clear benefit over those who do not, because we will create new ways to express our thoughts and ideas creatively.

1.14.1 Evolution of Generative AI Development on GitHub

Figure 1.7 highlights the rapid growth and increasing mainstream adoption of generative AI projects on GitHub in 2023. It underscores the transition from specialized research to practical application, as well as the expectation for continued innovation and expansion in the field of generative AI.

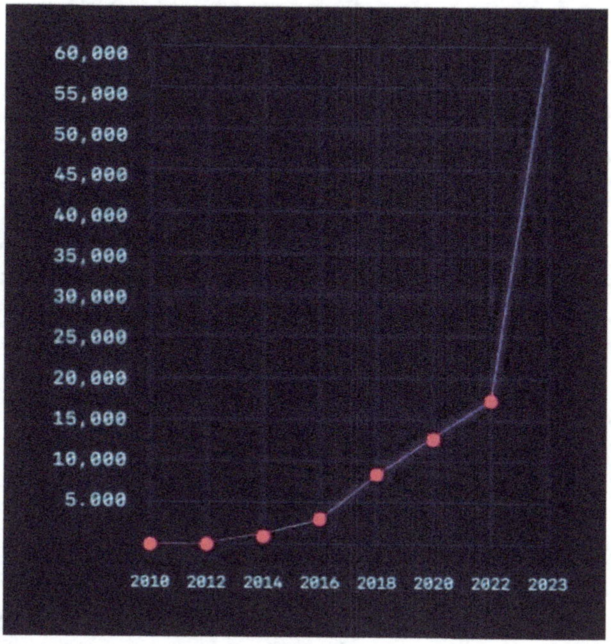

Fig. 1.7 Graph of open source and rise of AI in 2023

1.14.1.1 Generative AI Development Background

From this, we understand the trend of GitHub's generational AI development, which is on the rise, and it appears that it will draw significant attention in 2023, but it has been interesting to developers for several years. This indicates a gradual evolution rather than a sudden phenomenon.

1.14.1.2 Transition to Mass Implementation

This indicates the transition from specialized work and research to the mainstream adoption of generative AI projects. Developers are increasingly using pre-trained models and APIs for practical implementation, reflecting the transition from theoretical exploration to real-world application (Bellovin, 2019).

1.14.1.3 Significant Increase in Projects

The text notes a substantial rise in the number of AI projects created on GitHub in 2023 compared to previous years. Specifically, it mentions that by the middle of the year, the number of AI projects had already doubled compared to the entirety of 2022, indicating rapid growth and heightened interest in the field.

1.14.1.4 Expectations for Further Growth

GitHub reported a significant increase in the number of artificial intelligence projects created by 2023 compared to previous years. In particular, it pointed out that the number of artificial intelligence projects had doubled between mid-2016 and 2022, indicating rapid growth and increase in interest in the field.

Further growth expectations: The text expresses confidence in the possibility of continued growth of generative AI projects. It points out that the growth of activity is just the beginning, indicating significant potential that has not been exploited in the field. With more developers working with generative AI technology, it is expected to bring an innovation and integration into mainstream of software development.

1.14.2 Tools for Using Generative AI for Style Transfer

1.14.2.1 Style Transfer

Style transfer refers to the process of applying one image's visual style to another image while maintaining the original image's content and structure. For example, a city photo can look like a Van Gogh painting, or a person's portrait can look like a cartoon. Style transfer can be used for artistic expression, personalization, or enhancement of the aesthetics of web design.

1.14.2.2 Working of Style Transfer

Style transfer is based on a computer model based on a neural network to learn data and perform tasks such as recognition, classification, or generation. The neural network is composed of a layer of nodes that processes information and transmits it to the next layer. To carry out style transfer, we need two types of neural networks: content networks and style networks. The content network extracts the features and structures of the original image and the style network captures the colour, texture, and pattern of the style image. Thirdly, the network called the transmission network combines the output of the content network and the style network and generates a new image with both the content and style of the input image.

1.14.2.3 Challenges of Style Transfer

Style transfers are not simple tasks and involve several challenges and compromises. One of the main challenges is to balance output image content and style without losing detail or coherence. Another challenge is to deal with different resolutions, orientations, and scales of input images and to adapt the style transfer accordingly. The third challenge is to optimize the speed and quality of style transmission, because neural networks are expensive to calculate and time-consuming.

1.14.2.4 Tools for Style Transfer

Many tools and platforms allow you to transfer styles online or offline. Popular online tools are DeepArt, Artisto, and Prisma, which allow you to upload your own images or choose styles from a style gallery and apply style transfers in seconds. The offline tools TensorFlow, PyTorch, and Keras are frameworks for creating, training, and adapting parameters and options for building and training their own neural networks. You can also find tutorials and style transfer examples on websites such as Medium, GitHub, and YouTube.

1.14.2.5 Deepfakes

Deepfake is a synthetic video or audio that manipulates a person's appearance or voice to say or do something that has never been said or said. For example, a politician may make a video in which he speaks, or a famous person who supports products he never supports. Deepfakes can be used for entertainment, satire, education, as well as for deception, fraud, or propaganda illustrates in Fig. 1.8.

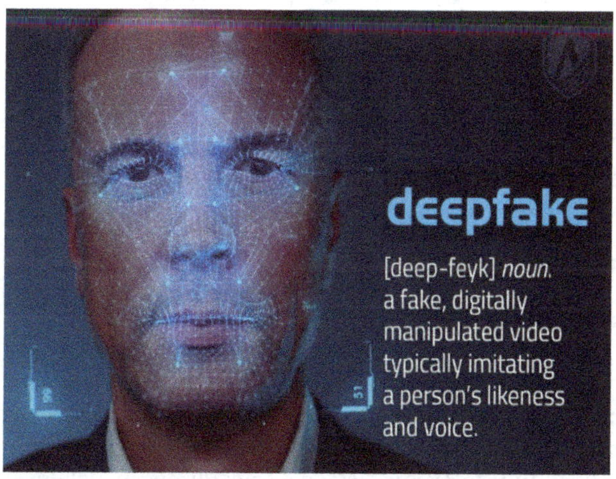

Fig. 1.8 Deep fake

1.14.2.6 Working of Deepfakes

Deepfake is a type of neural network that competes with each other to create real content using a technique called generative adversarial networks (GANs). GAN is composed of two networks: generators and discriminators. Generators try to produce fake content that looks like real content, but discriminators try to distinguish between fake and real content. Generators and discriminators learn from each other and improve their performance over time until generators deceived the discriminator and produced convincing content.

1.14.2.7 Challenges of Deep-Fakes

Deepfakes are not only difficult to create, but also difficult to detect and prevent. One of the main challenges is to ensure the authenticity and realism of deep falsification without introducing artifacts, distortions, or inconsistencies. Another challenge is to address ethical and legal issues such as consent, privacy, copyright, and information misuse. The third challenge is to develop and implement effective methods and tools for verifying the authenticity and source of content and to identify and eliminate harmful or malicious fakes.

1.14.2.8 Tools for Deep Fakes

There are many tools and platforms that allow you to create or watch deep-fake online or offline. Popular online tools include Deepfakes Web, Face, and MyHeritage, which allow you to upload your own photos and videos or choose from celebrities' libraries and apply Deepfakes in a few minutes. Offline tools such as DeepFaceLab depicts in Fig. 1.9, Faceswap, and FakeApp are software for building, editing, and controlling parameters and options. You can also find deep-fake tutorials and examples on sites such as Reddit, GitHub, and YouTube.

1.14.3 How Can You Assess the Quality and Accuracy of the Text Generated by Transformers and GPT-3 Models?

1.14.3.1 Transformers Overview

Transformers are neural network architectures that encode and decode information through listening mechanisms. Sequencing data such as text, language, and image can be processed without repeating or recursing layering, thereby increasing their efficiency and flexibility. GPT-3 is a special transformer model developed using an approach called self-attention to process text from large web texts. It can generate

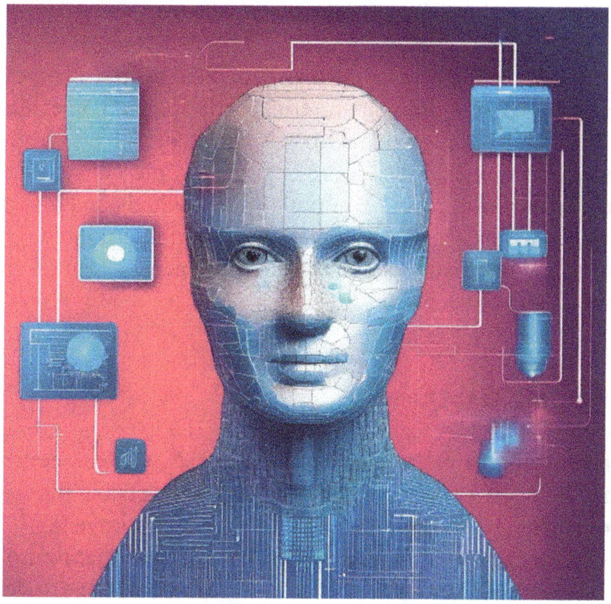

Fig. 1.9 DeepFaceLab

coherent and varied texts on almost any topic and can be inserted in a few words or sentences.

1.14.3.2 Text Generation Process

Text generation using transformers and gpt-3 is based on the language modeling concept, which is the task of predicting the next word or token, taking into account previous words. The model analyses large quantities of text data to learn the probability of different words or symbols occurring in different contexts. In order to generate text, the model uses decoding strategies, such as greedy, beam, or topk samples, to select the most likely or different words or tokens to continue the sequence until a predefined length or stop token is reached.

1.14.3.3 Advantages of Text Generation

Text production using transformers and gpt-3 has many potential benefits and applications, such as content creation, summary, translation, dialogue, questioning, etc. It can help automate and improve various tasks that require the understanding and generation of natural languages and provide new ways of interacting with information and knowledge. However, text generation also poses challenges and risks, such as ethical, social and legal consequences, quality and accuracy, data and model biases.

1.14.3.4 Quality and Accuracy Evaluation

Evaluating the quality and accuracy of text generated by transformers and gpt-3 is not easy because they do not have any metric or criteria. Human evaluation involves requiring experts and users to evaluate texts in different aspects, which is more reliable, more comprehensive, but also more costly and time-consuming. On the other hand, automatic assessment, using computational methods and algorithms to compare text to reference texts and data, is more effective, scalable, but limited and biased. Examples of automatic evaluation metrics include n-gram overlap, confusion, rouge, blue, and bertscore.

1.14.3.5 Enhancing Text Quality

Continuous improvements in the quality and accuracy of texts generated by transformers and gpt-3 are an active research area. Data expansion, model optimization, and text improvement are some of the possible ways to improve texts. Data enhancement involves adding multiple data sources and fields to the training or adjustment of models to increase their diversity and coverage of themes and styles. Model optimization involves the modification or adaptation of the model's architecture or parameters to increase its capacity and performance. Improve text using postprocessing technologies and methods to correct errors or problems such as grammar, spelling, punctuation, or style, etc.

Transformers are an essential neural network architecture for natural language processing (NLP) that efficiently captures long-range data dependency by using attention mechanisms. OpenAI's GPT-3 (Generative Pre-Trained Transformer3) is a well-known implementation of 175 billion parameters. It has achieved great success in generating coherent and contextually relevant texts, having been trained in various Internet texts. Both the transformers and the GPT-3 have revolutionized language processing and enabled advanced applications such as text completion and language translation in various contexts. Their impact extends to industry and demonstrates the transformational potential of advanced NLP models.

1.14.3.6 Moral Points to Remember

Multimodal generative AI is showing us the ethical issues about privacy, justice, and prejudices, as with any other artificial intelligence technology.

Bias and Fairness: Multimodal generative models may involuntarily reinforce existing biases in the training program and lead to unfair or discriminatory results (Rai et al., 2020).

Careful assessments of information assets, model designs, and assessment measures are essential to ensure fairness and mitigate biases.

Privacy: When operating with sensitive or non-public data, the growing real-time artificial data provides a boost to privacy concerns. One of the main ethical enterprises is to create appropriate and valuable records for AI programs while respecting the privacy of individuals.

Misuse: Multimodal generative AI can be abused for very bad purposes, such as creating false information or spreading false information. The creation of legal guidelines and ethical standards are required for us so that we can stop abuse and inspire moral AI improvement. To implement these type of moral recommendations, guidelines and laws which can balance the innovation with social values and norms, information and strategies should be developed that include policy makers, researchers, corporate stakeholders, and civil society, so that moral concerns can be addressed in multi-modal generative AI.

Efficiency and Scalability: When many intelligent design processes process large, high-dimensional data from many variables, scalability and efficiency issues arise:

Processing Complexity: Due to the high-dimensional representation of the interaction model across multiple domains modalities is a difficult task, and training multimodal generative models requires a significant workforce. Virtual environment and interactive multimedia systems: This might require us to use algorithms and other frameworks which can be optimized, increase performance, and produce real results; entry of data, which can include text, images, audio, and video.

Real-Time Generation: This content generation method is required for many applications, including virtual environments and interactive multimedia systems. In Fig. 1.10 it is clearly shown about the real-time generation of images using KERA.AI. It is important to develop algorithms and frameworks that enable quality preservation, efficiency, and real-time generation (Anton, 2017).

Fig. 1.10 Real-time generation using AI

Data Handling: Scalable techniques and infrastructure for data storage, retrieval, and processing are required for the effective processing and management of multimodal datasets, which may contain text, images, audio, and video.

Solving the problems of efficiency and scalability requires developments in hardware acceleration, distributed computing, optimization strategies customized for multimodal generative AI, and algorithmic efficiency.

Ongoing Education: In multimodal generative AI, the term "continuous learning" refers to the model's ability to adapt to the flow of data over time and acquire new knowledge, allowing the model to evolve and operate in a favourable environment. Issues with continuous learning include:

Catastrophic Forgetting: When multitasking models are exposed to new information, they tend to forget what they have previously learned, resulting in reduced work.

It is very hard to develop strategies for stopping the memory loss and encouraging full lifelong learning. It is typical to develop adaptive learning algorithms and systems that can learn and grow (Goodfellow et al., 2014).

Adaptability: Over time, multimodal generative models must adapt to changes in the environment, task requirements, or data distribution. It is very important to create required or adaptive learning algorithms and also the systems which can learn and adjust them incrementally.

Data Efficiency: To maintain model performance and avoid overfitting or underfitting, continuous learning frequently involves learning from sparse or streaming data. This calls for the efficient use of resources and data (Sharma et al., 2022).

In order to enable multimodal generative models to continuously improve and adapt to changing settings, research in online learning algorithms, transfer learning techniques, memory-based approaches, and model consolidation methods is necessary to address issues in ongoing learning.

1.15 Learnings from the Early Days of Generative AI

A strong engineering philosophy for the construction of new technologies: At the beginning, we placed an engineering philosophy based on exploration above the construction of a mature final product. However, with the passage of time, we hope to build the maturity of the right characteristics and experiences, and encourage exploration by placing the technology of cognitive artificial intelligence into the hands of interested engineers and product managers. This includes tools such as LinkedIn's internal generative AI playground, allowing engineers to explore LinkedIn data using advanced OpenAI and other AI models. Based on what we saw on the playground, we brought together LinkedIn's largest internal hackathon engineers, with thousands of participants, who brought new perspectives and ideas. Initially, we wanted to embrace exploration rather than push for something perfect. Access and tools are fundamental: The interest in generative AI is very high, but technological exploration requires thoughtful access and the appropriate tools. We are committed to creating tools that inspire our engineers to explore and quickly identify ideas with strong potential for our members and customers.

1.15.1 Directions for Implementing Generative AI

There's no question around it: Generative AI is a cathartic innovation—a progressive device that will change how we are working. Be that as it may, embracing any innovation requires cautious thought. It's not fair a case of "futz around and see".

From all these years, if there's one thing that we learn is it's that change never begins with the innovation itself. We can't tell everybody that in your trade to begin testing with ChatGPT is absent since there are exceptionally genuine challenges and confinements. (For case, we can't have your deals group uploading client information to a device like ChatGPT since that seems to possibly uncover people's individual data.)

So, yes, we need individuals to be utilizing these kinds of instruments as rapidly as conceivable, but astutely. Keenly and with all the bolster they require to produce the best out of unused innovation. With that in intellect, few tips for effectively actualizing generative AI in your business.

1.15.1.1 Keep in Mind That, Generative AI Is Apparatus (Not a Substitute for Humans)

Generative AI will not be doing employments for us. Instep, we will utilize AI to do our employments successfully. To computerize more tedious and ordinary assignments. And to free up time for more valuable errands. As such, gen AI will not supplant the requirement for exceptionally human abilities like imagination, problem-solving, and building relationships but it will ideally simplify work and is better for people.

1.15.1.2 Develop a Good Mindset and a Rightful Culture

Embracing generative AI effectively requires a change in culture and attitude. It requires proper mentality that grasps interest, lowliness, versatility, and collaboration.

This generative attitude toward AI must start proactively and spread throughout the organization. This means you need an organizational culture where people continually challenge the status quo, embrace change (and disappointment), don't seek out challenges, and are open to learning new things. I'm talking about a culture where people are always asking questions like "How can we create more respect for our customers?" and "How can we create more respect for the world?" and "…how can we use innovation to achieve this goal?"

1.15.1.3 Contribute in the Right Abilities and Talent

In generative AI, hidden variable models are invisible conductors, creating different data structures and revealing new insights. Like sculptors, they discover hidden forms in data, pushing the limits of imagination and innovation. Embrace their transformative power and embark on a journey of discovery in the field of generative AI. Recent advances in latent variable models for generative AI include improving variables automatic encoders (VAEs), improving normalization flows, and the search for unconnected representations. Hybrid models, such as the combination of VAE and the generation of adversarial networks (GANs), show promising results. Applications include image generation, data expansion, drug discovery, anomaly detection and text generation in tasks such as dialogue generation and linguistic translation (Baidoo-Anu 2023). These advances lead to a more stable training, improved sample quality, and improved model interpretability. The late variable model plays an important role in the creation of artificial intelligence (AI) by capturing the basic structure and pattern of data. These models introduce hidden variables, inductive variables derived from observed data, so that the model can learn a more compact and expressive representation. The hidden variable model continues to evolve and its application to artificial intelligence applications spans various areas. These models allow for more effective representation learning, real-world sample generation, and enhanced interpretation, contributing to the development of artificial intelligence in various industries.

1.15.2 Generative AI Powering Innovation and Personalization

Generative AI, is a branch of artificial intelligence that focuses on creating entirely new content, rather than predicting future events. While predictive AI excels at forecasting trends and behaviours, generative AI takes a creative leap, generating novel text, images, music, and even 3D models.

Generative AI is transforming various fields.

Enhanced Content Creation:

Generative AI empowers content creators by automating repetitive tasks and generating fresh ideas. For instance, it can create product descriptions, social media posts, or even scripts based on specified styles and tones.

Personalized Experiences:

Similar to predictive AI, generative AI personalizes user experiences. Imagine a streaming service that curates a movie trailer specifically tailored to your interests, or a marketing campaign that generates personalized product recommendations—both are achievable with generative AI.

Revolutionizing Design:

Generative AI assists with graphic design, generating logos, mock-ups, or creating variations of existing designs based on user input. 3D modeling can also be accelerated with generative AI, allowing for the creation of intricate objects or environments.

The Future of Generative AI.

Generative AI is still evolving, but its potential is vast. As the technology matures, expect to see even more creative and innovative applications emerge across various industries.

1.16 Conclusion

In conclusion, to fully realize the promise of multimodal generative AI and guarantee its positive effects across a range of fields, interdisciplinary cooperation, creative research, and responsible AI development processes are needed to solve these issues and determine the field's future paths.

References

Ali, S. R. (2024). Constructing dreams using generative AI. *Proceedings of the AAAI Conference on Artificial Intelligence, 38*, 23268–23275.

Anton, D. K. (2017). Real-time communication for Kinect-based telerehabilitation. *Future Generation Computer Systems, 75*, 72–81.

Baidoo-Anu, D. (2023). Education in the era of generative artificial intelligence (AI): Understanding the potential benefits of ChatGPT in promoting teaching and learning. *Journal of AI, 7*(1), 52–62.

Bellovin, S. M. (2019). Privacy and synthetic datasets. *Stanford Technology Law Review, 22*, 1.

Dwivedi, Y. K. (2023). "So what if ChatGPT wrote it?" multidisciplinary perspectives on opportunities, challenges and implications of generative conversational AI for research, practice and policy. *International Journal of Information Management, 71*, 102642.

Fui-Hoon Nah, F. Z. (2023). Generative AI and ChatGPT: Applications, challenges, and AI-human collaboration. *Journal of Information Technology Case and Application Research, 25*, 277–304.

Gao, J. L. (2020). A survey on deep learning for multimodal data fusion. *Neural Computation, 32*, 829–864.

Goodfellow, I. J., Pouget-Abadie, J., Mirza, M., et al. (2014). Generative adversarial nets. In *Proceedings of the 27th International Conference on neural information processing systems* (Vol. 2, pp. 2672–2680).

Harshvardhan, G. M. (2020). A comprehensive survey and analysis of generative models in machine learning. *Computer Science Review, 38*, 100285.

Huzaifah, M., et al. (2021). Deep generative models for musical audio synthesis. In *Handbook of artificial intelligence for music: Foundations, advanced approaches, and developments for creativity* (pp. 639–678).

Kola, R. S. (2019). *Generation of synthetic plant images using deep learning architecture*. Blekinge Institute of Technology.

Rai, A. K., et al. (2020). Landsat 8 OLI satellite image classification using convolutional neural network. *Procedia Computer Science, 167*, 987–993.

Saxena, D., et al. (2021). Generative adversarial networks (GANs) challenges, solutions, and future directions. *ACM Computing Surveys (CSUR), 54*, 1–42.

Sharma, P., Singh, A., Singh, K. K., & Dhull, A. (2022). Vehicle identification using modified region based convolution network for intelligent transportation system. *Multimedia Tools and Applications, 81*(24), 34893–34917.

Singh, K. K., Mehrotra, A., Nigam, M. J., & Pal, K. (2013, April). Unsupervised change detection from remote sensing images using hybrid genetic FCM. In *2013 Students Conference on engineering and systems (SCES)* (pp. 1–5). IEEE.

Singh, A., Dhull, A., & Singh, K. K. (Eds.). (2024). *Blockchain and deep learning for smart healthcare*. Wiley.

Sohn, K. S. (2020). Artificial intelligence in the fashion industry: Consumer responses to generative adversarial network (GAN) technology. *International Journal of Retail & Distribution Management, 49*(1), 61–80.

Souza, V. M. (2020). Challenges in benchmarking stream learning algorithms with real-world data. *Data Mining and Knowledge Discovery, 34*(6), 1805–1858.

Chapter 2
ChatGPT and BERT: Comparative Analysis of Various Natural Language Processing Applications

Saranya M and Amutha B

Abstract In natural language processing (NLP), transformer-based models have grown in popularity over the past few years. In light of the fact that these models have demonstrated promising outcomes across a number of metrics for example. For the purpose of comparing and contrasting the GPT-4 and BERT language models in a variety of settings, this research makes use of a wide variety of natural language processing applications within their respective categories. This research makes use of a battery of classification tasks in order to investigate the architecture of the GPT-4 and BERT language models as well as their performance in a variety of different environments.

Keywords Sentiment analysis · Speech recognition · Question answering · Spam filtering · BERT · ChatGPT · Natural language processing

2.1 Introduction

Based on text, it may be able to understand human language. ChatGPT could search a sizable database to reply to questions from humans. Understanding users' unstructured, natural-language opinions and feelings about current events across a variety of platforms is the aim of "sentiment analysis" algorithms created in the data science field (Liu, 2010). One needs to read works authored by people with particular opinions in order to gain perspectives. Certain individuals may hold neutral, favorable, or unfavorable opinions. Finding this information requires sentiment analysis. Sentiment analysis uses text data—which can be gleaned from a variety of media sources, including social media—to determine public sentiment. Twitter is just one website that may provide information useful for sentiment analysis; there are many

Saranya M (✉) · Amutha B
Department of Computing Technologies, School of Computing, SRM Institute of Science and Technology, Kattankulathur, Chennai, Tamil Nadu, India
e-mail: sm2317@srmist.edu.in; amuthab@srmist.edu.in

A. Singh, K. K. Singh (eds.), *Multimodal Generative AI*,
https://doi.org/10.1007/978-981-96-2355-6_2

more. Tweets encompass a wide range of topics and internet users. This makes it possible to observe a subject and, through analysis of the gathered tweets, gain more precise knowledge about that problem. Since many people use Twitter to express their feelings and viewpoints, it is a data source for discussing ideas and opinions on a variety of topics (Bholane Savita & Gore, 2016). In addition, the Twitter network gives sentiment analysis access to a huge quantity of information. Thus, the data used in this investigation is sourced from Twitter. You could gather information about topics discussed on Twitter by using a crawling technique. A machine learning model that aims to identify sentiment is fed the datasets produced by the crawling process. In the field of natural language processing, BERT is used to represent words for sentiment analysis tasks. A multilingual-cased model based on BERT is used in several studies. Sentiment analysis is employed in these studies. In order to categorize attitudes into various buckets, this study used the BERT (Bidirectional Encoder Representations from Transformers) paradigm. The term frequency-inverse document frequency (TFIDF) model is used to explain the topics. The results of the investigation showed that classification performance was pretty good. To compare the translations, textual analysis was used. The best choice is number four. The study discovers that even though the translations use very different styles and languages, they basically send the same message when looking at emotional and semantic similarities.

Transformer is a well-studied neural network model. As it happens, it is (Han et al., 2021). ChatGPT has gone popular because of its powerful asking and answering features. A kind of pre-trained transformer that powers it is called generative pre-trained transformer (Qadir, 2022). Transformer building may use artificial intelligence from the fields of natural language processing and cybernetics (Jia et al., 2022). Training the dataset comes after pre-processing and BERT model modifications. Using three state-of-the-art benchmark techniques—CNN, TF-IDF, and support vector machine (SVM)—we assessed our BERT model. At our company, we use many different test procedures. Among these measures include accuracy, precision, recall, and f1-score. A question generator is one of such tools. Generate different kinds of questions on demand with ease using the automated creation of questions service. This kind of questioning includes "wh-" questions, matching, multiple-choice, and "true or false" ones. These context-dependent questions are produced by means of natural language processing and transformer models. Natural language processing is the study and understanding of text by computers. Text files and papers contain this kind of data, also referred to as linguistic data. Natural language processing (NLP) builds a model using the gathered linguistic data. Here a syntactic or semantic structure may be at work. Whereas the semantic structure clarifies the meaning that the text conveys, the syntactic structure clarifies the grammatical connections between the text. As defined by Verspoor and Cohen (Wei & Zou, 2019), a natural language processing system is a collection of linked components that take text as input and carry out different processing operations on it. As so is generated the right output text. Within the fields of neural networks and natural language processing, the transformer model generates an output sequence from an input sequence. Modern approach to handle natural language is the transformer model. When encoders modify input data to extract information, which is then sent on to the decoder, data transformations are made feasible.

Within a transformer is a multi-level configuration of decoders and encoders. These models can also understand the relationships between data points and use this knowledge to predict the intended outcome with accuracy. The Bidirectional Encoder Representations from Transformers (BERT) model is one encoder-based transformer language model that uses text embedding for sentence classification, extraction, and recognition (Bholane Savita & Gore, 2016). Many more transformer models exist, each with a unique way of handling input data. Using transformers, the Generative Pertained Transformer (GPT) model (Akkaradamrongrat et al., 2019) can generate, finish, and translate language. Through an evaluation of the connections between two sets of input text, this text-generation system forecasts the next word or phrase. This ability includes paragraphs and other longer textual sections. Finding the contextual connections between the two texts is how it does this. Expanding sentences is one of GPT models' primary objectives. Built on transformers, this language model is capable of handling jobs like translation, summarizing, and question answering. As said in (Francisca et al., 2021), this model is called the Text-To-Text Transfer Transformer (T5). T5 outperforms GPT and BERT for any sequence-to-sequence (seq2seq) operation since it can combine the encoder and decoder designs. That's the reason it outperforms those two algorithms. Teaching T5 to rebuild sentences starts with teaching it to identify missing or broken tokens.

As a use case, we have here a system that answers questions. Generative big language models may preserve some factual information and provide responses that seem natural, but they have their limitations. First of all, these algorithms produce tractable responses, which mean that it is impossible to easily ascertain their source. This is so because the models satisfied their goal of word prediction by using all the data at their disposal. This weakness may cause the model to go through "hallucination," in which it provides answers that appear logical but actually are incorrect. Second, although expensive, these models may be quickly trained to satisfy the needs of a particular place.

We presented an architecture that allows semi-automatic production of the data set needed to train a conventional quality assurance system. The system used a generative language model to increase the efficiency of creating the questions.

One could use the framework to create a company-specific data collection and then use these parameters to refine a Reader model. The experiment shows that using an already-tuned open-domain quality assurance approach produces better results than starting from scratch and fine-tuning. They show that, without modifying the neural network model, domain-specific, factually-grounded answers can be produced by combining the output of the Reader model with a generating language model using the prompting strategy. Another fascinating aspect is that the comments' naturalness was partially verified by human assessors.

Additionally available is voice recognition software. Speech recognition has advanced recently with BERT and connectionist temporal classification (CTC)-based ASR. By means of these transformers, feature extraction, audio modeling, language modeling, and decoding can all be carried out in a single network thanks to the self-attention mechanism. Conversely, the transformer architecture is a neural network model that gives relevant context top priority while processing data sequentially. Transformers have also demonstrated promising results in ASR, and it is expected that they will significantly influence next advancements in this area. The authors of (Benyamin & Ali, 2020) look into a number of text-based emotion identification methods using BERT and its derivatives. Alongside their thorough assessment of their approaches, contributions, and precise results, they draw attention to the flaws in their models. We mostly discuss ASR, even though BERT-based emotion detection is intriguing. The authors augment their work with models like ALBERT and ELECTRA even though they focus their study mostly on the BERT variants (XLM, Roberta, Distill BERT, BERT base, and BERT large). The work described in (Salman et al., 2022) mostly seeks to give a general introduction and tutorial on transformers, BERT, GPT, and the attention mechanism. This article covers transformer components, the attention mechanism, and the mechanism itself. All the same, the main focus of our thorough review is on BERT and CTC transformer applications, especially in ASR. Researchers examined the performance of transformer models including BERT, GPT, Roberta, and T5 on machine translation, question answering, and text categorization among other language tasks in (Qadir, 2022). Additionally discussed in the article are potential uses of these techniques in computer vision. Our study differs from the others, though, in that it concentrates on the components of the ASR domain—BERT and CTC transformers—instead of the other domains evaluated in the survey that was already described. Among the many speech-related fields covered in (Siddique et al., 2023) are automated speech recognition (ASR), Para-linguistics, augmentation, voice synthesis, and translation. As the survey authors explain, transformers face challenges in every one of these areas. We start our assessment with the same issues raised in the survey, but we also include more transformers—ELECTRA, ALBERT, and CTC Transformers—especially those that are part of the ASR framework. The next use, though, is scam filtering. Among other possible content kinds in these mass emails could be advertisements, phishing attempts, and money manipulation scams. Alarm has been raised by this issue for the most part of a century. Substance-, origin-, and rule-based spam filtering is one of the detection techniques proposed by researchers to address this issue. One such technique that can add or remove the sender's IP or email address from a whitelist or greylist to differentiate between spam and legitimate communications is origin-based spam filtering. While machine learning (ML) models are used to learn from data and detect spam, content-based filtering searches email content for trends. Application of machine learning (ML) models to content-based spam email filtering is the primary goal of this work. This class contains the models naive Bayes classifier, random forest, and support vector machine. The training dataset provides these models with information on particular spam characteristics from which they draw their conclusions. The naive Bayes classifier, for instance, searches for both spammy terms like "free" and "gift" and common phrases like "earn extra money." Spammers' methods of producing spam have developed concurrently with spam filters. To avoid detection, spammers are making their emails less feature-rich. As was mentioned in (Cheng et al., 2022), blocking "magic" spam words, substituting common spam phrases with their equivalents or purposefully using grammatical errors are some ways to get past content-based filters. Crafting excellent, spam-free emails is another approach to get past spam filters.

Recent advances in generative AI have made it feasible for computers to produce prose that is nearly human-quality, such as transformer-based large language models. These are the generative pre-trained transformer (GPT) models, which learn to generate new text from massive datasets given a signal. Author R. I. Karanjai looked into how well the GPT-2 and GPT-3 language models could generate phishing emails. Attackers could create excellent spam emails with these language models, but doing so would require a deep understanding of generative models and programming. Anyone can access generative models since OpenAI's ChatGPT chatbot was made available (OpenAI, 2023a). It makes use of the large language models GPT-4 and GPT3.5 from OpenAI. This has significantly improved artificial intelligence. Regardless of degree of coding experience, both groups quickly embraced ChatGPT. The chatbot can compose emails and provide nearly precise answers to user questions in any subject. It is also capable of continuing conversations because of this feature. Even with answers that violate moral principles, prompt engineering can still fool ChatGPT. The purpose of this work is to draw attention to one of the many shortcomings of generative AI models: the generation of harmful data. Having said that, they carefully investigate the potential abuses of ChatGPT, including the use of it for fast engineering to produce spam emails. Considering the current state of machine learning-based spam filters, there's a good chance that the generated emails will be recognized by at least one of them. They therefore use adversarial prompt engineering techniques, which involve synthesizing and rewriting pre-existing spam emails, to get past spam filters. By means of experimental data analysis on two standard email datasets and standard spam filter algorithms, they demonstrate how adversarial prompt engineering diminishes the efficacy of conventional machine learning models. They also provide research on possible defenses, like employing AI-generated text detection methods, like ChatGPT's rewriting of emails to the training set, and so on.

2.2 Literature Survey

2.2.1 Sentiment Analysis

The BERT model is preprocessed before it is fine-tuned. By comparing our BERT model to four state-of-the-art benchmark algorithms—SVM, TF-IDF, CNN, and others—we assess its efficiency. They make use of f1-score, accuracy, precision, and recall among other test metrics. Notable algorithms that use TF-IDF are Support Vector Machines (SVMs). An algorithm for classifying objects according to the linear distance between their labels is the support vector machine (SVM) (Ameliasari et al., 2021). Assume for the sake of argument that the two labels cannot be separated linearly. Consider sentiment analysis as one classification method that often yields outstanding results in text analysis case studies (Ghobakhloo & Ghobakhloo, 2022).The first algorithm lays out the training process for the SVM and TF-IDF models, beginning with the pre-processing stage. The Remove Stop words (Text) method removes words that are very similar. Words can also be simplified by stemming or lemmatizing. Lemmatizing is the process of returning a word to its root form; stemming removes affixes. One way to think of the procedure is as lemmatize. Convolutional neural networks (CNNs) are one form of deep learning in which convolutional kernels are used to learn features across the entire dataset (Erwin et al., 2021). Our models are evaluated by means of precision, accuracy, recall, and f1-score metrics. It takes only comparing all of the predictions to all of the guesses to determine the accuracy. You can find out how accurate a forecast is by picking the best mix of all the possible outcomes. The details of the real data are now being looked at by recall to find out what makes a prediction correct. "f1-score" is the square root of the sum of the accuracy and recalls scores in a certain situation.

2.2.2 *Text Summarization*

A text summary is the process of removing, condensing, or polishing the most significant information from a work to determine its main ideas or general meaning. Text summarizing is mostly done in two ways: extractive and abstractive. This method uses self-attention and CNNs to filter the global encoding of text, so resolving the problem of misalignment between the source text and the target summary. Pre-training and fine-tuning were introduced into GPT in what is generally regarded as a watershed in the field of natural language processing, thanks to the groundbreaking work of BERT (Kenton & Toutanova, 2019). Trans ABS (Liu & Lapata, 2019) applied a two-stage tuning approach to use generative summarization with the Bert SUM (Liu, 2019). We achieved optimality with our approach on three distinct datasets. CAVC (Song et al., 2020) applied a Mask Language Modeling (MLM) approach based on the BERT model to obtain these remarkable results. As shown in (Karn et al., 2022), one instance is the training of a bidirectional LSTM-based model for summary extraction using the Multi-agent Reinforcement Learning approach. After BERT was developed, BioBERT (Lee et al., 2020) used extensive biomedical corpora to pre-train it. Among the downstream medical tasks it outperformed, earlier methods were named entity recognition, association extraction, and question answering. The possible application of pre-trained language models in the biomedical field was examined in this work. Radiological diagnosis: Similar job is done by BERT (Rezayi et al., 2022). We pre-trained a BERT model and proposed a knowledge-infused few-shot learning (KI-FSL) method. Using subject expertise, this method interprets radiation clinical reports. ChestXrayBERT (Cai et al., 2021) summarized diagnostic reports by using a Transformer decoder and pre-trained BERT on a corpus related to radiology.

Pre-training and fine-tuning methods have been applied in an increasing number of Natural Language Processing (NLP) research projects since the

Transformer-based BERT paradigm was presented. The results demonstrate that the method that first trains on a lot of unlabeled data and then refines on a tiny fraction of tagged data works better. Consider the 120 million parameter GPT1 (Liu et al., 2018) model, trained by combining supervised fine-tuning with self-supervised pre-training prior to the use of BERT. Excellent results were obtained on question-and-answer and natural language inference tasks by direct use of the Transformer decoder. Next came other developments sparked by Google's revolutionary BERT paradigm, such as the Transformer encoder. To enhance performance even more, mask language modeling and next sentence prediction were also applied during the pre-training phase. Presently, 350 million parameters are in BERT-Large. Not long after BERT was published, the GPT-2 (Radford et al., 2019) model was also made public. This model, built atop the GPT-1 model, had much larger parameters and training dataset. Extra Large model GPT-2 has 1.8 billion parameters. The fact that the extended training dataset spared the researchers from fine-tuning allowed them to achieve exceptional performance in downstream tasks using a big language model. Constructed upon GPT-2, GPT3 (Brown et al., 2020) considerably expanded the data and parameters. With an 185 billion maximum parameter, it did noticeably better on downstream jobs. These results came from using GPT3. Furthermore, they suggested an unsupervised pre-training paradigm for few-shot cue training.

Larger LLMs are more generalizable than smaller PLMs. Without modifications, they reliably learn possible input text properties and effectively complete a range of downstream activities. Built atop the GPT-3.5 model, ChatGPT (Brown et al., 2020) is one well-known big language model core model example. ChatGPT allows simple human-machine interaction using conversational training data. Just two of the several industries that have greatly benefited from ChatGPT are education and healthcare. Excellent results have been obtained with the tool for many natural language processing tasks, such as text categorization, data expansion, summarizing, and more (OpenAI, 2023b; Liu et al., 2023a, 2023b; Shi et al., 2023a; Dai et al., 2023). Though it struggles with more complicated tasks, ChatGPT performs admirably most of the time. Thus, we propose Impression GPT, an iterative optimization method, to assist ChatGPT in its job of summarizing radiological reports. Timely engineering is a controversial and fast expanding area of study that may enhance LLM performance. One of the main ideas of prompt engineering is programming LLMs using prompts. Effective interaction with models such as ChatGPT requires this proficiency (Liu et al., 2023c). As demonstrated in the current work, prompts that point the model towards important input characteristics may enable more precise and consistent outputs. This is very important in tasks like language translation and text summarizing when the output quality is crucial. Many times, prompt engineering is regarded as a paradigm change in natural language processing. It shows promise for efficient prompt patterns even though it is still in its infancy (Tang et al., 2023). These patterns provide many of ideas, thus it is important to create prompts that provide value beyond text or code creation.

Developing questions that complement the models, though, might be difficult. Any little change to the instructions could have a significant effect on the model's accuracy. That is the reason selecting the appropriate stimulation is still rather challenging. There will usually be both automated template versions and prompts created by humans. First up, there's the handcrafted prompt, designed especially to point LLMs in the direction of input. These reminders instruct the model in how to approach the current task and which data to prioritize (Wang et al., 2023a).

Human prompts function best when the input data is well specified, as is the format or structure of the intended output. Use of hand-crafted signals to direct the model to concentrate on particular areas of the entering data is common practice in the medical industry, for instance, where interpretability is essential. Handwritten instructions were used, for example, by medical text security specialists to teach the model how to recognize and extract sensitive information from medical records. Our aim was to give serious attention to the moral problems with medical data (Liu et al., 2023c). Usually, human recommendations are required to enhance the models' performance in several domains. This is so because they give the model a more efficient framework to approach the present work methodically and focused.

Automated Prompt Operated by Templates could never write off manual prompts because they have so many applications. For instance, writing prompts requires work and a certain amount of skill. Little changes in the prompts could significantly affect the model's predictions. For individual users, this is especially important because it could be difficult to provide sufficient manual guidance for intricate tasks (Jiang et al., 2020). Many methods to automate prompt design have been developed by researchers in an effort to solve these problems. Numerous signal types could be used to teach language models to finish jobs more successfully.

2.2.3 Question Generation

Natural language processing (NLP) can automatically turn text into questions. Answering each type of question required many algorithms and methods. Khokhlov and Reznik (Rodriguez-Torrealba et al., 2022) used LSTM neural networks. They answered "Wh-" questions, true/false questions, and fill-in-the-gaps questions. We grouped phrases by nouns, verbs, and adverbs first. This lets us learn the question's premise and create a tree object. The four steps matched the fill-in-the-blanks question. Pre-made and gap queries were used to train an LSTM model. The second step used natural language processing to extract nouns, verbs, and pronouns from the tree object. A Stanford parser removed noun and adjective phrases. These became blanks afterward. Then stop words were removed. Filling in numbers and blanks was the last priority. Three steps were needed to make true/false questions: Replace all sentences on the list with lies except one. In sentences, change negative words to positive ones and vice versa. Third, remove numbers from the phrase. Last, the "Wh-" question had three steps. Goutham (Devlin et al., 2018a) suggested that true or false questions change depending on the situation. OpenAI's GPT-2, BERT, and Berkeley consistency parsing are used. The original sentence was changed to remove falsehoods and keep the truth. Write down what you saw or heard before lying. Consistency parsing separated each sentence next. Then, the GPT-2 completion model was given the broken sentence and told to form a new phrase to join the parts. Make a list of GPT-2 model flaws. BERT is then used to compare false and true statements. Each false claim is scored by how similar it sounds. Then, the false claims are ranked by likelihood and the most likely one is chosen. It was also better because the questions used six models and the GPT-2 transformer model. Rodriguez-Torre alba and others (Androutsopoulos, 2023) set up a 10-option MCQ system using the T5 transformer model. You can ask a multi-answer question five ways. First, Wikipedia article parts were assembled. They were then changed to make useful statements. Questions will be based on this. After making phrases, question and answer sheets were made. The question and keyword were created by trained T5 models. Now it was time to list obstacles and incorrect answers. We reinstated the questions, answers, and background text. The distractions were scored. Compare the cosine of each word vector to the key response to get scores. The answers to each stage were combined to create an MCQ.

2.2.4 *Automatic Speech Detection*

The transform encoder-based BERT model is used for NLP operations before training. Devlin and colleagues created it in 2019 (AbdulNabi & Yaseen, 2021). First, learn the language basics, then improve. The second type is better for assessing emotions, answering questions, and summarizing. BERT pre-trains using several methods. NSP and masked language modeling are examples. To train for MLM, hide words in phrases and use their training context to reassemble them. The NSP checks if the second phrase comes after the first so BERT can understand how they fit together. BERT was trained on 13 TB of English Wikipedia and book corpus text data. Changing the BERT output layers after pre-training fine-tunes the model for a task. Fine-tuning works better here because it only learns model parameters that don't change the outcome. The BERT package has two versions: base and large. BERT-base has 135 million parameters. It has 12 transformer encoder blocks, 768 hidden layers, and 12 head self-attention layers. BERT-large has over 360 million parameters and 28 transformer encoder blocks with 32 head self-attention layers. BERT-large has better accuracy than BERT base but requires more computer power. BERT has several major issues. Because it was designed for monolingual classifications, it works best with one language. BERT could be modified to perform multilingual tasks, but it may not be as effective.

2.2.5 *Spam Filtering*

Generally speaking, there are two types of data used by the existing spam classification algorithms: The first group employs postal characteristics for identification purposes. Using supervised classifiers, such the SVM model, for classification, this technique takes mail title features and applies it (Google, 2023). Some studies have also included the cumulative relative frequency of each spam feature to enhance the performance of supervised classifiers (ZeroGPT Website, 2023). A number of classification methods are well-suited to the feature-based spam detection dataset due to its multidimensional structure. Researchers have therefore compared the efficacy of several machine learning models in spam identification tasks (Writer.com Website, 2023), including models such as support vector machines (SVMs), deep neural networks (DNNs), and k-nearest neighbor (KNN) among them. There is a second group whose primary focus is email content identification. Some examples of effective methods for filtering and classifying email content are the relaxed online SVM model [54] and naive Bayes classifiers (Mujahid et al., 2021). Over the past few years, spam detection has increasingly made use of deep learning algorithms. One area where CNNs have demonstrated promise is in spam detection. The use of BERT model transfer learning for spam categorization has shown very accurate results. Machine learning classification approaches such as random forest and support vector machines have not been able to match the performance of Long Short-Term Memory (LSTM) models on short messages (SMS) datasets (Tran et al., 2021). Based on Vaswani's 2017 Transformer model, the GPT model is one of the most influential models in the field. A large number of individuals are already using the anticipated GPT2 and GPT3. They did an excellent job with everything from content creation and question and answer engagement to translation, summarization, and conversation. In 2022, ChatGPT was introduced as an upgraded variant of OpenAI. The GPT-3.5 model, fine-tuned for conversational tasks, is its basis. Control over chat policies, conversation history input/output, and other features are all a part of it. Natural and fluid conversations with humans are also within ChatGPT's capabilities. Chat jobs are where it really shines. One method of text classification is spam recognition. Researchers have demonstrated that the model works well on specific text categorization tasks, and the general public agrees that ChatGPT does a good job overall. It is possible to use GPT to aid in classification decision-making by using ChatGPT to extract refined and structured knowledge from the knowledge graph, which is a feature of classification jobs. Furthermore, with the right prompt design, it may be used directly for text classification, such in agricultural, business news, or medical conversations.

ChatGPT can classify text as true or false, which could help find spam, according to research. ChatGPT can correctly identify and group textual emotions, feelings, and ideas, according to recent studies. GPT also did a great job determining how people felt about diseases in publications and texts (Muneshwara et al., 2022), suggesting that ChatGPT can handle many sentiment analysis tasks in many different areas.

To get the most consistent and useful results when using ChatGPT to categorize text, carefully develop the suggestions and algorithms. Because big language model results are uncertain. If necessary, samples could be added to the model to guide it. Chain-of-Thought (CoT), One-shot, and Few-shot prompting are examples (Parimala et al., 2021).

2.3 Methodology

2.3.1 ChatGPT Based Sentiment Analysis

The ChatGPT language model, created by OpenAI, can compose text that is very similar to human speech. People's native tongue. As an added bonus, it can engage in genuine conversations with humans (Aslam et al., 2022a). The ChatGPT platform has a huge advantage in consumer sentiment analysis because of its capacity to understand and handle human language efficiently (Aslam et al., 2022b). The capacity to assess consumer mood is one such benefit. A wide and extensive dataset was used to train the model so that it can understand the contextual, emotional, and complicated elements buried in customer contacts. According to research in the field of sentiment analysis, ChatGPT can tell if a consumer is being positive, negative, or neutral when they submit words (Araujo et al., 2022). Thanks to this function, ChatGPT can tell the difference between neutral, positive, and negative emotions. As a consequence, companies may use ChatGPT as an analytical tool to find significant outcomes from customer data. Companies can make important discoveries because of this.

Correct implementation of a number of sequential tasks is necessary to include ChatGPT into consumer sentiment research. Companies must, for example, first compile and organize relevant customer data. Possible formats for this data include product reviews, social media comments, surveys gauging customer satisfaction, and audio recordings of customer service agent conversations. Better results from sentiment analysis can be obtained by compiling a more varied data. After data collection, step three is training the ChatGPT model with the appropriate training dataset. Every sample of a customer text that is given to the model during training has a label that designates its mood—positive, negative, or neutral. Following its learning of the underlying patterns in the supplied dataset, the algorithmic model will establish a link between textual content and the sentiment that corresponds to it. The ChatGPT model could begin easily analyzing customer sentiment after the training is over. Businesses can feed textual data into computational models to extract consumer sentiment from it. Figure 2.1 shows the working principle. When a customer writes a review expressing great satisfaction with the good or service, an algorithm could be able to identify the positive message they are sending. By comparison, the algorithm will see a bad attitude as a sign of dissatisfaction from the client.

2.3.2 Transformer Based Sentiment Analysis

It takes accurate execution of several sequential tasks to incorporate ChatGPT into studies on consumer sentiment. Companies have to, for instance, first gather and arrange pertinent client information. Product reviews, remarks on social media, customer satisfaction surveys, and audio recordings of customer service agent

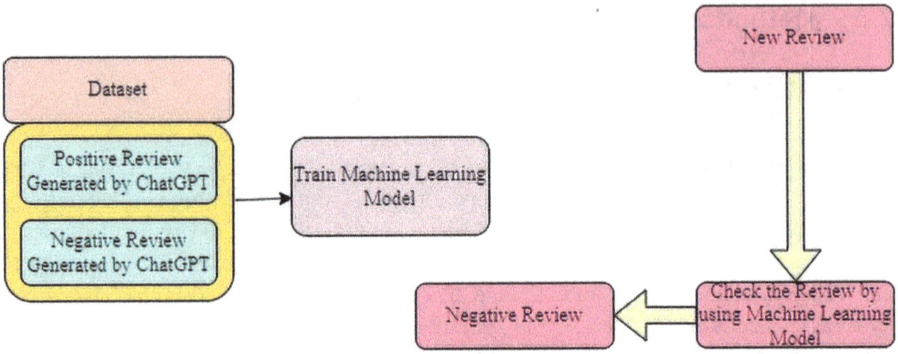

Fig. 2.1 Working principle of sentiment analysis by ChatGPT

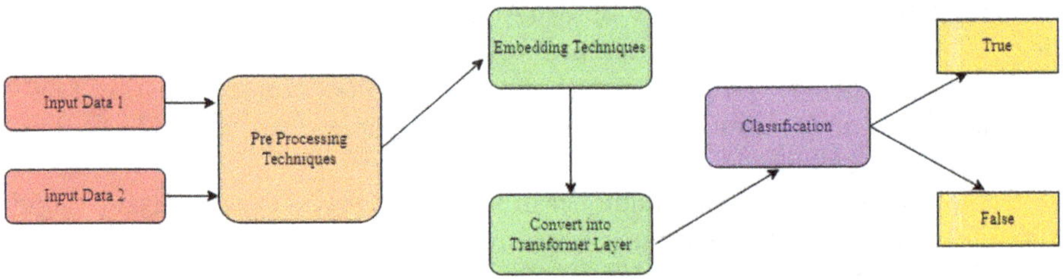

Fig. 2.2 Working principle of BERT transformer model in sentiment analysis

conversations are some of the possible formats for this data. Building up a more varied set of data will yield better sentiment analysis results. Training the ChatGPT model with the suitable training dataset comes after data collecting. Every customer text sample the model receives during training has a label indicating whether it is positive, negative, or neutral. The algorithmic model will link textual content and the sentiment that goes along with it after learning the underlying patterns in the provided dataset. After the training, the ChatGPT model could start easily analyzing consumer sentiment. Companies can use textual data to feed computational models in order to get customer sentiment. Figure 2.1 illustrates the practical idea. An algorithm could be able to recognize the positive message a customer is sending when they write a review expressing their great satisfaction with the good or service. By comparison, the algorithm will see a bad attitude as a sign of dissatisfaction from the client (Fig. 2.2).

An extensive pre-trained vocabulary is used by the BERT to produce input ids, or numerical values of the text being entered. The process gives each token a unique identifier starting with a set of tokens created from the whole input text tokens. All that input ids are numerical representations of the processed text. Much as an attention method, BERT's input mask keeps input text tokens apart from padding. The model can ascertain which tokens in the input sequence are evaluated and which are not by use of the input mask. Section IDs are tokens used to distinguish between several sentences. It is concatenated once linked to the BERT Keras layer.XLNet, designed like BERT, was made public by Ashish Vaswani. Among the auto-encoder models are the BERT and the autoregressor models are the XLNet (Zhang et al., 2018). Dependencies amongst tokens in a sentence will be ignored by the BERT model. By using permutation-based training goals instead of mask-based ones, XLNet overcomes this issue. The permutation-based objective allows XLNet to define the dependencies using each token in a paragraph. Transformer-based Robustly Optimal BERT pretraining (RoBERTa) is a paradigm found in several natural language processing applications (Mandl et al., 2020). Development on it started in 2019. One BERT model variant that helps to get beyond its limitations is called RoBERTa. Whereas 160 billion words have been taught to RoBERTa, just 3.3 billion words have been taught to BERT.

2.3.3 Question Generator

Ids, or numerical values of text, are generated by the BERT using a large pre-trained vocabulary. Starting with the entire input text tokens, the process assigns tokens unique identifiers. Input ids are numerical representations of processed text. Like attention, BERT's input mask separates text tokens from padding. The input mask lets the model decide which tokens in the sequence are evaluated. Section ID tokens identify sentences. Concatenation occurs when it joins BERT Keras. Ashish Vaswani launched XLNet with a BERT-like design. Auto-encoder models include BERT and autoregressor models like XLNet (Zhang et al., 2018). BERT ignores sentence token dependencies. XLNet avoids this by using permutation-based training objectives instead of masks. XLNet can define dependencies using every token in a paragraph due to its permutation-based goal. Many natural language processing applications use transformer-based Robustly Optimal BERT pretraining (RoBERTa) (Mandl et al., 2020). Development began in 2019. RoBERTa helps BERT models overcome constraints. BERT learned 3.3 billion words, while RoBERTa learned 160 billion. RoBERTa is ideal for large data sets due to its large batch sizes and fast training. BERT masks statically, but RoBERTa masks dynamically.

2.3.4 Chat GPT in Question Answering System

Cleaning and preprocessing the data corpus first guarantees consistency and eliminates superfluous information. With a pre-trained language model such as ChatGPT, every corpus document or text passage is embedded. Similarity searches are best done with generated embeddings in a vector database. In query processing, a trained language model transforms a user query into an embedding, or string of numbers. Vector database similarity searches and query embedding can find the best matches.

Prompt engineering is the process of making a prompt that includes both the user's query and the content that was returned. Figure 2.3 shows how to answer a

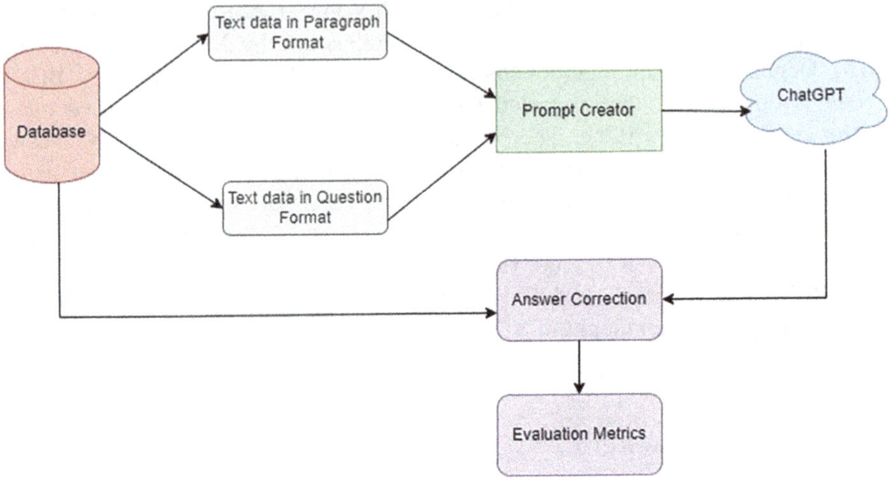

Fig. 2.3 Question answering system based prompt engineering approach

prompt-based question. If you give ChatGPT this instruction, it might come up with a good answer. The answer is given by ChatGPT when the generated prompt is fed into it. This is the last step in the process of making a response.

2.3.5 BERT in Question Answering System

With the input question and passage as its two parameters, the BERT algorithm creates a single packed sequence for the Question Answering System software. When the embeddings for tokens and segments are added together, the input embedding may be determined. The question and paragraph both end with a [SEP] token, and a [CLS] token is appended to the input word tokens at the beginning of the question for token embeddings. Those two tokens are tucked within the paragraph's last paragraph. When using segment embeddings, you can tell if a token represents Sentence A or Sentence B by adding a marker to it. This allows the model to recognize distinct sentences. Illustration of the BERT Architecture in Fig. 2.4. The following example shows that the question has all of the tokens indicated with the letter A, but the paragraph contains all of the tokens marked with the letter B.

2.3.6 ChatGPT in Text Summarization

With input of a short text or a lengthy document, ChatGPT can produce an understandable and succinct summary that summarizes the key ideas of the original material. We have gathered an overview of the key ideas of texts to assist you in

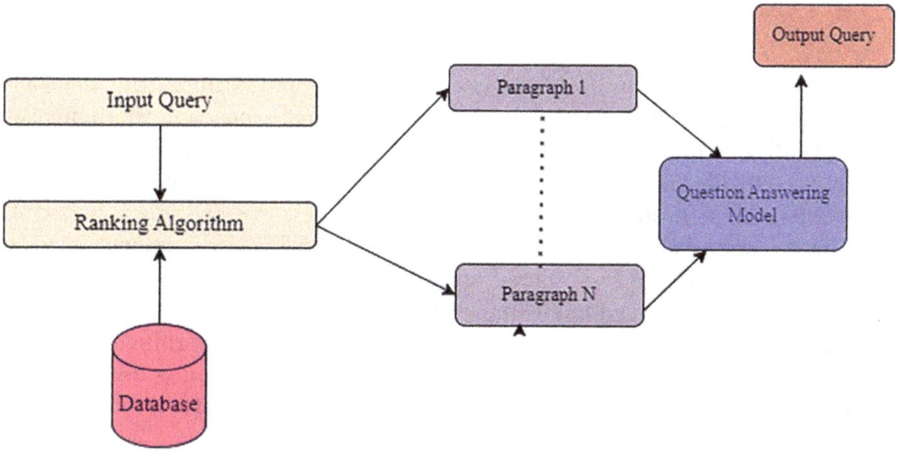

Fig. 2.4 Architecture diagram of BERT question answering system

summarizing them. One can efficiently manage long-form material by using ChatGPT to examine and extract intricate information from big papers. It is thus an excellent tool for professionals, students, and researchers who have to handle a lot of material. Moreover, ChatGPT can produce logical, understandable summaries so you can be sure that the key ideas will be made very evident. Encourages time and effort saving: Because ChatGPT can automate the summary process, time and effort are saved. Long documents will not need to be read and summarized by hand any longer; ChatGPT can do it quickly.

2.3.7 BERT in Text Summarization

This work tries to investigate DistilBERT, a compressed version of BERT (Vaswani et al., 2017) that is used to produce a summary while preserving the performance of the model and to reduce the computation required for training an extractive summarizer so that it can be deployed on devices with limited resources. The idea is to retrain the model, log the ROUGE scores, and update the BERT Sum design such that the DistilBERT encoder replaces the BERT-base encoder. Utilizing the GPU capacity on the Google Colab laptop, this experiment was carried out. DistilBERT aims to demonstrate that, by applying knowledge distillation in the pre-training stage, a BERT model can be 60% faster and 40% smaller while retaining 97% of its language understanding capabilities. Bucila et al. (Radford et al., 2018; Devlin et al., 2018b; Brown et al., 2020) first proposed knowledge distillation, a compression method. Here a bigger model known as "the teacher" gives a smaller one known as "the student" instructions on how to act.

2.3.8 Speech Recognition

The paper previously mentioned proposes a two-stage architecture using the CNN and BERT encoder. The proposed approach looks for hate speech in Hindi that can lower the dignity of the impacted people or groups. More especially, Hasoc 2021 (Liu et al., 2023d) and 2020 (Ouyang et al., 2022), to give an example, this work trains and tests the proposed model using Hindi datasets to make sure it is flexible to a larger range of situations. With the pre-trained BERT (Bidirectional Encoder Representations from Transformers) encoder model, the embeddings were built from the textual data to capture textual properties. Furthermore, this convolutional neural network fed these embeddings. A dropout layer gives learning stability to the model. Lastly, a sigmoid layer is used to assist in determining the degree of hatefulness of the individual. Among the several significant achievements of this work are:

We developed a hate speech detection Hindi language binary classification model using the BERT encoder and a convolutional neural network. The data were improved by the oversampling technique to level the playing field between the majority and minority classes. Results from the proposed model were contrasted with those from the transformer-based BERT classifier model for the balanced and unbalanced versions of the Hasoc 2020 and 2021 Hindi task datasets.

2.3.9 Spam Filtering in ChatGPT

Among the foundational models to emerge from Vaswani's 2017 Transformer model (Shi et al., 2023b) is the GPT model (Zhao et al., 2023). Another example is the BERT model proposed by (Lopez-Lira & Tang, 2023). The anticipated GPT2 and GPT3 are already being used by many (Wang et al., 2023b; Susnjak, 2023). Everything from creating content and participating in Q&A to translating, summarizing, and having conversations went really well.

2022 saw the release of ChatGPT (Sharma et al., 2023), an improved version of OpenAI. Its foundation is the conversational task-tuned GPT-3.5 model. Among its features are control over chat policies, input/output of conversation history, and other. Additionally possible with ChatGPT are natural and flowing conversations with people. That really comes through in chat jobs.

Recognition of spam is one approach to text classification. Everyone concurs that ChatGPT is a strong text categorization model, and research has demonstrated its efficacy on a number of tasks (Brown et al., 2020). A characteristic of classification jobs, ChatGPT can be used to extract structured and refined knowledge from the knowledge graph, so supporting classification decision-making. There is a reference at (Borji, 2023). More precisely, with the appropriate prompt design, it can be applied for instant text categorization in agriculture (Reiss, 2023), business news (Korini & Bizer, 2023), and medical discourse (Zhao et al., 2021).

The few studies investigating the use of ChatGPT for text binary classification could be one source for spam detection. An accurate identification and categorization of emotions, feelings, and ideas contained in text has been demonstrated by a number of recent studies. Furthermore, GPT did well in evaluating the emotions in tweets and medical articles (Liu et al., 2023e), indicating that ChatGPT is capable of handling a wide variety of sentiment analysis jobs for texts in many fields.

Carefully developing the recommendations and related algorithms is necessary to get the best consistent and efficient results when using ChatGPT for text categorization tasks. Reason being, the most precise results might not always come from massive language models. If necessary, the model might include a few examples to provide direction. Among the formats are One-shot Prompting, Few-shot Prompting, and (Singh et al., 2024a) Chain-of-Thought (CoT).

2.3.10 Spam Filtering Using BERT

Tokenizing vocabulary files are then used to convert all input tokens into integer IDs. These IDs and a numerical matrix are next fed to the BERT model. The BERT model transforms each token input into a vector of size equal to the BERT hidden size, in which the matching element in the related matrix is set to one for real inputs and zero for padded inputs. From our work with BERT (Singh et al., 2024b), we are aware that every encoder has an attention layer that symbolizes the token in context. We next feed the classification layer, which decides whether or not the arriving email was spam, this final hidden state.

2.4 Result and Discussion

Using standard sentiment analysis criteria, we put both models through their paces on a held-out test set. A number of natural language processing systems use the Squad 2.0 dataset to get results, and these metrics include accuracy, precision, recall, and F1-score. Designed specifically for tasks like sentiment analysis, BERT is a transformer-based model. To successfully gather contextual information, attention techniques are employed. ChatGPT, in contrast, is mostly designed for text generation duties; nevertheless, it can also do sentiment analysis adequately with few tweaks. This system's architecture isn't sentiment analysis-optimal, in contrast to BERT's. BERT's training takes a lot more time and computational resources than ChatGPT's because of its wider architecture and the nature of its pre-training objectives. Because ChatGPT is not a heavy model, it requires less computing power and can be trained relatively faster.

Considering that text production is its primary objective, ChatGPT does rather well, even if BERT typically does much better in tasks requiring sentiment analysis. The capacity of ChatGPT to understand context and generate logical text may be the reason for its excellent sentiment recognition performance in conversational settings. Better interpretability is made feasible by BERT's attention mechanism (Rai et al., 2020), which makes it obvious which sections of the input text are more crucial for sentiment prediction. Because there isn't this obvious attention mechanism, ChatGPT makes less obvious decisions than it ought. 89.2% accuracy was achieved by BERT, 87.5% by ChatGPT. Both of these results, one could contend, make sense. The accuracy rates with BERT were, for the positive, negative, and neutral conditions, 84.2%, 84.5%, and 84.6%, respectively. 85.3% of positive sentiment, 89.4% of negative sentiment, and 83.6% of neutral sentiment were remembered by the BERT method. For positive sentiment, ChatGPT was accurate to 88.4%; for negative sentiment, to 83.8%; and for neutral sentiment, to 81.7%. With relation to ChatGPT, recall rates for neutral mood were 80.2%, for negative mood 84.9%, and for good mood 85.3%. The F1-score put neutral mood at 83.2%, negative sentiment at 87.9%, and positive sentiment at 89.3%. With scores of 81.2% positive, 83.4% negative, and 80.6% neutral, ChatGPT handled the sentiment analysis rather well.

For training and evaluation, we used a sample dataset (the Stanford Question Answering Dataset) including conventional question-answering questions. The compilation covers many topics and includes questions and answers about those extracts. The passages and questions could be tokenized so that BERT and ChatGPT could use the input forms. This was one preparation method among others. We used the processed data when the time came to fine-tune BERT and ChatGPT for the question-answering task. Whereas BERT learned from a span prediction head, ChatGPT learned from its most recent unknown state using a linear layer. Both approaches (Raghunath et al., 2022) were applied in model training. We took care to train the models with same hyper parameters and settings in order to enable an objective comparison. Comparison of the results became feasible as a result. We assessed both models according to the correctness and consistency between predicted and real answers using industry-standard question-answering metrics, such F1-score and Exact Match (EM). The exact match (EM) results showed ChatGPT scored 75.2% and BERT scored 82.5%. We find that ChatGPT and BERT had, respectively, F1-scores of 82.8% and 89.7% . The BERT model based on transformers was inspired by these kinds of queries. One method it uses to effectively log the relationships between the textual words and the query is self-attention. ChatGPT may not be very good at answering questions because it is mostly trained for text generating jobs and cannot clearly show the relationship between the given passage and the question. Generally speaking, BERT is the best option even though it requires a lot of processing power for applications that highly value accurate query results. When computing resources are limited or a model that can provide consistent answers across multiple jobs is required, ChatGPT can still be a viable choice even if its performance is a little worse. If you want to replace BERT or ChatGPT speech recognition, look elsewhere. Work involving text is more appropriate for their designs than audio processing.

We evaluated the speech recognition flexibility of BERT and ChatGPT by converting audio data into token sequences. Concurrently optimizing BERT and ChatGPT, we used a sequence-to-sequence (Seq2Seq) architecture combining attention processes. To present an objective analysis, we trained both models with the same hyper parameters and settings. Using commonly accepted speech recognition metrics such as Accuracy, Character Error Rate (CER), and Word Error Rate (WER), we assessed the two models.

WERs generated by BERT were 18.5% and those by ChatGPT 22.8%. These are both shortened to WER. By contrast, BERT had a CER of 8.2% and ChatGPT of 10.5%. At 8.2%, BERT met its goal. The accuracy ratings for ChatGPT were 77.2% and for BERT they were 81.5%. The original goal behind the development of BERT and ChatGPT was not voice recognition tasks. Even if they could be modified for tasks like these by seeing audio pieces as sequences of tokens, their designs most likely don't make full use of the temporal correlations and acoustic characteristics in voice signals. Speech recognition is one of its many uses; its bidirectional self-attention mechanism helps to collect contextual information that could aid in word detection inside sentences. Being mostly trained on text generation, ChatGPT may struggle to comprehend the subtleties of context and long-range relationships found in spoken signals. One of the several goals of the text data pre-training of BERT and ChatGPT is next sentence prediction, another is masked language modeling. The auditory characteristics of spoken language could not be clearly connected to these goals. Their performance on speech recognition tasks is probably going to suffer since the pre-training objectives and the properties of speech signals are not in line. Particularly when trained on speech recognition tasks, BERT and ChatGPT demand a lot of processing power. Unfortunately, these architectures perform less well in processing audio data than specific models made for speech recognition. Performance and efficiency may therefore both suffer.

In voice recognition tasks, BERT and ChatGPT may yield results that commensurate with specialized models like deep neural networks (DNNs) or convolutional neural networks (Singh et al., 2013) (CNNs) created especially for audio processing. Still, they could be sensible options if there are few computing resources available or if a single model that can process text and speech is required.

The emails and the labels indicating whether or not they are spam are part of a dataset that our group has compiled. One of the preprocessing steps to prepare the dataset for BERT and ChatGPT input was to tokenize the emails and extract text features. We optimized ChatGPT and BERT for the spam filtering task using the preprocessed dataset. The BERT model used a classification head; the ChatGPT model was trained over its most recent hidden state using a linear layer. With the same hyperparameters and settings, we trained both models to allow an objective comparison. On a hidden test set, the two models were evaluated using F1-score, accuracy, precision, and recall—traditional spam filtering metrics.

At accuracy of 96.5% and 93.8%, respectively, BERT and ChatGPT both attained high performance levels. For the detection of spam emails, BERT achieved a 97.2% accuracy rate and for non-spam emails a 95.7% accuracy rate. Results: BERT achieved a recall of 96.8% for spam emails and 94.5% for non-spam emails; ChatGPT a precision of 92.3% for non-spam emails and 94.5% for spam emails.

ChatGPT found spam emails 91.5% of the time and non-spam 92.3%. BERT scored 96.9% and 95.1%, better at distinguishing spam from non-spam emails.

ChatGPT was an effective email filter, detecting 91.9% non-spam and 93.4% spam. BERT and ChatGPT have different goals and designs before training. BERT is a transformer-based model for categorization, while ChatGPT is trained for text production. They both filter spam well, but BERT does better than ChatGPT. BERT uses its bidirectional self-attention mechanism to gather contextual data to understand spam emails. ChatGPT can generate comprehensible language, but it may not understand spam emails' complex patterns and context-specific components. Thus, its performance may suffer.

BERT and ChatGPT pre-trained on large text corpora, but their goals differed. Training with masked language modeling and next sentence prediction gives BERT contextual representations for spam filtering and other downstream tasks. Therefore, BERT was pre-trained. ChatGPT's pre-training focuses mostly on language modeling, which may not help identify spam email traits. ChatGPT pre-trains language modeling.

BERT usually outperforms ChatGPT at spam filtering, even though its larger architecture and pre-training objectives require more training time and computing resources. Because ChatGPT is lighter, it can be trained faster and with less computing power, despite slightly worse performance.

BERT, though CPU-intensive, may be best for spam filtering applications that need high accuracy. ChatGPT is still a good choice with respectable performance when computing resources are limited or a lighter model is desired. We managed a dataset of full and condensed lengthy articles. The dataset preparation ended with tokenization and conversion of documents and summaries into BERT and ChatGPT input forms.

We improved BERT and ChatGPT for text summarization using the preprocessed dataset. BERT was trained using Sequ2Seq architecture with attention mechanisms. A similar training architecture included a linear layer above ChatGPT's latest hidden state. For an objective comparison, we trained both models with the same hyper parameters and settings.

Model performance was evaluated using standard text summarizing criteria. They included the ROUGE (Recall-Oriented Understudy for Gusting Evaluation) scores—ROUGE-1, ROUGE-2, and ROUGE-L. BERT scored 0.42 on the ROUGE-1 scale, outperforming ChatGPT's 0.38. ChatGPT scored 0.18 and BERT 0.22 on ROUGE-2. BERT scored 0.40, ChatGPT 0.36 on ROUGE-L. BERT and ChatGPT were trained with Seq2Seq designs with attention mechanisms to summarize text. In contrast, BERT's bidirectional self-attention mechanism provides more contextual information about arriving content. Implementing this feature may make it a better summary generator than ChatGPT.

BERT contextualizes text with bidirectional attention. This may clarify the paper's main points and sentence structure. Longer articles with contextual subtleties and long-range dependencies are harder for ChatGPT to summarize. Yes, even though ChatGPT has good sentence structure. Though pre-trained on large text corpora, BERT and ChatGPT had different training goals. Next sentence prediction and masked language modeling train contextual representations for text summarization.

Machine learning algorithm BERT. Document structure and key information may not be covered in ChatGPT's pre-training on language modeling. BERT summarizes text better than ChatGPT despite its time- and computer-intensive architecture and pre-training objectives. Though less performant, ChatGPT can be trained faster and with fewer computing resources due to its lighter model.

BERT produces high-quality summaries but is CPU-intensive. ChatGPT could replace limited computing resources or a lighter model. Question pairs represent dataset paragraphs or context phrases. Context paragraphs and phrases were tokenized and converted into BERT and ChatGPT-compatible input formats to improve question-generating. BERT learned its Sequ2Seq architecture with attention, while ChatGPT added a linear layer above its latest hidden state. All designs in this collection employ attention processes. For an objective comparison, we trained both models with the same hyper parameters and setup. BLEU scores were used with other metrics to evaluate performance. BERT had 0.52 BLEU and 0.48 ChatGPT. Attention-process-incorporating Seq2Seq designs improved BERT and ChatGPT question-generating. BERT gets a more complete input context from its bidirectional self-attention mechanism. Instead of ChatGPT, this may generate observational and situational questions. Its bidirectional attention mechanism gives BERT contextual information from input text in both directions. Understanding statement-context relationships is possible. BERT and ChatGPT were pre-trained on large text corpora, but their training objectives were different, so ChatGPT may struggle to understand context and generate less relevant or coherent queries. BERT can learn contextual representations for question generation by pre-training with masked language modeling and next sentence prediction. ChatGPT's pre-training only models language, so it may not help you understand context paragraph structure and key information. ChatGPT courses use language modeling. BERT generates questions better than ChatGPT, but its larger architecture and pre-training objectives make training it more time- and resource-intensive. A lighter model, ChatGPT, trains faster with less computing power despite performing worse. BERT may be best for precise, contextually relevant queries, but it requires a lot of processing power. ChatGPT may work well with limited computing resources or a lighter model.

2.5 Conclusion

Our main objective is to compare ChatGPT's application results to the updated BERT model. Researchers may read faster with ChatGPT. Investigation concludes. Data is converted into readable text and sentence-level measurement is similar to the refined BERT model. It may succeed. Worse, other platforms may be more innovative, robust, and effective at spreading knowledge than ChatGPT. This is especially true for data dissemination.

References

AbdulNabi, I., & Yaseen, Q. (2021). Spam email detection using deep learning techniques. *Procedia Computer Science, 184*, 853–858. The 12th International Conference on Ambient Systems, Networks and Technologies (ANT) / The 4th International Conference on Emerging Data and Industry 4.0 (EDI40) / Affiliated Workshops.

Akkaradamrongrat, S., Kachamas, P., & Sinthupinyo, S. (2019). Text generation for imbalanced text classification. In *2019 16th international joint conference on computer science and software engineering (JCSSE)* (pp. 181–186). IEEE.

Ameliasari, M., Putrada, A. G., & Pahlevi, R. R. (2021). An evaluation of svm in hand gesture detection using imu-based smart watches for smart lighting control. *JURNAL INFOTEL, 13*(2), 47–53.

Androutsopoulos, I. (2023). Ling Email. Retrieved April 15, 2023, from https://www.kaggle.com/datasets/mandygu/lingspam-dataset

Araujo, A. F., Gôlo, M. P., & Marcacini, R. M. (2022). Opinion mining for app reviews: An analysis of textual representation and predictive models. *Automated Software Engineering, 29*, 1–30.

Aslam, N., Rustam, F., Lee, E., Washington, P. B., & Ashraf, I. (2022a). Sentiment analysis and emotion detection on cryptocurrency related tweets using ensemble LSTM-GRU model. *IEEE Access, 10*, 39313–39324.

Aslam, N., Xia, K., Rustam, F., Lee, E., & Ashraf, I. (2022b). Self-voting classification model for online meeting app review sentiment analysis and topic modeling. *PeerJ Computer Science, 8*, e1141.

Benyamin, G., & Ali, G. (2020). *Attention mechanism, transformers, BERT, and GPT: Tutorial and survey*. OSF preprint.

Bholane Savita, D., & Gore, D. (2016). Sentiment analysis on twitter data using support vector machine. *International Journal of Computer Science Trends and Technology (IJCST), 4*(3), 831–837.

Borji, A. (2023). *A categorical archive of ChatGPT failures*. arXiv (Cornell University). https://doi.org/10.48550/arxiv.2302.03494

Brown, T., Mann, B., Ryder, N., Subbiah, M., Kaplan, J. D., Dhariwal, P., Neelakantan, A., Shyam, P., Sastry, G., Askell, A., et al. (2020). Language models are few-shot learners. *Advances in Neural Information Processing Systems, 33*, 1877–1901.

Cai, X., Liu, S., Han, J., Yang, L., Liu, Z., & Liu, T. (2021). Chestxraybert: A pretrained language model for chest radiology report summarization. *IEEE Transactions on Multimedia, 25*, 845–855.

Cheng, Q., Xu, A., Li, X., & Ding, L. (2022). Adversarial email generation against spam detection models through feature perturbation. In *2022 IEEE International Conference on Assured Autonomy (ICAA)* (pp. 83–92).

Dai, H., Liu, Z., Liao, W., Huang, X., Cao, Y., Wu, Z., Zhao, L., Xu, S., Liu, W., Liu, N., et al. (2023). AugGPT: Leveraging ChatGPT for text data augmentation. In *arXiv preprint arXiv:2302.13007*.

Devlin, J., Chang, M.-W., Lee, K., & Toutanova, K. (2018a). Bert: Pre-training of deep bidirectional transformers for language understanding. In *arXiv preprint arXiv:1810.04805*.

Devlin, J., Chang, M.-W., Lee, K., & Toutanova, K. (2018b). Bert: Pre-training of deep bidirectional transformers for language understanding. In *arXiv preprint arXiv: 1810.04805*.

Erwin, M. R. S., Putrada, A. G., & Triawan, M. A. (2021). Deteksi hama ulat pada tanaman selada berbasis aquaponic menggunakan cnn (convolutional neural network). *eProceedings of Engineering, 8*(5).

Francisca, A. A., Henry, N.-M., & Wenyu, C. H. (2021). Transformer models for text-based emotion detection: A review of bert-based approaches. *Artificial Intelligence Review, 54*, 5789–5829.

Ghobakhloo, M., & Ghobakhloo, M. (2022). Design of a personalized recommender system using sentiment analysis in social media (case study: Banking system). *Social Network Analysis and Mining, 12*(1), 84.

Google. (2023). *Google colaboratory*. https://colab.research.google.com

Han, K., Xiao, A., Wu, E., Guo, J., Xu, C., & Wang, Y. (2021). Transformer in transformer. *Advances in Neural Information Processing Systems, 34*, 15908–15919.

Jia, J., Chen, X., Yang, A., He, Q., Dai, P., & Liu, M. (2022). Link of transformers in CV and NLP: A brief survey. In *5th International Conference on Pattern Recognition*.

Jiang, Z., Xu, F. F., Araki, J., & Neubig, G. (2020). How can we know what language models know? *Transactions of the Association for Computational Linguistics, 8*, 423–438.

Karn, S. K., Liu, N., Schutze, H., & Farri, O. (2022). Differentiable multi-agent ¨ actor-critic for multi-step radiology report summarization. In *Proceedings of the 60th Annual Meeting of the Association for Computational Linguistics. Vol. 1: Long Papers* (pp. 1542–1553).

Kenton, J. D. M.-W. C., & Toutanova, L. K. (2019). Bert: Pre-training of deep bidirectional transformers for language understanding. In *Proceedings of NAACL-HLT* (pp. 4171–4186).

K. Korini, & C. Bizer, "Column type annotation using ChatGPT," 2023.

Lee, J., Yoon, W., Kim, S., Kim, D., Kim, S., So, C. H., & Kang, J. (2020). Biobert: A pre-trained biomedical language representation model for biomedical text mining. *Bioinformatics, 36*(4), 1234–1240.

Liu, B. (2010). Sentiment analysis and subjectivity. *Handbook of Natural Language Processing, 2*(2010), 627–666.

Liu, Y. (2019). Fine-tune bert for extractive summarization. *arXiv preprint arXiv:1903.10318*.

Liu, Y., & Lapata, M. (2019). Text summarization with pretrained encoders. In *Proceedings of the 2019 Conference on Empirical Methods in Natural Language Processing and the 9th International Joint Conference on Natural Language Processing (EMNLP-IJCNLP)* (pp. 3730–3740).

Liu, P. J., Saleh, M., Pot, E., Goodrich, B., Sepassi, R., Kaiser, L., & Shazeer, N. (2018). Generating Wikipedia by summarizing long sequences. *arXiv preprint arXiv:1801.10198*.

Liu, Z., Zhong, A., Li, Y., Yang, L., Ju, C., Wu, Z., Ma, C., Shu, P., Chen, C., Kim, S., et al. (2023a). Tailoring large language models to radiology: A preliminary approach to llm adaptation for a highly specialized domain. In *International workshop on machine learning in medical imaging* (pp. 464–473). Springer.

Liu, Y., Han, T., Ma, S., Zhang, J., Yang, Y., Tian, J., He, H., Li, A., He, M., Liu, Z., et al. (2023b). Summary of ChatGPT/Gpt-4 research and perspective towards the future of large language models. In *arXiv preprint arXiv:2304.01852*.

Liu, Z., Yu, X., Zhang, L., Wu, Z., Cao, C., Dai, H., Zhao, L., Liu, W., Shen, D., Li, Q., et al. (2023c). DeID-GPT: Zero-shot medical text de-identification by GPT-4. In *arXiv preprint arXiv:2303.11032*.

Liu, Y., Han, T., Ma, S., Zhang, J., Yang, Y., Tian, J., He, H., Li, A., He, M., Liu, Z., Wu, Z., Zhu, D., Li, X., Qiang, N., Shen, D., Liu, T., & Ge, B. (2023d). Summary of ChatGPT/Gpt-4 research and perspective towards the future of large language models. *Computation and Language, 1*, 100017.

Liu, Y., Iter, D., Xu, Y., Wang, S., Xu, R., & Zhu, C. (2023e). GPTeval: NLG evaluation using gpt-4 with better human alignment. In *arXiv preprint arXiv:2303.16634*.

Lopez-Lira, A., & Tang, Y. (2023). Can chatgpt forecast stock price movements? return predictability and large language models. In *arXiv preprint arXiv: 2304.07619*.

Mandl, T., et al. (2020). Overview of the HASOC track at FIRE 2020: Hate speech and offensive content identification in Indo-European languages. In *CEUR Workshop Proceedings* (Vol. 2826, pp. 87–111).

Mujahid, M., Lee, E., Rustam, F., Washington, P. B., Ullah, S., Reshi, A. A., & Ashraf, I. (2021). Sentiment analysis and topic modeling on tweets about online education during COVID-19. *Applied Sciences, 11*, 8438.

Muneshwara, M., Swetha, M., Rohidekar, M. P., & Pranove, A. B. (2022). Implementation of therapy bot for potential users with depression during Covid-19 using sentiment analysis. *Journal of Positive School Psychology, 6*, 7816–7826.

OpenAI. (2023a). *ChatGPT: OpenAI Language Model*. OpenAI.

OpenAI. (2023b). Introducing ChatGPT—openai.com. Retrieved August 28, 2023, from https://openai.com/blog/chatgpt.

Ouyang, L., Wu, J., Jiang, X., Almeida, D., Wainwright, C., Mishkin, P., Zhang, C., Agarwal, S., Slama, K., Ray, A., et al. (2022). Training language models to follow instructions with human feedback. *Advances in Neural Information Processing Systems, 35*, 27730–27744.

Parimala, M., Swarna Priya, R., Praveen Kumar Reddy, M., Lal Chowdhary, C., Kumar Poluru, R., & Khan, S. (2021). Spatiotemporal-based sentiment analysis on tweets for risk assessment of event using deep learning approach. *Software: Practice and Experience, 51*, 550–570.

Qadir, J. (2022). *Engineering education in the era of ChatGPT: Promise and pitfalls of generative AI for education*. IEEE.

A. Radford, K. Narasimhan, T. Salimans, I. Sutskever, et al., "Improving language understanding by generative pre-training," 2018.

Radford, A., Wu, J., Child, R., Luan, D., Amodei, D., Sutskever, I., et al. (2019). Language models are unsupervised multitask learners. *OpenAI Blog, 1*(8), 9.

Raghunath, K. K., Kumar, V. V., Venkatesan, M., Singh, K. K., Mahesh, T. R., & Singh, A. (2022). XGBoost regression classifier (XRC) model for cyber-attack detection and classification using inception v4. *Journal of Web Engineering, 21*(4), 1295–1322.

Rai, A. K., et al. (2020). Landsat 8 OLI satellite image classification using convolutional neural network. *Procedia Computer Science, 167*, 987–993.

M. V. Reiss, "Testing the reliability of chatgpt for text annotation and classification: A cautionary remark," 2023.

Rezayi, S., Dai, H., Liu, Z., Wu, Z., Hebbar, A., Burns, A. H., Zhao, L., Zhu, D., Li, Q., Liu, W., et al. (2022). Clinicalradiobert: Knowledge-infused few shot learning for clinical notes named entity recognition. In *Machine Learning in Medical Imaging: 13th International Workshop, MLMI 2022, Held in Conjunction with MICCAI 2022, Singapore, September 18, 2022, Proceedings* (pp. 269–278). Springer.

Rodriguez-Torrealba, R., Garcia-Lopez, E., & Garcia-Cabot, A. (2022). End-to-end generation of multiple-choice questions using text-to-text transfer transformer models. *Expert Systems with Applications, 208*, 118258. https://doi.org/10.1016/j.eswa.2022.118258

Salman, K., Muzammal, N., Munawar, H., Syed, W. Z., & Fahad, S. K. (2022). Transformers in vision: A survey. *ACM Computing Surveys, 54*, 1–41.

Sharma, S., Aggarwal, R., & Kumar, M. (2023). Mining twitter for insights into chatgpt sentiment: A machine learning approach. In *2023 international conference on distributed computing and electrical circuits and electronics (ICDCECE)* (pp. 1–6). IEEE.

Shi, Y., Xu, S., Liu, Z., Liu, T., Li, X., & Liu, N. (2023a). Mededit: Model editing for medical question answering with external knowledge bases. In *arXiv preprint arXiv:2309.16035*.

Y. Shi, H. Ma, W. Zhong, G. Mai, X. Li, T. Liu, and J. Huang, "Chat graph: Interpretable text classification by converting chatgpt knowledge to graphs," 2023b.

Siddique, L., Aun, Z., Heriberto, C., Fahad, S., Moazzam, S., & Junaid, Q. (2023). Transformers in speech processing: A survey. *arXiv preprint arXiv:2303.11607*.

Singh, K. K., Mehrotra, A., Nigam, M. J., & Pal, K. (2013, April). Unsupervised change detection from remote sensing images using hybrid genetic FCM. In *2013 Students Conference on Engineering and Systems (SCES)* (pp. 1–5). IEEE.

Singh, K. K., Rho, S., Singh, A., & Sergei, C. (2024a). Big data analytics and knowledge discovery for urban computing and intelligence. *Complex & Intelligent Systems, 10*(1), 1–2.

Singh, A., Dhull, A., & Singh, K. K. (Eds.). (2024b). *Block chain and deep learning for smart healthcare*. Wiley.

Song, K., Wang, B., Feng, Z., Liu, R., & Liu, F. (2020). Controlling the amount of verbatim copying in abstractive summarization. In *Proceedings of the AAAI Conference on Artificial Intelligence* (Vol. 34, pp. 8902–8909).

T. Susnjak, "Applying bert and chatgpt for sentiment analysis of Lyme disease in scientific literature," 2023.

Tang, C., Liu, Z., Ma, C., Wu, Z., Li, Y., Liu, W., Zhu, D., Li, Q., Li, X., Liu, T., et al. (2023). PolicyGPT: Automated analysis of privacy policies with large language models. In *arXiv preprint arXiv:2309.10238.*

Tran, A. D., Pallant, J. I., & Johnson, L. W. (2021). Exploring the impact of chatbots on consumer sentiment and expectations in retail. *Journal of Retailing and Consumer Services, 63*, 102718.

Vaswani, A., Shazeer, N., Parmar, N., Uszkoreit, J., Jones, L., Gomez, A. N., Kaiser, Ł., & Polosukhin, I. (2017). Attention is all you need. *Advances in Neural Information Processing Systems, 30.*

Wang, J., Shi, E., Yu, S., Wu, Z., Ma, C., Dai, H., Yang, Q., Kang, Y., Wu, J., Hu, H., et al. (2023a). Prompt engineering for healthcare: Methodologies and applications. In *arXiv preprint arXiv:2304.14670.*

Z. Wang, Q. Xie, Z. Ding, Y. Feng, and R. Xia, "Is ChatGPT a good sentiment analyzer? A preliminary study," 2023b.

Wei, J., & Zou, K. (2019). EDA: Easy data augmentation techniques for boosting performance on text classification tasks. *arXiv preprint arXiv:1901.11196.*

Writer.com Website. (2023). *AI content detector*. https://writer.com/aicontent-detector/

ZeroGPT Website. (2023). *ZeroGPT: AI text detector*. https://www.zerogpt.com/

Zhang, X., Zhou, X., Lin, M., & Sun, J. (2018). Shuffle Net: An extremely efficient convolutional neural network for mobile devices. In *Proceedings of the 2018 Conference on computer vision and pattern recognition (CVPR), Salt Lake City, UH, USA, 18–22 June 2018.*

Zhao, Z., Wallace, E., Feng, S., Klein, D., & Singh, S. (2021). Calibrate before use: Improving few-shot performance of language models. In *International conference on machine learning* (pp. 12697–12706). PMLR.

Zhao, B., Jin, W., Ser, J. D., & Yang, G. (2023). ChatAgri: Exploring potentials of chatGPT on cross-linguistic agricultural text classification. *Neurocomputing, 557*, 126708.

Chapter 3
Large Language Model on Multi-Modal Data

Avi Aneja, Anuradha Dhull, Akansha Singh, and Krishna Kant Singh

Abstract Large Language Models (LLMs) are used as the brain to do multimodal tasks in the emerging field of multimodal Large Language Model (MLLM) represented by GPT-4V. Surprisingly, MLLM's emergent capabilities—like OCR-free math reasoning and the ability to write stories based on images—are uncommon in conventional multimodal approaches and point to a possible route towards artificial general intelligence. In an attempt to create MLLMs that are on scale with or even superior to GPT-4V, researchers in academia and industry have been working surprisingly quickly to push the boundaries of their field. This paper explores the potential of multimodal large language models, the core concepts of LLM, and the distinct design that sets the existing multimodal LLM apart. Additionally, it has outlined the difficulties and restrictions that the existing LLM faces, including model complexity and data heterogeneity.

Keywords Large Language Model · Methodologies · Framework · Challenges · Applications · Limitations

3.1 Introduction

Language Models (LMs) are designed to forecast the probability of words within a given language, that aims on generating new text, completing sentences, and predicting new ideas based on a given context (Hadi et al., 2023a). Large language

A. Aneja · A. Dhull
Computer Science Engineering (CSE), The NorthCap University, Gurugram, Haryana, India
e-mail: avi21csu247@ncuindia.edu; anuradha@ncuindia.edu

A. Singh (✉)
School of Computer Science Engineering and Technology, Bennett University, Greater Noida, Uttar Pradesh, India

K. K. Singh
Delhi Technical Campus, Greater Noida, Uttar Pradesh, India

A. Singh, K. K. Singh (eds.), *Multimodal Generative AI*,
https://doi.org/10.1007/978-981-96-2355-6_3

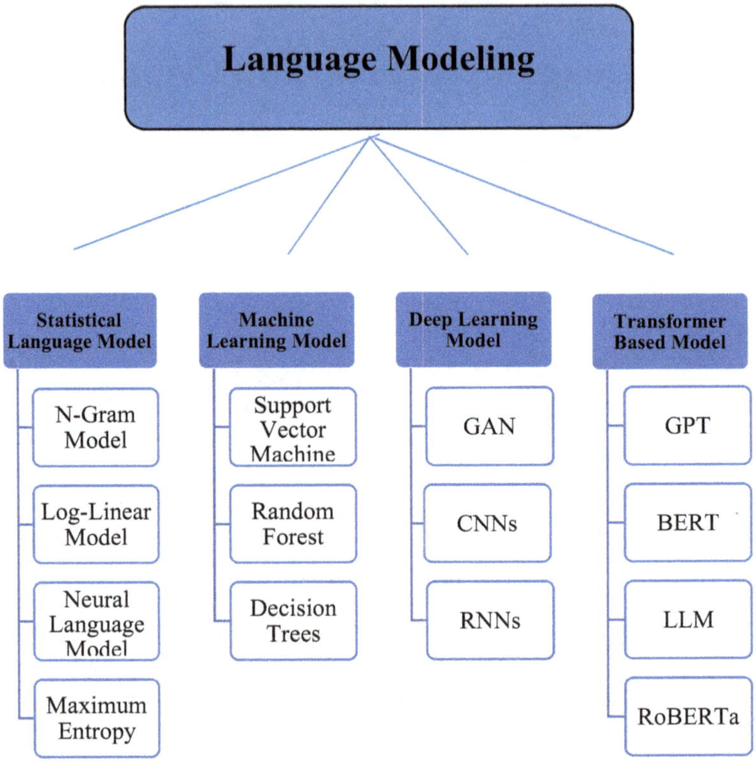

Fig. 3.1 Classification of Large Language Models (LLMs) divided into four main categories: Statistical language models, machine learning models, deep learning models, and transformer-based models

models (LLMs) can be broadly classified into different architectures, which include statistical methods, machine learning, and deep learning approaches as illustrated in Fig. 3.1. Deep learning architectures encompass techniques like generative adversarial networks (GANs), convolutional neural networks (CNNs), and transformer-based models such as GPT-2. Large Language models have, however, been made possibly by the growing need of deep learning in Natural Processing (NLP), the availability of robust computer equipment, and the quantity of publicly accessible datasets.

3.2 Overview of Multimodal Data

Multimodal machine learning is a dynamic interdisciplinary area of study that addresses core AI objectives through the integration and modeling of multiple forms of communication, including linguistic, acoustic, and visual signals. Initially centered on audio-visual speech recognition, recent advancements have expanded into language and vision projects like language-guided reinforcement learning, visual question answering, and labelling of images and videos (Morency et al., 2022). The rapid advancement in multimodal research has posed challenges in identifying the fundamental themes across historical and recent studies, as well as the primary unanswered questions in the field.

3.3 Overview of Large Language Models

Large language models, often called "next-generation" or "Innovative" language models, represent a major progress in Natural Language Processing (NLP) (Hadi et al., 2023b). They support deep learning techniques, specifically transformers, to use complex patterns within large amounts of language data, signifying a remarkable leap in NLP. Large volumes of unstructured text can be easily processed by LLMs, who are expert at identifying the semantic connections between words and phrases. They may also learn about correct links between many kinds of data, such as auditory, visual, multi-modal, and audiovisual data. As a result, LLMs have greatly enhanced machines' understandability and production of human-like language. LLMs have undergone various stages of development, there by growing in size and complexity.

The GPT series, like GPT-3 with its 1.7 trillion parameters, showcases how Large Language Models (LLMs) have grown in size and complexity. This has enabled a jump in their capacity to comprehend and generate human-like language.

Early language models particularly aimed to model for producing text data, whereas more recent language models (such as GPT-4) focus on solving complex tasks. Moving forward from language modeling to task solving shows a remarkable progress in scientific comprehension, essential for grasping the generation of language models over research history (Zhao et al., 2023a). In terms of task solving, the four generations of language models have shown varying levels of model capacities. Figure 3.2 illustrates the evolution of language models in relation to their task-solving capacity. Initially, statistical language models primarily aided specific tasks, such as retrieval or speech tasks, by leveraging predicted or estimated probabilities to enhance task-specific approaches. Subsequently, neural language models concentrated on learning task-agnostic representations, aiming to reduce the need for human feature engineering. Apart from this, pre-trained language models gained context-aware representations adaptable to downstream tasks. In the latest generation, LLMs have been advanced by exploring the scaling effect on model capacity, making them capable of solving general-purpose tasks.

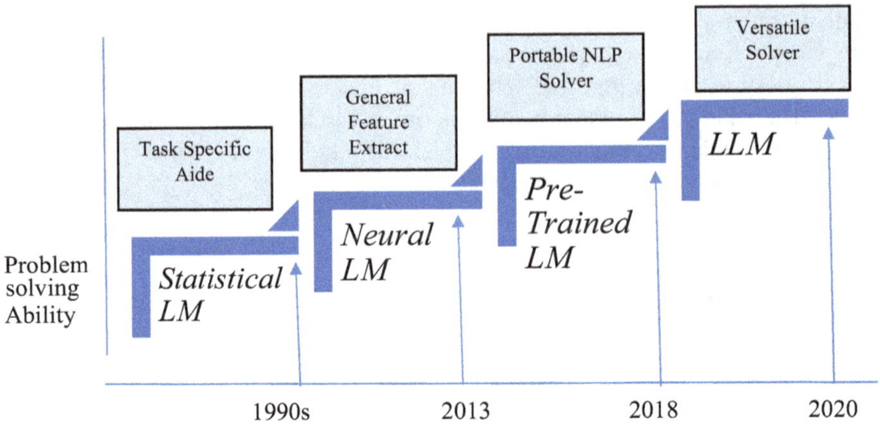

Fig. 3.2 A description of the progression of four generations of language models (LM) based on their ability to solve tasks

3.4 Overview of Multimodal Large Language Models

The development of LLMs has advanced remarkably in recent years. These LLMs increase remarkable emergent abilities by scaling up the size of the data and the model; common examples of these abilities are teaching based on InContext Learning (ICL) and Chain of Thought (CoT). Since LLMs can only comprehend discrete text, they are intrinsically "blind" to visual, despite their surprising fewshot reasoning performance on the majority of Natural Language Processing (NLP) tasks. At the same time, Large Vision Models (LVMs) have clear vision but they typically have slow reasoning (Liang et al., 2023). Because of their complementary nature, LLM and LVM operate the development of the new field of Multimodal Large Language Model (MLLM) toward one another. In formal terms, it describes an LLM-based model that can process, receive, and output multimodal data. Multimodality has been the subject of numerous publications prior to MLLM, which can be categorized into discriminative and generative paradigms. The foundation of MLLM is LLM with billionscale parameters, which was not possible in earlier models. To reach its maximum potential, MLLM leverages novel training paradigms. For example, multimodal instruction tuning is used to motivate the model to obey novel commands. With these two qualities, MLLM demonstrates new abilities including creating code for websites based on images, deciphering memes, and doing OCR-free mathematical reasoning.

3.5 Existing Work on Multimodal LLM

MM-LLMs use LLMs as a cognitive powerhouse to support a wide range of MM tasks. LLMs provide desirable qualities such strong language generation, zero-shot transfer, and In-Context Learning (ICL). Concurrently, foundation models in other modalities generate high-quality representations. MM-LLMs have trouble connecting pre-trained models from several modalities for collaborative inference. Better granularity support is one of the usage scenarios or features that have been expanded in subsequent developments. More control over user prompts is being developed to allow clicking on specific objects or locations via boxes.

Large language models (LLMs) have been the subject of current research, which demonstrates their significant influence and wide range of applications. Reif et al. (2021) explore using LLMs for few-Shot Text Style Transfer, by introducing a novel method called augmented zero-shot learning (Zhang et al., 2023). Kojima et al. (2022) discuss using LLMs for zero-shot reasoning and significantly improving performance by adding necessary prompts (Reynolds & McDonell, 2021). Du et al. (2022) review challenges in LLMs related to shortcut learning and find out methods to reduce them (Chen et al., 2021). Huang and Chang (2022) provide a comprehensive overview of reasoning in LLMs, covering and evaluating techniques, and future research directions (Du et al., 2022). Anil et al. (2022) investigate transformer-based LLMs' performance on length generalization tasks, highlighting issues with simple fine-tuning approaches (Austin et al., 2021). Morrison (2022) examined the use of LLMs in publicly available platforms, specifically in education, analysing the characteristics of generated text and suggesting adaptations for educators (Kojima et al., 2022). Vaithilingam et al. (2022) investigated how programmers utilize Copilot, an LLM based code generation tool, discovering that despite its occasional inability to increase task completion times or success rates, it is recommended for everyday activities because of its practicality and time saving features (Anil et al., 2022). Zhu et al. (2023) explored the intersection of LLMs and Information Retrieval (IR) systems, majorly targeting on components like query rewriters, retrievers, and readers, pointing on potential synergies (Reif et al., 2021). Jansen et al. (2023) discussed about the efficient use of LLMs in survey research, specifically in generating responses to survey items, offering new perspectives on survey design and analysis (Zhao et al., 2023b). Wang et al. (2023) provided insights into improving the calibration of LLMs with human-oriented tasks and expectations, emphasizing the need for better understanding and response to human nuances (Navigli et al., 2023). Thirunavukarasu et al. (2023) outlined the development of LLM applications in clinical settings, discussing about their strengths, limitations, and potential to increase efficiency in medicine (Morrison, 2022). Myers et al. (2023) delved into Foundation and Large Language Models (FLLMs), highlighting their major roles in training LLMs for various AI tasks, particularly in Natural Language Processing (NLP), underscoring their importance in order to advance LLM capabilities (Kalyan, 2023). Overall, these studies demonstrate the wide-ranging capabilities of LLMs and the ongoing efforts to enhance their performance in various domains.

3.5.1 Multimodal LLM Architecture

The model design, as shown in Fig. 3.3, consists of five components, each of which is covered in detail in this section. The first three components are usually found in Multimodal Large Language Models (MM-LLMs) that majorly focuses on multimodal (MM) understanding. In most of the cases, the Modality Generator, LLM Backbone, and Modality Encoder are set as static during training, and input and output projector optimization takes over the priority. MM-LLMs are benefited from lightweight projector designs, resulting in only 2% of their total parameters being trainable (Thirunavukarasu et al., 2023). The scale of the core LLM being utilized in the MM-LLMs gives the overall number of parameters (Zhang et al., 2024). However, MM-LLMs can be effectively taught for a range of multimodal activities.

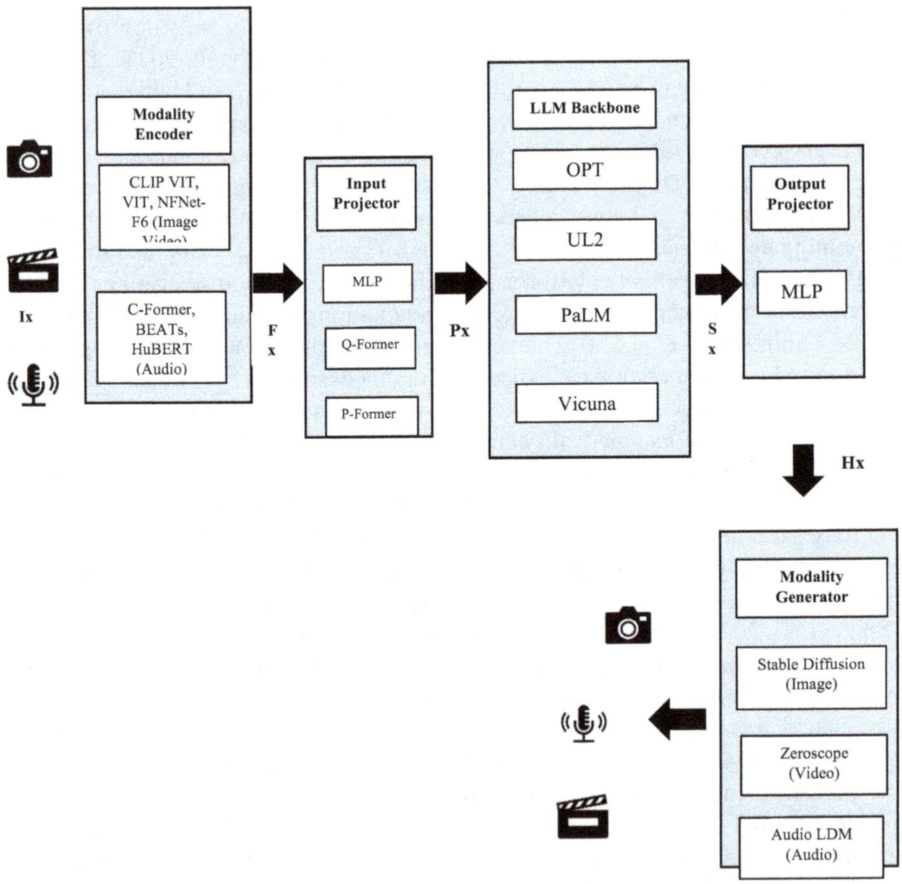

Fig. 3.3 The general architecture of multimodal LLM where the modality encoder, input projector, LLM backbone represents multimodal understanding and output projector, modality generator represents multimodal generation

3.5.1.1 Modality Encoder

The modality encoder is a major component in the architecture of a Multimodal Large Language Model (MM-LLM). It is used to process input data from different modalities, such as text, images, or audio, and put them further into an understandable structure for the model. In the context of MM-LLMs, the modality encoder is basically designed to extract relevant features from each modality while preserving the semantic information which is specific to that modality. These features, denoted as FX, are obtained through the operation MEX(IX), where MEX represents the specific pre-trained encoder used for each modality. FX = MEX(IX).

3.5.1.2 Input Projector

The Input Projector ($\Theta X \rightarrow T$) bridges the gap between encoded features from various modalities and the text feature space. Prompts are generated by this bridging procedure and mixed with textual features before being sent to the LLM Backbone for additional processing. It is still difficult to maintain high-quality text creation while integrating non-textual modalities. An input projector is used to close the gap between the text creation process and the non-textual features in order to address this. This projector can be as simple as a linear projector or as complex as a multi-layer perceptron (MLP), which can learn complicated non-linear correlations between the text space and the encoded properties. Even more sophisticated methods are available with advanced approaches such as Q-Former, P-Former, MQ-Former, and Cross-attention. Various approaches for aligning features are used in the system to generate less impact on the quality of text produced. Additionally, some advanced techniques may require initialization through a few pre-training stages.

3.5.1.3 LLM Backbone

LLM Backbone performs semantic comprehension, inference, and input-related decision-making when processing representations from several modalities. It generates signal tokens SX from other modalities and direct textual outputs t. These signal tokens serve as directives that tell the generator whether to make multimedia content or not. The aligned representations of other modalities PX can be seen of as soft prompt-tuning for the LLM, as in t, SX = LLM(PX,FT).

3.5.1.4 Modality Generator

The Modality Generator (MGX) works as an agent that allows the model to produce output in the form of audio video or even pictures. It incorporates well-known methods such as audio production with AudioLDM2, video creation with Zeroscope, and

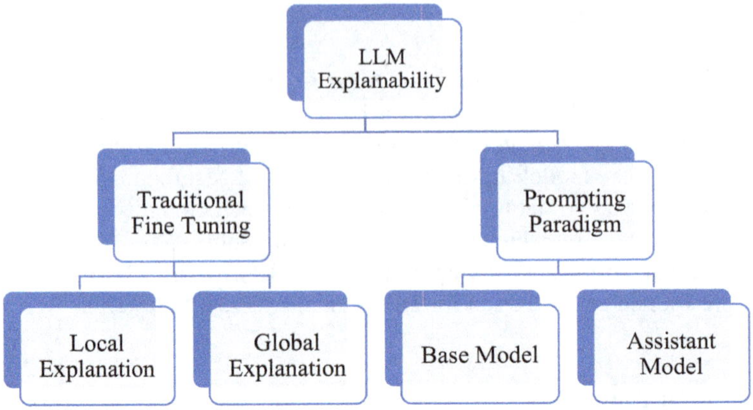

Fig. 3.4 Categorization of LLM explainability into 2 major paradigms along with various explainability techniques

image formation using Stable Diffusion. These models take features (HX) from the Output Projector as the input. During training, MGX begins with the actual content and converts it into a hidden feature. This feature is then altered by adding some random elements to create a noisy version. This noisy version is used to generate the final output.

3.6 Training Methodologies for Large-Scale Language Models

There are two main methods for specializing large language models (LLMs) to specific tasks: fine-tuning and prompting. These approaches differ significantly in how they achieve this adaptation. The substantial distinctions between these methods have led researchers to propose various explanations for how each works, as illustrated in Fig. 3.4.

3.6.1 Traditional Fine Tuning

Large language models are trained like language trainees. In the first stage where a trainee absorbing information from a vast library, the model goes under *pre-training* on a staggering amount of raw text data. This self-supervised learning allows it to grasp the underlying rules and structures of language, building a strong foundation in understanding how words interact and convey meaning. Imagine the trainee familiarizing themselves with grammar, vocabulary, and the nuances of different writing styles. Following this foundational training, the model moves to the *fine-tuning* stage, where the trainee being assigned specific tasks. Here, the model focuses on a particular domain or task by training on a smaller, but more targeted, dataset of labelled examples. Datasets like SST-2 (sentiment analysis) or MNLI (natural language inference) become the trainee's workshop, honing their skills in a specific area. This two-stage approach, used for models like BERT and RoBERTa, allows them to first develop a broad understanding of language and then specialize in tackling specific tasks effectively.

3.6.2 Prompting Paradigm

The prompting paradigm revolutionizes language model training by utilizing natural language prompts, like sentences with blanks, to enable few-shot or zero-shot learning without requiring further training data. This paradigm's model can be classified into two primary groups:

Base Models: These models, such as GPT-3 OPT, LLaMA-1, LLaMA-2, and Falcon, are characterized by their massive scale, often containing billions of parameters. Despite their size, they can perform impressively well in few-shot learning tasks through prompting. Base models are considered foundation models that can interact with users without specific alignment to human preferences (Myers et al., 2023). Explanations for base models focus on how these models leverage their pre-trained knowledge when prompted.

Assistant Models: While base models excel in many areas, they have limitations, including an inability to follow complex user instructions and a tendency to produce biased or toxic content. To overcome these limitations, base models undergo supervised fine-tuning to achieve human-like capabilities, particularly in open-domain dialogue. This process involves aligning the model's reaction with input from humans and their preferences, this is frequently done through Reinforcement Learning from Human Feedback (RLHF) and instruction tuning. Assistant models, such as GPT-3.5, GPT-4, Claude, LLaMA-2-Chat, Alpaca, and Vicuna, are trained to engage in complex, multi-turn conversations. Explanations for assistant models focus on how they acquire and refine their interactive abilities through conversations.

3.7 GPT-3 Family Large Language Model

OpenAI, a company founded in 2015, initially focused on creating generative models. They began with exploring Recurrent Neural Networks (RNNs) but later shifted to the transformer model, known for capturing long-term dependencies well. This led to the development of GPT-1117M parameter pretrained language model which was the first transformer-based model. The Large Model Training paradigm was coined by GPT-1 as a useful tool for developing models (Wang et al., 2023). Building

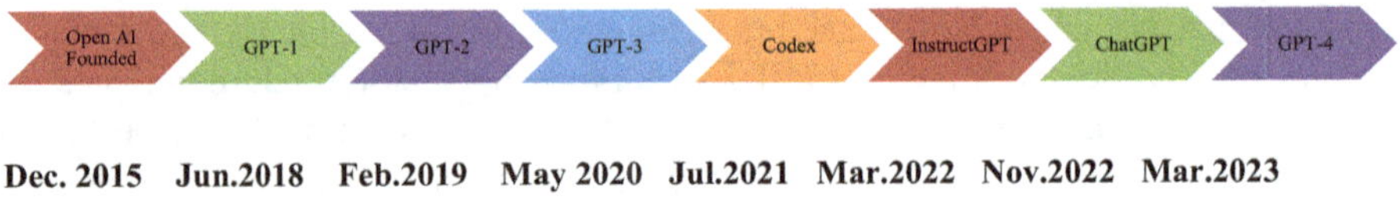

Fig. 3.5 OpenAI's LLM advancement from GPT-1 to GPT-4

on the foundation of GPT-1, OpenAI introduced GPT-2. This iteration boasted a larger dataset for pre-training and offered multiple versions with varying capabilities. The success of both GPT-1 and GPT-2 precipitated the way for the GPT-3 family of LLMs, which includes models like ChatGPT and culminates in the latest GPT-4. This progression from GPT-1 to GPT-4 showcases OpenAI's continuous efforts in pushing the boundaries of language models for natural language understanding and generation. Figures 3.5 and 3.6 illustrate this evolution, highlighting the journey from the initial GPT-1 to the most recent GPT-4, and detailing the expansion of the GPT-3 family.

The era of large language models (LLMs) commenced approximately with the advent of GPT-3. Subsequently, in the wake of GPT-3's achievements, OpenAI introduced successor Generative AI Tools such as Instruct GPT, Codex, ChatGPT, and GPT-4.

3.8 GPT-3 Models

The findings from experiments with GPT-2 demonstrated that expanding the model's dimensions leads to a decrease in perplexity, with larger models achieving better results compared to smaller ones. Inspired by the positive impact of GPT-2, OpenAI developed GPT-3, a model with 175 billion parameters, making it 100 times larger than its predecessor. Unlike GPT-1 and GPT-2, which were trained on text from specific sources, GPT-3 learned from a diverse mix of webpages, Wikipedia, and books. Unlike prior GPT models requiring supervised learning, GPT-3's wide scaling empowers it to handle new tasks without specific training. Rather, GPT-3 employs in-context learning without instruction.

3.8.1 GPT-3.5 Models

Due to the absence of code within its training data, GPT-3 exhibits limitations in tackling intricate reasoning tasks, such as solving mathematical problems. Additionally, potential biases inherent in its training data can influence its outputs and it sometimes struggles to follow user instructions, leading to the generation of harmful text. To address these issues, GPT-3.5 models have been introduced. These models, such as Codex, are specifically designed for coding tasks and are developed

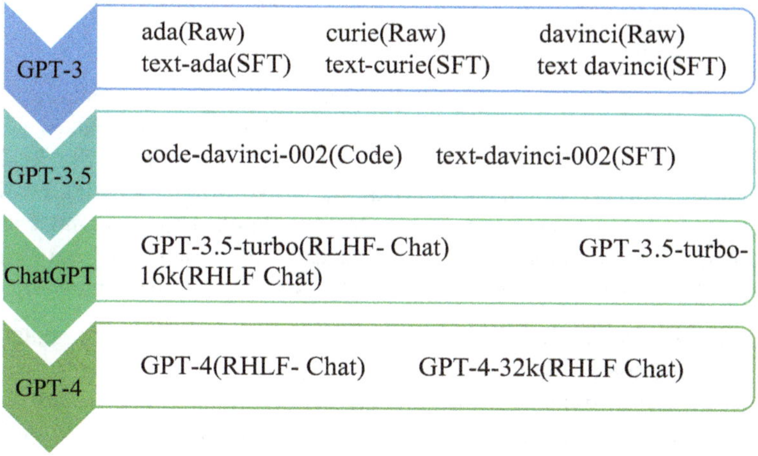

Fig. 3.6 Training paths for OpenAI's GPT-3 family (GPT-3 to GPT-4). Models can be fine-tuned from a general pre-trained state ("raw") to more specialized versions through techniques like supervised fine-tuning (SFT) or reinforcement learning with human feedback (RLHF), with "RLHF-Chat" specifically targeting chat applications

by fine-tuning GPT models with code data from sources like GitHub (Austin et al., 2021). Additionally, GPT-3.5 receives human feedback for supervised or reinforcement learning fine-tuning to enhance their capacity for reasoning and decrease the production of damaging text. This model improves the proficiency of the model in understanding and executing user instructions, while also mitigating the risks associated with generating undesirable content.

3.8.2 *Chat GPT and GPT-4*

While GPT-3.5 models can comprehend and produce code in addition to natural language, GPT-3 models can only understand and generate natural language. The GPT-3.5 and GPT-3 models, however, are not chat-optimized (Chen et al., 2021). The ChatGPT (GPT-3.5-turbo) and GPT-4 (OpenAI, 2023) versions overcome this problem. With its November 2022 launch, ChatGPT attracted millions of users due to its remarkable conversational skills. Later in March 2023, OpenAI unveiled the GPT-4 model, which was able to handle inputs as images and text. These models not only write as fluently as humans do, but they also excel in a variety of NLP applications.

3.9 Challenges and Future Directions

Large language models like GPT-4 have significantly advanced natural language processing but come with challenges. These include high computational costs, adversarial robustness, and interpretability issues (Zhang et al., 2024). Scaling up for complex tasks raises scalability, privacy, and real-time processing challenges. While foundational research delves into merging multi-modality and enhancing transfer learning, continuous learning introduces hurdles. Tackling these challenges is essential for maximizing the real-world effect of large language models.

- *Computational Cost*: Training large language models (LLMs) is computationally intensive, resulting in increased manufacturing costs and an adverse effect on the environment due to the large energy usage. Better performance can be achieved by applying more processing power to the task, but as the law of diminishing returns, these benefits get decreased with time.
- *Overfitting*: Large language models (LLMs) face the challenge of remembering peculiarities from their enormous training datasets. For LLMs to generate accurate responses, they need a proper balance between memorization and generalization. Memorization helps the model to store specific information for recall, whereas generalization allows it to adapt to new situations and tasks. It is essential to acquire the memorization-generalization equilibrium, because excessive memorization may lead to overfitting and can also hinder adaptability to novel inputs (Yin et al., 2023).
- *Cognitive Planning*: Certain reasoning and planning tasks, like sensible planning, which seems to be effortless for humans, are still challenging for current large language models (LLMs) to perform. This is not an unexpected struggle, because LLMs are optimized to find statistical pattern rather than robust reasoning. As a result, their ability to solve problems in a way that mimics human logic remains uncertain.

3.10 Limitations of Large Language Model

LLMs have contributed much to NLP, but they are not without drawbacks. This section outlines several of these drawbacks, such as biased data, an excessive dependence on surface-level patterns, a lack of common sense, a poor capacity for reasoning and interpreting feedback, etc. Furthermore, obstacles to its wider implementation include high resource requirements, restricted generalizability, interpretability problems, and difficulties with uncommon terms or strange syntax. Further limitations include susceptibility to manipulation, ethical considerations, struggles with ambiguous language (Guo et al., 2023). They also lack causality understanding, have low real-time capabilities, significant maintenance and training expenses, struggle with multimodal inputs, have a limited attention span, and offer restricted transfer learning capabilities. Furthermore, they have insufficient knowledge of the world beyond text, human behaviour, and struggle with long-form text generation, collaboration, ambiguity, cultural differences, incremental learning, structured data, and noise in input data. It is necessary for researchers and practitioners to recognize and mitigate LLM limitations for its effective use. Furthermore, developing new models that can excel these limitations is essential.

Generative Inconsistencies in LLMs: At times, GPT-4 might produce results that are completely imaginary or factually inaccurate because it is producing information that is not grounded in its training set. In LLMs, hallucinations frequently arise from the model's attempt to make assumptions based on patterns it has learnt during training in order to fill in context or knowledge gaps. Research on the cause of hallucinations in LLMs is ongoing. According to recent developments, the problem is complex and involves the dataset, architectural design, and training technique of the model. LLMs may have a bias toward producing more "interesting" or fluid outputs, which raises the likelihood of hallucinations.

Difficulties in Specific Application Domains Like Spelling Accuracy: GPT-4 may have trouble dealing with particular specific tasks because of their statistical character, such as identifying and correcting spelling mistakes. Counting errors are another example of this. When the model miscounts or misinterprets numerical numbers, counting error happens. Counting the words or characters in lengthy paragraphs is one example of how it could give inaccurate results or misplace decimal points while doing arithmetic operations.

Cognitive Biases: LLM may make mistakes in logical thinking due to ambiguities in prompts or limits in understanding complex operations. Also, LLMs lack the ability to plan, reason, and have a limited understanding of the physical world.

3.11 Use-Cases and Applications

This section covers various applications and use cases of Large Language Models (LLMs), highlighting their versatility and practical relevance in various fields.

LLM in Clinical Knowledge: As discussed by Zhu et al., 2023, ChatGPT has gained a lot of attention in the medical field for its ability to pass medical school exams and respond to patient inquiries on social media platforms with responses that are more sympathetic and correct than those from actual physicians. GPT-4 is seen as a major development in medical knowledge and reasoning, even though GPT-3.5 performance in more specialized tests has been unsatisfactory. Apart from this, Pathways Language Model 2 (PaLM 2) and Large Language Model Meta AI 2 are two more LLMs that are widely used.

LLM in Education: The research of Huang & Chang, 2022 showed that the application of LLMs to education has garnered a lot of interest lately, especially with regard to automated scoring. This research has concentrated on optimizing LLMs such as ChatGPT for automated scoring applications, showcasing their capacity to assess student answers with a high degree of accuracy. Additional insights into the possibilities of LLMs in educational contexts are provided by recent developments in AI for education, like the creation of Artificial General Intelligence (AGI) for educational reasons and chain-of-thought prompting approaches.

LLM in the Environment: Large Language Models (LLMs) must be applied in order to meet environmental issues, as discussed by Vaithilingam et al. (2022). The environment is expected to be directly impacted by other potentially transformative technological innovations like the metaverse, as they may lead to increased energy consumption and consequent resource consumption and carbon dioxide production. This is obviously also a problem for LLMs, as training LLMs and making inferences need a lot of energy, which almost always calls for algorithmic efficiency. Depending on the energy used and the carbon intensity of the energy source, the carbon footprint will change. Apart from carbon dioxide emissions, the computing facilities could also have other environmental effects like water consumption, soil pollution, or sealing, which could have wider consequences for the quality of the environment.

LLM Hallucinations: Large Language Models (LLMs) have been crucial in enhancing Artificial Intelligence and Natural Language Processing (NLP), as discussed by Wang et al., 2023. Hallucinations in LLMs have been the subject of an increasing amount of NLP literature. Some of the studies that have been conducted include automating factuality detection for real-world claims, investigating the source of hallucinations from a training data and putting automated feedback mechanisms in place to correct LLM-generated content.

LLM for Software Engineering: The usage of Large Language Models (LLMs) in Software Engineering is examined by Jansen et al., 2023. Any application where the software products or procedures are based on the use of Large Language Models is referred to as LLM-based Software Engineering. With their strong language representation and contextual awareness capabilities, pre-trained models and LLMs may use a variety of training data and, by rapid engineering, transfer learning, and fine-tuning, adjust to creative tasks. These benefits make them useful instruments in generative tasks and they have shown outstanding performance in the software engineering domain.

3.12 Conclusion

The application of NLP to solve problems and deliver practical services has seen a radical transformation in the last several years. We are now witnessing advanced conversational AI courtesy to the transformer architecture and, consequently, LLMs that demonstrate strong logic and problem-solving abilities. Due to this, the field of computer vision has crossed over, and as a result, we are now seeing what are known as MM-LLMs. Therefore, this paper has provided a brief overview about Large Language Models (LLM), their implications and usage. Also, it provided a brief overview of LLMs on multimodal data. Further, we have discussed about the architecture of MM-LLMs and its training methodologies. This study also highlighted the use of AI technologies like ChatGPT.

Future research should concentrate on enhancing the performance and accuracy of these models as the area of MM-LLMs develops and advances, resolving their shortcomings and investigating fresh applications for them. Researchers and practitioners may guarantee that MM-LLMs are utilized responsibly and benefitively by embracing the guidelines provided in this survey and helping to further the continued advancement of LLMs.

References

Anil, C., et al. (2022). Exploring length generalization in large language models. *Advances in Neural Information Processing Systems, 35*, 38546–38556.

Austin, J., et al. (2021). Program synthesis with large language models. *arXiv preprint arXiv:2108.07732*.

Chen, M., et al. (2021). Evaluating large language models trained on code. *arXiv preprint arXiv:2107.03374*.

Du, M., et al. (2022). Shortcut learning of large language models in natural language understanding: A survey. *arXiv preprint arXiv:2208.11857*.

Guo, Z., et al. (2023). Evaluating large language models: A comprehensive survey. *arXiv preprint arXiv:2310.19736*.

Hadi, M. U., et al. (2023a). Large language models: A comprehensive survey of its applications, challenges, limitations, and future prospects. *Authorea Preprints*.

Hadi, M. U., et al. (2023b). A survey on large language models: Applications, challenges, limitations, and practical usage. *Authorea Preprints*.

Huang, J., & Chang K. C. (2022). Towards reasoning in large language models: A survey. *arXiv preprint arXiv:2212.10403*.

Jansen, B. J., Jung, S.-G., & Salminen, J. (2023). Employing large language models in survey research. *Natural Language Processing Journal, 4*, 100020.

Kalyan, K. S. (2023). A survey of GPT-3 family large language models including ChatGPT and GPT-4. *Natural Language Processing Journal, 6*, 100048.

Kojima, T., et al. (2022). Large language models are zero-shot reasoners. *Advances in Neural Information Processing Systems, 35*, 22199–22213.

Liang, W., et al. (2023). Can large language models provide useful feedback on research papers? A large-scale empirical analysis. *arXiv preprint arXiv:2310.01783*.

Morency, L.-P., Liang, P. P., & Zadeh, A. (2022). Tutorial on multimodal machine learning. In *Proceedings of the 2022 Conference of the North American Chapter of the Association for Computational Linguistics: Human Language Technologies: Tutorial Abstracts*.

Morrison, R. (2022). Large language models and text generators: An overview for educators. Online Submission.

Myers, D., et al. (2023). Foundation and large language models: Fundamentals, challenges, opportunities, and social impacts. *Cluster Computing, 27*, 1–26.

Navigli, R., Conia, S., & Ross, B. (2023). Biases in large language models: Origins, inventory and discussion. *ACM Journal of Data and Information Quality, 15*, 1.

OpenAI. (2023). About OpenAI. Retrieved from https://www.openai.com.

Reif, E., et al. (2021). A recipe for arbitrary text style transfer with large language models. *arXiv preprint arXiv:2109.03910*.

Reynolds, L., & McDonell, K. (2021). Prompt programming for large language models: Beyond the few-shot paradigm. In *Extended Abstracts of the 2021 CHI Conference on Human Factors in Computing Systems*.

Thirunavukarasu, A. J., et al. (2023). Large language models in medicine. *Nature Medicine, 29*(8), 1930–1940.

Vaithilingam, P., Zhang, T., & Glassman, E. L. (2022). Expectation vs. experience: Evaluating the usability of code generation tools powered by large language models. In *Chi conference on human factors in computing systems extended abstracts*.

Wang, Y., et al. (2023). Aligning large language models with human: A survey. *arXiv preprint arXiv:2307.12966*.

Yin, S., et al. (2023). A survey on multimodal large language models. *arXiv preprint arXiv:2306.13549*.

Zhang, S., et al. (2023). Instruction tuning for large language models: A survey. *arXiv preprint arXiv:2308.10792*.

Zhang, D., et al. (2024). Mm-llms: Recent advances in multimodal large language models. *arXiv preprint arXiv:2401.13601*.

Zhao, W. X., et al. (2023a). A survey of large language models. *arXiv preprint arXiv:2303.18223*.

Zhao, H., et al. (2023b). Explainability for large language models: A survey. *ACM Transactions on Intelligent Systems and Technology, 15*, 1–38.

Zhu, Y., et al. (2023). Large language models for information retrieval: A survey. *arXiv preprint arXiv:2308.07107*.

Chapter 4
Adaptive Learning Technologies: Navigating the Road from Hype to Reality

S. Valai Ganesh, M. Gomathy Nayagam, V. Suresh, S. Rajakarunakaran, and B. Bensujin

Abstract As AI techniques such as machine learning, neural networks, and natural language processing continue to improve at a very fast rate, they will be increasingly used in education. This will lead to more flexible and fair learning experiences that are tailored to each student's strengths and needs. This paper looks at the main AI features that power some of the most important education technology innovations, such as smart tutoring systems, personalized virtual tutors, automated content creation, and smart assessment analytics. After looking at results and examples from a number of important areas, such as simulated environments, data-driven ideas, and conversation-based tutoring agents, the main pros and cons of these new sociotechnical systems are discussed. AI has the ability to improve teaching and learning in ways that people cannot. However, as the field grows, it is important to think about ethics, responsibility, and including everyone. A study of adding teachers versus replacing them is also performed, and framework principles for fair learning progress are proposed. The summary is easy to understand and discusses past successes, current skills, current trends, and possible future developments. It tells many people about both good and bad outcomes and stresses how important it is to

---

S. V. Ganesh (✉)
Department of Mechanical Engineering, Ramco Institute of Technology, Rajapalayam, India
e-mail: valaiganesh@ritrjpm.ac.in

M. G. Nayagam
Department of Computer Science and Business Systems, Ramco Institute of Technology, Rajapalayam, India

V. Suresh
Department of Electronics and Communication Engineering, National Engineering College, Kovilpatti, India

S. Rajakarunakaran
Department of Mechanical Engineering, Ramco Institute of Technology, Rajapalayam, India

B. Bensujin
College of Engineering and Technology, University of Technology and Applied Sciences, Muscat, Sultanate of Oman

A. Singh, K. K. Singh (eds.), *Multimodal Generative AI*,
https://doi.org/10.1007/978-981-96-2355-6_4

have freedom, choice, and design that is based on results for systems that are moral and help students learn.

Keywords Personalized learning · Smart tutoring systems · Virtual tutors · Automated content creation · Smart assessment analytics · Simulated learning environments · Data-driven educational technology · Conversation-based tutoring agents · Sociotechnical systems in education

4.1 Introduction

In the past 10 years, the use of artificial intelligence (AI) in education and learning has rapidly increased. HolonIQ's (2020) market report states that the world market for AI in education will increase from $1.7 billion in 2019 to over $10 billion by 2025. Improvements in natural language processing, machine learning, and the sharing of educational data are some of the main factors that have led to this.

Specifically, within e-learning, AI has opened doors to more adaptive and personalized online education experiences (Khan & Ponzanelli, 2021). According to an analysis by Adams Becker et al. (2017), applying learning analytics powered by AI is one of the most significant trends in the e-learning industry. By analysing different learner modalities, including texts, speech data, video conferencing interactions, and assessment performance, AI e-learning systems can offer customized and dynamic content at the individual level (Shah, 2020).

Current real-world implementations of multimodal AI in online education range from intelligent tutoring systems that tailor lesson plans (VanLehn, 2011) to AI-driven plaque platforms such as anthropic platforms that respond to verbal and visual student inputs (Taggart, 2022) to next-generation virtual environments using embodied AI agents as teachers and facilitators (Zawacki-Richter et al., 2019).

4.1.1 Brief History of AI in Education

There is a long history of using artificial intelligence (AI) in schooling that dates back to the 1970s. The SCHOLAR system, created by Jaime Carbonell in 1970, is one of the first examples. It used AI to construct a dialogue-based, interactive curriculum to instruct geography (Nwana, 1990).

Smart tutoring systems (ITS) became popular in the 1980s. The goal of these methods was to give each student individualized instruction by showing what they already knew and changing how they taught based on that. The SOPHIE system (Brown et al., 1982) for fixing problems with electronics and the LISP Tutor (Anderson & Reiser, 1985) for teaching computing are two well-known examples.

Pedagogical bots and customisable hypermedia systems were introduced in the 1990s. Adaptive hypermedia systems change how online learning materials are presented based on how well and what the user likes (Brusilovsky, 2001). Pedagogical agents have moving characters that lead students through educational content.

The spread of internet-based learning and the easy access to increasingly more educational data in the early 2000s led to the creation of data mining for education and learning analytics methods (Chen & Zhu, 2019). These methods that use AI attempt to identify trends in learner data to help people make better decisions based on the data (Romero & Ventura, 2010).

Recently, improvements in machine learning, especially deep learning, have made it possible to develop AI systems for education that are smarter and more useful. Adaptive learning platforms change students' learning paths all the time based on how well they do, and AI-powered tools grade essays automatically, provide personalized feedback, and act as smart tutors (Luckin et al., 2016).

As AI technologies continue to improve, it becomes increasingly clear that they can completely change education and solve long-standing problems in teaching and learning. However, putting AI to use in schools also raises important moral, social, and teaching issues that need to be addressed to ensure that its use is responsible and effective (Singh & Saravanan, 2024).

4.1.2 Scope and Structure of the Chapter

This chapter provides an in-depth look at how artificial intelligence (AI) is used in education and online learning. It talks about the present state of AI-powered tools for learning, how they could change the way we teach and learn, and the problems and moral issues that arise when we use them.

The first part of the chapter discusses the main ideas and terms used in AI in education. These include personalized learning, adaptable structures, and mixed data analysis. After that, it talks about various ways that AI is being used in schooling, such as:

1. Personalized and adaptive learning systems that tailor instruction to individual learner needs and preferences.
2. Multimodal content generation techniques that create engaging and interactive learning materials.
3. Virtual teachers, facilitators, and peers who provide intelligent support and guidance to learners.
4. Automated assessment tools that evaluate student work and provide timely feedback.

This chapter discusses the basic AI technologies, how they are currently used, and study results on how well they work in each of these areas. Using real-life instances and stories, it also discusses the pros and cons of each method.

The next part of the chapter discusses the greater effects of AI on education, such as how it might make learning more accessible, fair, and efficient. It also discusses the moral and social issues that arise when AI is used in education, such as data privacy, computer bias, and how teachers' jobs are changing.

The chapter stresses how important it is for AI to be developed and used in education in a way that is responsible and focused on people. Considerable attention is given to the fact that educators, researchers, lawmakers, and tech developers all need to work together to get the most out of AI for learning while also minimizing its risks.

At the end of the chapter, the main points are summed, and suggestions are made for those who are involved in creating, using, and controlling AI in education. It also discusses where future research should go and how new ideas can be used in this area that is changing so quickly.

The goal of this chapter is to provide a thorough and critical look at AI in education so that teachers, scholars, legislators, and technology developers can learn how to use AI to improve learning and instruction while also addressing the many problems and moral issues that arise when it is used in schools.

4.2 Key Terms

Multimodal: The use of multiple modes of data input and output by an AI system, including text, speech, images, video, sensory data, and more (Baltrušaitis et al., 2018). In the context of e-learning, a multimodal AI tutoring system can analyse a student's verbal response, facial expressions, and written work (Whitehill et al., 2014).

Generative AI: AI models that can automatically generate new content, such as text, images, video, and 3D objects (Panwar et al., 2023). Key techniques include generative adversarial networks (GANs) (Singh et al., 2024), variational autoencoders (VAEs), diffusion models and transformer language models (Bommasani et al., 2021). Generative AI can generate personalized e-learning materials on the fly.

E-learning: Electronic sources, most often the internet, are used for learning. Some important types of learning are online courses, webinars, simulations, and platforms that change based on student needs and are driven by educational data mining and learning analytics (Rasheed et al., 2020).

Personalized/adaptive learning: Systems for education that change and adapt lessons and tasks based on each student's strengths, weaknesses, and preferences (Lee & Hannafin, 2016). Different types of learning analytics, systems for feedback, and learning reinforcement are some of the examples in which it works (Fig. 4.1).

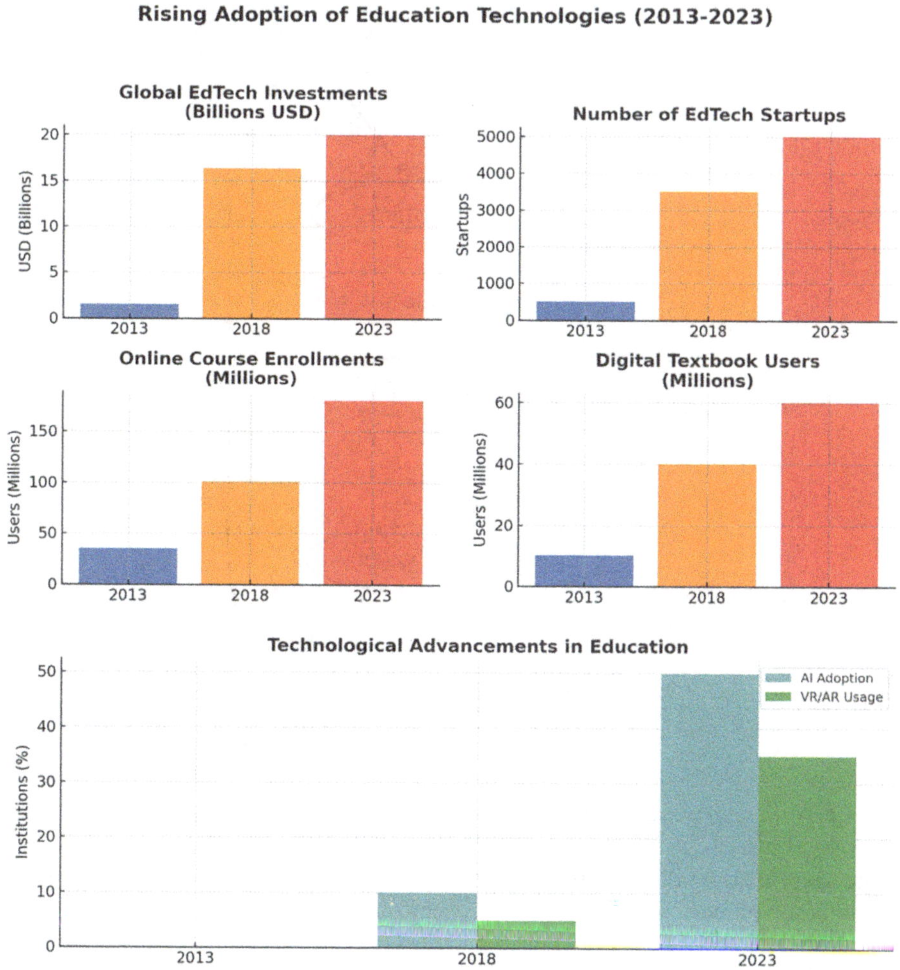

Fig. 4.1 Rise adoption of education technologies (2013–2023)

4.3 AI-Powered Personalized and Adaptive Learning

Multimodal AI can read many different types of student data, such as written content, spoken answers, test results, discussions among educators and students, and body expressions during home lessons. Emotional examination can identify expressions of anger or confidence (Aroyo et al., 2019), and speech recognition programs can write down what a student says when they answer a math problem (Fig. 4.2).

An intelligent teaching method or a personalized e-learning tool is built around a model of what each student knows, which is fed by these large datasets. When different types of learning are linked, the system can see more about each student's strengths and flaws (Worsley & Blikstein, 2014). Therefore, lessons can be made

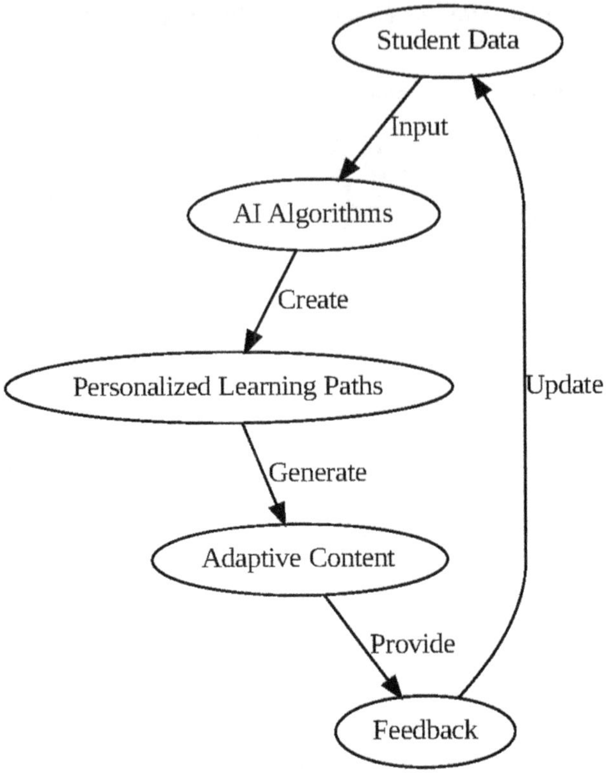

Fig. 4.2 AI-based personalized learning

more specific, right down to the subject, classroom idea, or instructional technique stage, based on how well they work for every learner (Tseng et al., 2008a).

If a student needs help with their vocabulary, an interdisciplinary intelligent writing teacher might look at the different words they use in their articles, the stops and tone of voice they use when they talk, and the way their eyes move while they read. After that, it can make vocabulary tests, clarifications, and instances of usage based on how the student discovers best, such as by using elements from games or listening to talks for a student who learns best by hearing (Echeverria et al., 2018).

Over time, tracking behaviour, conversations, and work build a long-term learner profile that allows for very specific customization that boosts engagement (Fig. 4.3).

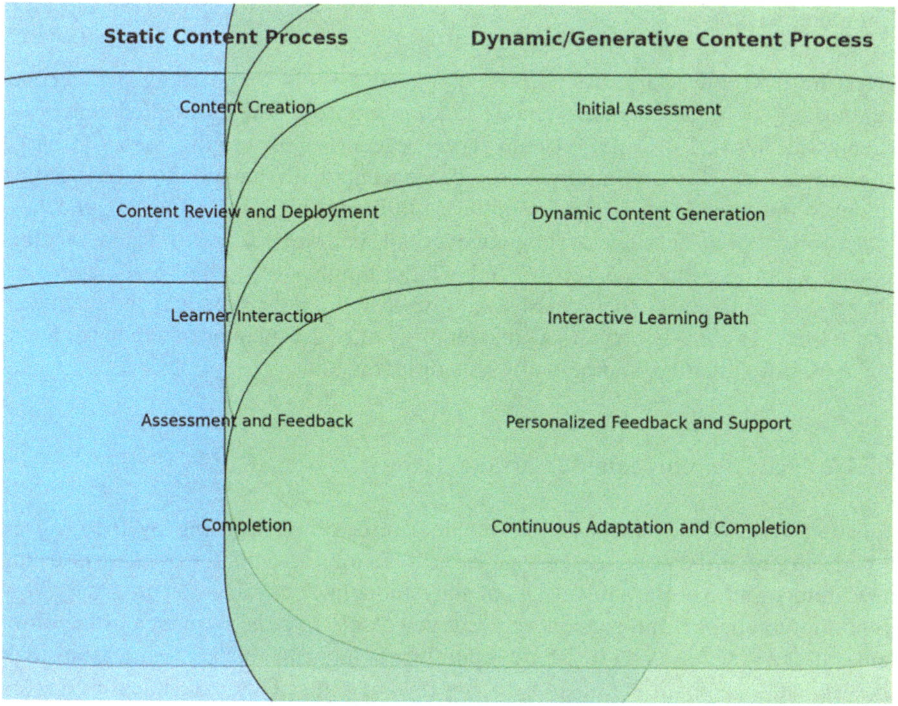

Fig. 4.3 Static versus dynamic (generative) content processes

4.3.1 Natural Language Processing, Computer Vision, Reinforcement Learning, and Generative Models in Education and E-Learning

4.3.1.1 NLP: Natural Language Processing

Because of changes in technology in schools, there are now more ways for students to be active and learn better. Liu et al. (2019) studied how useful it is to use written and spoken notes to measure mood analysis as a way to observe how learners behave as well as how involved they are. Language processing techniques help teachers learn a great deal about how learners feel. They can then adapt how they teach to fit those feelings. Siddharthan and Angrosh's (2014) methods for text simplicity also allow educational materials to be changed automatically to different reading levels. This makes them more accessible and open to everyone. Kumar et al. (2020a) additionally discuss how integrating intelligence dialogue agents can make teaching and chats more personalized, which helps students obtain more help and encourages them to learn in their own way. Now that these changes have occurred, schools will be able to adapt and respond to the requirements of each of their learners better, which will improve learning.

Computer Vision

Over the past few years, the use of cutting edge tools for learning has become increasingly popular. Kang and Liew (2020) proposed a new idea called generative adversarial networks (GANs). Could serve to generate newer quiz questions based on training ideas This new method provides teachers with many options for keeping students interested and pushed. Similarly, Pullin et al. (2021) investigated how transformer-based models of language could help people who teach writing. Teachers can use these models to find a large number of words, parts, and even entirety books to add to their classes and give learners more personalized feedback. These kinds of ideas show how AI may entirely alter the manner in which we learn, make us more creative, and help all types of learners.

4.3.1.2 RL: Reinforcement Learning

Schatten et al. (2020) checked fresh concepts regarding learning by focusing on how to improve the way teachers perform their jobs, especially by always trying new things and asking students what they thought. They learned that it is very important to change things about the way you teach, like the sequence of the problems, all the time based on real-time input. Instructors can change their lessons to fit the way students learn and taste by using reward signs. This makes learning more fun and effective. The emphasis on dynamic educational optimization shows a move towards teaching methods that are more adaptable and put the student first. The ultimate objective is to help students learn more and better understand things that are difficult for them to grasp.

Generative Models

Many more people are currently using cutting edge technology than they were a few years ago. GANs can now be used in a new way by Kang and Liew (2020) to automatically make quiz questions based on ideas that have already been taught. This new method provides teachers with many options for keeping their students interested and challenged. Another study by Pullin et al. (2021) investigated how transformer-based language models help teach writing. Teachers can obtain an immense collection of sentences, chapters, and sometimes whole books through these models. They can use these tools to supplement their instruction and provide learners with more specific comments. These findings show that AI can change the way instructions work, spark people's imaginations, and meet a wide range of learning needs.

4.3.2 *Virtual Instructors, Smart Tutoring Tools, and Simulated Environments*

4.3.2.1 Virtual Instructor Systems

Technology in schools has come a long way, and now there are ways to give each child personalized help. Ma et al. (2014) made systems that can change lessons based on the needs of each student and keep track of their unique skills and weaknesses. Customized learning is possible for each student because these systems monitor their growth and performance all the time. This keeps them interested and helps them comprehend better. Ritter et al. (2007) were also the first to use thinking tutors to help students with mathematics inquiries by giving them personalized notes, hints, and answers. Teachers who use complicated code and adaptive learning to determine where learners have trouble and help them in a way that makes sense to them. This helps them understand maths better. These new methods are a step towards making learning more personalized for each student. Technology plays a large role in making these possible and boosting learning outcomes.

4.3.2.2 Smart Tutoring Tools

Because of progress in artificial intelligence (AI), it is now possible to make virtual figures that are controlled by AI and can teach better than humans. These avatars were made by Zawacki-Richter et al. (2019) to help with talks and make them more fun. The incorporation of fun into classwork was made possible by this approach. When these teachers teach online, they can see how excited each student is, talk about difficult topics, and maintain learners' motivation while they learn. These avatars can change how they interact with students based on their needs and how they learn best. This makes the learning experience more personalized and immersive (Singh et al., 2024). This new method not only allows students to become more involved but can also be used on a large scale to adapt to the changing needs of modern education.

4.3.2.3 Simulated Environments

The idea of immersive venues, such as virtual labs and patient simulators, was created by Merchant et al. (2014). These will change clinical teaching by adapting to what students are doing in real time. These places give students the chance to learn by doing in a safe and controlled setting, where they can practice important skills without the dangers that come with real-life situations. Learners can also move through simulated situations with the help of AI bots while receiving personalized feedback and helping to improve their skills. By using immersive technologies and AI-powered help, teachers can provide students with more realistic training that is more like the problems they will face in the real world. This will better prepare them for future jobs in healthcare. This method not only helps people who want to work in healthcare learn new skills but also helps them feel more confident and competent.

In addition to generative neural networks, these apps use knowledge tracking, processing of natural languages, reinforcement learning, and other types of data and AI to make learning fun and easy for each person.

4.3.3 The Benefits and Downsides of Personalized and Adaptive Learning

Implementing personalized and adaptive learning methods driven by AI in schools has many benefits, but it may also come with some problems that need to be solved for programmes to work well and help students learn.

4.3.3.1 Advantages

1. *Customized learning*: AI algorithms can look at much information about each learner, such as their skills, weaknesses, learning styles, and level of engagement, to make learning paths that are just right for them. This kind of personalization can make people more interested and motivated, which can lead to better learning results.
2. Making good use of resources: AI-driven approaches can help teachers focus on more important tasks, such as one-on-one help and teaching students how to think critically, by automating parts of the teaching and learning process, such as delivering content, grading, and giving feedback.
3. *Improvement all the time*: AI systems can keep making the customized lessons they give better as they understand more regarding the way learners perform and how involved they are in learning. This process is repeated to ensure that the content and teaching methods used remain relevant and useful over time.
4. Moreover, personalized and adaptable learning tools can help learners from all walks of life and with different learning styles and skills perform better in school. People may acquire knowledge at their own rate and in a way that works for them with these AI-driven methods, which may close gaps in educational opportunities.

4.3.3.2 Disadvantages

1. *Privacy and security of information*: People are worried about privacy and security of information when private student information is collected and studied. To protect the privacy of students' data and the public's faith in the system as a whole, schools need to set up firm data control tools that follow these rules.
2. *Algorithm bias*: Artificial intelligence (AI) algorithms can unintentionally make mistakes in their training information or assumptions that were made to make them stronger. When developing AI systems, it is important to think about many different points of view and to ensure that they are fair and include everyone on a regular basis.
3. *Relying too much on technology*: AI-powered personalized learning may be very beneficial, but it is important to find a mix of technology-driven teachers and people you can talk to in person. Students could miss important social and mental skills that are good for their overall growth if they rely excessively on AI systems.
4. *Infrastructure and resources*: To use customized and adaptable learning methods powered by AI, people need to spend a great deal of money on technology infrastructure, staff training, and on-going upkeep. It might be difficult for schools, especially those that lack much money, to obtain and maintain these important parts.
5. *Not willing to change*: People who have adapted to the way those things are currently conducted, such as teachers, managers, or even students, may be unwilling to use innovative instruments or teaching methods. For change management methods to work and for professionals to grow, it is important that modifications are put into action well and are accepted by many.

4.3.4 *Case Studies of Successful Implementations*

To show what AI-powered personalized and flexible learning systems can do, it helps to look at examples of how they have been used successfully in the real world. The examples below show how these technologies can improve learning results, student engagement, and classroom efficiency in different settings.

Case Study 1: Cognitive Tutor for Mathematics
The cognitive tutor from Carnegie Learning is an intelligent teaching device that provides students with individualized math lessons. The system constantly checks students' information using the cognitive method of learning to change the way they are taught based on what they determine. Ritter et al. (2007) performed a study that showed that students who used Cognitive Tutor for Algebra I performed much better on standardized tests than did students who received regular teaching.

Case Study 2: Knewton Alta for Higher Education
Knewton Alta is an educational tool that makes math, chemistry, and economics lessons more relevant to each student. The tool looks at data on student achievement to identify knowledge gaps and provide specific help. Arizona State University performed a case study and found that students who took developmental math classes using Knewton Alta did better in school and had higher pass rates than did students who took regular classes (Knewton, 2018).

Case Study 3: DreamBox Learning for K-8 Mathematics
DreamBox Learning is a customizable learning platform for math for grades K-8. It changes how students are taught based on how they interact with it and how well they do. The platform helps students learn by giving them personalized comments, hints, and building blocks. The Centre for Education Policy Research at Harvard University performed a study that showed that students who used DreamBox Learning for 1 year performed much better in math than students who did not use DreamBox Learning (DreamBox Learning, 2016).

Case Study 4: Duolingo for Language Learning
Duolingo is a popular app for learning languages. It uses AI to tailor lessons to each student's needs. Machine learning algorithms are used by the app to make lessons harder, provide personalized feedback, and suggest learning paths. According to a study by Vesselinov and Grego (2012), people who used Duolingo were much better at speaking and understanding languages. They found that 34 h of Duolingo lessons were the same as a full term of college-level language classes.

These case studies show how personalized and adaptive educational platforms driven by AI can help students in a range of subjects and grade levels. These technologies can greatly improve learning results, engagement, and efficiency by adapting lessons to each student's specific needs and providing individualized support. However, it is important to remember that the success of these implementations relies on aspects such as the quality of the AI models that are used, how well they are integrated with the curriculum and teaching methods, and how well teachers and students are supported.

4.4 Measuring the Effectiveness of Personalized and Adaptive Learning

As personalized and flexible learning systems become more popular in schools, it is important to develop reliable ways to test how well they work. It is important to determine how these AI-powered technologies affect learning results, engagement, and efficiency so that they can be developed, used, and improved. This part talks about some important things to think about and ways to determine how well personalized and adaptable learning works.

4.4.1 Defining Metrics and Indicators

To determine how well personalized and adaptive learning works, we must first develop clear metrics and signs that match the learning goals. Some of these are:

1. *Learning gains*: Tracking how much students' information, skills, and abilities have grown over time.
2. *Engagement and motivation*: Checking how much students participate, how long they stay on tasks, and how they feel.
3. *Efficiency and success rates*: determining how much time and money students need to reach their learning goals.
4. *Learner satisfaction*: Finding out how learners feel about the adaptive learning method and what they have done with it.

4.4.2 Comparative Studies

Researchers often compare the results of customized educational methods in traditional classroom instruction or nonadaptive e-learning to determine how well personalized and personalized learning environments work. The plans of these studies could be experimental or quasi-experimental, and learning gains could be measured before and after the studies. For instance, Nakic et al. (2015) examined the differences between an adaptive e-learning system and a traditional e-learning method. They found that students who used the adaptive system learned more and were happier with it.

4.4.3 Learning Analytics and Data Mining

Personalized and personalized educational systems collect much information about how students connect with each other, how well they do, and how far they have come. Knowing data analysis and mining methods can be used to find patterns and insights in these data, which can then be used to judge how well the system works. For example, looking at data on pupil achievement across various educational paths or educational strategies can help individuals find the best ways to teach different types of students (Romero & Ventura, 2020).

4.4.4 Longitudinal Studies and Long-Term Impact

To understand how well personalized and adaptive learning support lifelong learning and job success, it is important to know how they affect people in the long term. Researchers who keep track of how students do over long periods of time can learn much about how these tools help students in the long term through longitudinal studies. For instance, Tseng et al. (2008b) examined the long-term benefits of an adaptable educational system for vocabulary retrieval and observed that learners who used the system retained much more of what they learned than did students who received traditional instruction.

4.4.5 Qualitative Feedback and User Experience

Along with quantitative measures, it is important to obtain qualitative comments from students and teachers to fully understand how users feel and find ways to make such comments better. Focus groups, surveys, and conversations can provide useful information about how learners feel about personalized and adaptive learning systems, the problems they face, and the things they would like to see improved. This input can help improve the layout, user interface, and teaching methods of the system so that it better meets the needs of learners.

To determine how well personalized and adaptive learning works, a variety of methods that blend qualitative and quantitative information from different sources need to be used. This is how researchers and teachers can fully understand how these technologies affect learning outcomes and experiences by using strict study designs, analytics for learning, longitudinal research, and user feedback. This method based on proof is very important for ensuring that personalized and adaptive educational platforms are developed and used in a way that is responsible for and helps students.

4.5 Multimodal Content Generation

4.5.1 Automatic Content Generation

There is a brand-new way to make teaching content on the spot using generating artificial intelligence (AI) techniques such as variational autoencoders, adversarial generative networks (GANs), and neural networks for language generation (NLGs). With these suggestions, you may create a lot for different kinds of learning tools that can be used by any student and fit their needs. It is possible to make learning materials such as clarifications, illustrations, and tasks that are unique to a subject or learning goal in real time with NLGs. Iterative learning is used by neural networks and variational autoencoders to make accurate and unique material that helps them improve this ability. In regard to being able to customize and expand, dynamically generated lessons are better than standard lessons. They can help students with a wide range of learning styles and skills. In this way, they can ensure that each learner has an individual experience that assists them in understanding more and being more interested. This type of AI can also quickly provide a large amount of information about a wide range of topics because it is flexible. An environment that has become increasingly digital meets the need for high-quality educational tools that are easy to find (Fig. 4.4).

Types of Generated Content

- Text-based tools such as passages, stories, and quiz questions that are written for each student.
- Conversational agents who provide details and answers through dialogue.
- Images, videos, and models to make multimodal courseware more useful.

Educational Considerations

- The importance of smartly integrating user-generated material into the curriculum to meet learning goals.
- There may be a need for human review before delivery to ensure quality.
- Relevance increases motivation and involvement.

Evaluation and Optimization

- Testing the comprehension, retention, and perceived quality of generated content.
- Gathering both qualitative and quantitative metrics.
- Using student feedback and responses to continuously improve output.

Limitations and Ethics

- Current limitations in factual accuracy, logical coherence, and reasoning abilities.
- Risk of perpetuating biases, need for oversight and transparency.

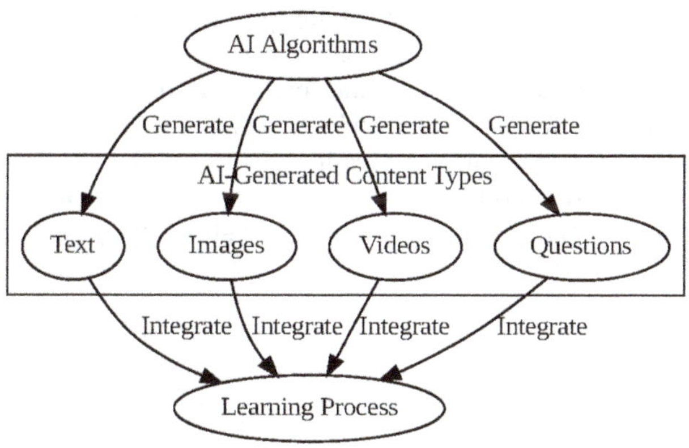

Fig. 4.4 AI-generated content in the learning process

- Following moral rules in regard to data protection, inclusion, and personalization.

4.5.2 Generative Language Models

A new study in education has shown that the latest algorithms can be utilized to help learners better understand and use what they learn. GPT-3 is a cutting-edge model for making language. Kumar et al. (2020b) used it to make assessments automatically based on reads. It is an excellent method to see the information that people learn and get them to engage with the material more closely. Similarly, Martin et al. (2020) used transformer models to make books easier so that they would be appropriate for every learner's reading level. The above techniques make it easy for all individuals to comprehend and employ, especially learners with different needs, because they change the level of challenge of the work to fit every learner's skill. When these projects collaborate, they demonstrate how AI-powered tools can be used to tailor learning and help students understand what they read better in schools. Different kinds of people can learn together in schools, and teachers can perform their jobs better in many places with these tools.

4.5.3 Generative Adversarial Networks (GANs)

Generative adversarial networks (GANs), a new type of artificial intelligence, have changed how educational materials are made and how they are reviewed. Wang et al. (2021) showed that GANs may generate word problems in mathematics on the spot, with tools that may be altered based on the theme, the amount of challenge, and other things. Teachers can use this new invention to perform many different kinds of tests, and it also provides students with specific jobs that can help these individuals become more proficient in solving problems. Zhang et al. (2022) went further into the way GANs might be used to create novel chemical examination inquiries that concentrate on subjects that learners found difficult to understand. Many students have different ways in which they learn best, and these made-up questions that use machine learning permit them to better understand and master hard topics. All of these modifications show that tests are becoming more flexible and useful and will help instructors learn and do well in all areas.

4.5.4 *Variational Autoencoders (VAEs)*

Variational autoencoders (VAEs) were suggested by Vahdat et al. (2021) as a good way to give each student a personalized lesson review. VAEs learn the hidden meanings of educational ideas, which helps them quickly understand the most significant elements of tough subjects. By examining how every learner learns and interacts with others, VAEs can construct course outlines that fit the style and needs of each student. As a result, learners know and recall things better because they are shown in an approach that fits how they think. VAE summaries also tell teachers useful information about how learners behave as well as how much they understand. This allows them to use targeted help and solutions. In general, VAEs at educational institutions are a good way to give each student a unique learning experience and help them improve their education.

4.5.5 *Multimodal Generative Models*

Zhu et al. (2021) were the first to use generative models in education. They did this by making movies with virtual teachers explaining ideas. These models use computer-generated voices, facial emotions, and gestures to make learning fun and useful. These methods change the way people usually learn by making practice questions, explanations, summaries, and other teaching materials that are dynamically tailored to each student. This customized approach ensures that every student obtains information that fits their unique learning style and abilities, which eventually helps them understand and remember more. Additionally, because generative models are flexible, teachers can change lesson plans instantly to meet the changing needs of their students and help them learn better. Therefore, using these new methods could make learning much more effective and interesting in a wide range of settings.

4.6 Benefits

Reduced costs: Costs are lower because generative models reduce the amount of work that teachers, instructional designers, and others have to do to create material by hand. Additionally, the cost of changes can be lowered by changing materials on the fly instead of redesigning static materials.

Widened access: Because it is personalized and there is an endless supply of created content, each student can obtain materials that are just right for them. This amount of customization has never been seen before, and it makes learning easier for people with different needs.

Rapid content updates: AI systems can quickly update training data and produce revised content, which is different from how curriculums are usually created. This allows the material to be updated and released more quickly to keep up with changing educational standards.

Additionally, automatically created content can have other benefits, such as the following:

- Getting teachers less work to do so, they can work on more important things.
- Making it possible to test answers and practice problems continuously using A/B.
- Supporting scenario-based teaching by allowing for endless changes.

Overall, these benefits can improve student engagement, motivation, and results due to timely, low-cost, and easy-to-find educational resources.

4.7 Challenges

Generative artificial intelligence has much potential to change how personalized and educational content is made, but it also has some issues that have to be solved before it can be used safely.

Alignment of instruction and quality assurance: One of the toughest parts of using generative AI for educational content is ensuring that the content it creates fits with the syllabus, objectives for learning, and standards for teaching. In cases where there is not enough quality control and management, people might make things that are wrong, do not make sense, or do not work to be a teaching tool. Subject matter experts and teaching designers should be part of strict review processes for educational materials made by AI to ensure that they remain high quality and honest.

Fairness and bias in algorithms: Generative AI models might support or boost biases found in the data used for training. These biases could be social norms or views that are prejudiced against certain groups. Someone who has been taught historical books might, for example, make work that fails to represent enough or inappropriate types of people to certain groups. To reduce these risks and promote equitable representation of students in instructional resources, ensuring that the information used for instruction varies and includes everyone is essential. Additionally, content that is made should be regularly checked for bias.

The right to ownership intellectual property and copyright: When creative AI is used, there are concerns about rights to intellectual property and copyright theft. These models learn from a lot of current content, so they may make content that sounds a lot like stolen works with no giving credit or approval. Make clear rules regarding when to employ artificial intelligence (AI) to make teaching content. This will help defend property rights and keep people from getting into trouble with the law. These rules should explain how to obtain rights and properly cite sources.

Privacy and data safety-generating AI models need to be able to access large datasets, which could include private information about students, such as their likes and dislikes, academic success, and learning histories. People need to believe the school system and follow certain rules, such as the General Data Protection Regulation (GDPR) in the EU or the Family and Educational Rights and Privacy Act (FERPA) in the USA. This information needs to be kept secret and safe. Many privacy rules and methods of anonymization should be used to protect the privacy of students.

Since generative AI models are complex, determining how certain results are made can be tricky. We need to be responsible and able to explain! This makes me worried about who is responsible and how open things are. It is important for both learners and educators to understand how the things they are studying are used in educational institutions, particularly in regard to testing and receiving feedback. Artificial intelligence techniques can be described, and the boundaries and possible flaws of algorithms for machine learning can be identified. This will help people trust AI and make smart decisions.

Being in charge and working with people: Generative AI can make the procedure of making material easier and faster, but people have to remain involved and always keep a watchful eye on things. It is very important for AI systems to have human teachers who can guide their development, choose the material they make, and provide thorough feedback that takes each student's needs into account. It is important to find the right mix between how useful AI-made content is and how much humans know to make educational opportunities that are both useful and fun.

AI experts, lawmakers, teachers, and anyone else who cares about these issues need to keep together to develop best practices, guidelines, and rules for using generative AI in schools in a reasonable and moral way. If we know about these problems ahead of time and work to fix them, we can use artificial intelligence (AI) to make teaching materials better while lowering risks and problems that were not meant to happen.

4.8 Simulating Virtual Teachers and Peers

4.8.1 AI Beings That Act as Virtual Teachers, Guides, and Classmates

Moreover, artificial intelligence (AI) agents are being employed in e-learning for a range of roles, such as virtual teachers, guides, and fellow learners. They AI-powered things can give you specific help, get your team to work together, and generally make learning better.

Virtual teachers: Virtual teachers who are driven by AI can give each student a customized lesson, adapt to various methods of learning, and make notes. Georgia Tech made an AI called Jill Watson to help teach computer science. Watson helped students take an online course by answering their inquiries, providing them with guidance, and offering these individuals customized assistance (Goel & Polepeddi, 2016). Teaching agents such as AutoTutor are another example. These talk to learners in natural languages and provide them with answers, hints, and notes, similar to what an actual instructor might (Graesser et al., 2005).

Virtual tutors: AI models can also help learners with talks and group projects by taking on the role of virtual tutors. One project called Bazaar made artificially intelligent agents that help with the web group work. These agents watch how students interact, look for problems, and step in to help learners work together better (Adamson et al., 2014). Using AI-powered robots as teachers in massive open online courses (MOOCs) has also led to hope for attracting more interested students and providing personalized help to each student on a large scale (Pérez et al., 2020).

Agents with artificial intelligence (AI) can even behave like virtual friends. This allows students to interact with and learn from each other. The SimStudent project, for example, made AI figures that learn with real students. They can work together, fight, and teach each other (Matsuda et al., 2013). Another example is social machines such as EMYS, which are driven by AI. These robots can work with children on schoolwork and be their friends to help them feel better (Kanda et al., 2012) (Fig. 4.5).

Some examples of how to use it in e-learning settings:

1. *Customized help*: AI online teachers can offer one-on-one help classes, ensuring that their lessons and feedback are based on the needs and progress of each student.
2. *Help with group projects*: AI tutors can guide learners via collaborative tasks by observing how they operate together, giving them comments, and suggesting ways for them to work together better.
3. *Learning language*: Artificial intelligence *(AI) agents* can have conversations with people who are trying to learn a language. This gives them a chance to work in real life and obtain feedback in real time on their speech and grammar.
4. *Adaptive courseware*: AI-powered courseware—can change the pace, material, and level of difficulty based on how well learners are doing and how interested they are in learning. This ensures that students have the most beneficial learning experience possible.
5. *Virtual labs and simulations*: As students work through online labs and simulations, AI bots can help them learn by pointing them in the right direction and providing direct feedback.

The use of AI agents such as fake educators, trainers, and peers in e-learning settings can make such experiences more fun, useful, and tailored to each student. However, we should consider the moral problems and possible limits associated with such AI-driven jobs. We should ensure that they are developed and utilized in

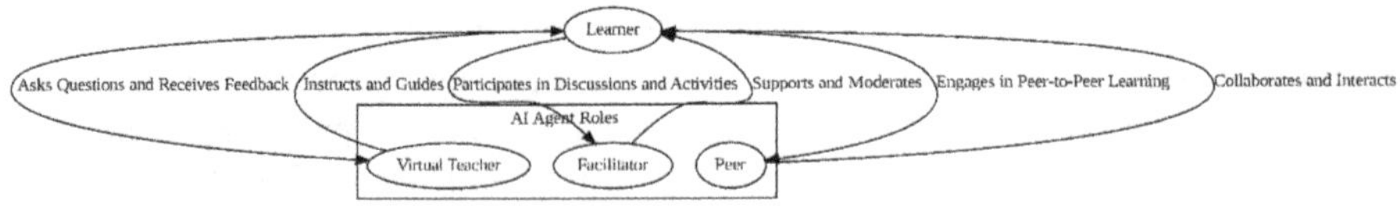

Fig. 4.5 Various roles of AI agents (virtual teachers, facilitators, and peers)

a manner that puts students' needs first and complement teachers rather than replacing them.

4.8.2 Conversations in Natural Words

Natural language conversations: Kim et al. (2020) created contextual dialogue agents who are good at talking about school topics by using recognition of speech, conversational generation, and information retrieval techniques. Processing language signals from peers to keep conversations on track and logical.

Emotional intelligence is the ability to read feelings from body language, facial expressions, and tone of voice (Zhou et al., 2020). Responding with the right amount of understanding, laughter, and support based on how the student is feeling.

Personalized interaction-adaptable dialogue managers create a series of queries and explanations that are specific to each learner's needs based on a review of past talks (Chi et al., 2010). Making sure conversations go smoothly by using student profiles to determine their backgrounds, hobbies, and so on.

These AI capabilities allow virtual agents to foster personalized connections with learners critical for social-emotional engagement, motivation and, ultimately, learning outcomes.

4.8.3 Example Systems and Evaluation Results

Jill Watson (Georgia Tech)

- IBM Watson-based Chabot acts as a teaching assistant in online courses.
- Approximately 40% of the student questions were answered accurately and provided helpful resource recommendations (Goel & Polepeddi, 2016).

SimCoL (University of Sydney)

- Playing the role of a learner participating in collaborative dialogue.
- Modelled different personalities (extravert vs. introvert).
- Qualitative feedback indicated that it improved students' collaboration skills (Zagal & Bruckman, 2005).

Auto Tutor (Memphis University)

- Conversational intelligent tutoring system across multiple domains.
- Dialogues are 35% more effective than reading textbooks.
- Improves learning gains up to 0.8 standard deviations (Graesser et al., 2005).

Overall, these evaluations show that AI-powered virtual agents increase learner motivation, enjoyment, and information retention compared to self-study agents by providing persuasive, social, and adaptive guidance.

4.8.4 Ethical Considerations and Best Practices for Using Virtual Agents Powered by AI

As virtual agents driven by AI become more common in e-learning settings, it is important to discuss ethical issues and establish guidelines for the responsible creation and use of these technologies.

Openness and skill to explain: Learners should be conscious when they are dealing with a computerized agent and be given clear details about what the agent can and cannot do. Developers should try to make artificial intelligence systems that are clear and easy to understand so that teachers and students can know how the agent makes choices or suggestions. This openness helps build faith and gives users the information they need to choose how to connect with the AI helper.

Privacy and data safety: In e-learning settings that use fictional characters with AI, those characters collect and use private information regarding learners, such as the way they acquire knowledge, the amount of homework they do, and personal information. Strict means of data security and protection must be implemented to prevent learners' information from being misused, hacked, or viewed without permission. Data protection rules, such as the FERPA and GDPR, should always be followed. Likewise, learners should be able to decide what information is collected about them and ask for this to be deleted.

Being fair and not favouring anyone: AI-powered virtual beings—could be made and trained to be equal, impartial, and not favouring anyone. People who use algorithms need to ensure that the information used for training is highly accurate and that algorithms do not support or make more severe opinions among humans based on race, gender, or social class. Often, checks and tests should be performed on artificial intelligence (AI) systems to identify and correct any errors they might have. Additionally, any unfair results should be found and fixed all the time.

Oversight and accountability by people: People should keep an eye on and answer to artificial beings that are controlled by AI. These agents may be very useful in e-learning environments, yet they should not be considered a replacement for real teachers. While these tools are being used, it is essential to keep people in charge and responsible. Teachers need to be active in making AI bots, putting them to use, while maintaining an eye on them to ensure that they are following their goals and the best ways to teach. It is important to make clear rules and steps regarding what to do when an AI agent's actions or ideas are harmful or wrong.

Accepting and making AI-powered virtual agents available to everyone: It is important that these agents be able to satisfy the requirements of every learner, including those who have learning difficulties or who prefer different ways to learn. They need to be able to talk to the AI client in a manner that works for them, so the user interface and ways of talking to the AI agent should be flexible and fluid. People who make apps should follow the rules for access and include kids with a variety of needs in the planning and testing stages.

Reviewing and improving all the time: AI-powered virtual assistants should be made and used in online classrooms all the time, and they should be reviewed and improved over time. It is important for developers to make it easy for users to rate how helpful, efficient, and good the AI agent is for studying. Comments and suggestions for how to improve tasks should be provided by both students and teachers. The AI system should always be updated and improved based on this input.

By following these moral standards and guidelines, developers and schools may be sure that AI virtual agents are employed in online education in a helpful and responsible way. Focusing on transparency, confidentiality, fairness, human control, usefulness, and ongoing enhancement will assist students in getting the most from these advancements while reducing any risks or negative impacts they may have.

4.9 AI for Automated Assessment

Unlike traditional multiple choice tests, open-ended assignments such as essays, verbal explanations, diagrams, and presentations allow students to demonstrate higher-order skills such as logical reasoning, creative problem solving, and scientific communication. However, providing detailed evaluation and feedback places a massive grading burden on human teachers (Ke & Ng, 2019).

Emerging AI techniques offer a potential solution through automated assessment at scale. Algorithms can evaluate writing structure, argument quality, vocabulary sophistication, factual accuracy, and other aspects that align with grading rubrics (Burrows et al., 2015). Wang et al. (2020) demonstrated how speech recognition captures verbal responses for textual analysis, while computer vision categorizes visual data.

By comparing open-ended responses to expert reference materials across various criteria, AI promises to lower grading costs, accelerate feedback turnaround times, enable continuous personalized improvement advice, and support overall scalability. However, the effectiveness of these methods depends on training data quality and fairness considerations (Riordan et al., 2017).

While questions remain about assessing higher-order skills, multimodal AI assessment offers data-driven opportunities to enhance open-ended learning activities through rapid analysis and recommendations grounded in real student work (Fig. 4.6).

4.9.1 Speech Recognition, NLP and Computer Vision for Automated Grading and Feedback

Speech recognition involves converting verbal student responses to short answer or essay questions into text through speech-to-text transcription. The assessment entails analysing the vocabulary, concepts, and discourse markers within the transcript to gauge its quality. Compares words used for ideal responses to identify gaps.

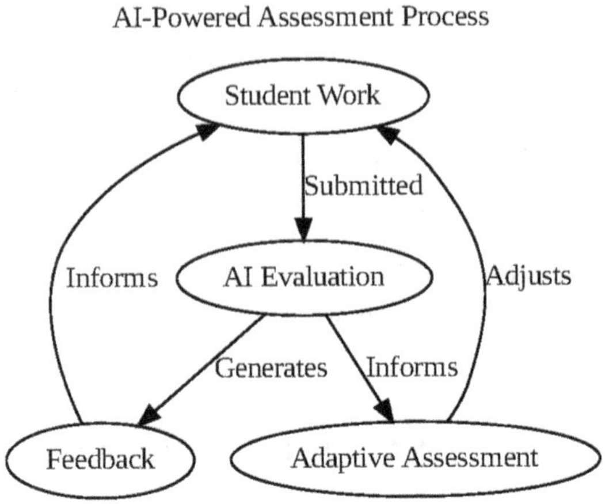

Fig. 4.6 AI-powered automated assessment

NLP: Assesses writing structure, coherence, conciseness, and evidence usage via NLP. Plagiarism checks are conducted by comparing the transcript to an existing corpus. It provides personalized feedback on grammar, readability, and stylistic improvements.

Computer vision: Evaluates visual elements such as charts and diagrams based on expected features. Additionally, it draws outlines around missing labels, incorrect axes, and unsuitable visual comparisons. Finally, simplicity, clarity, and interpretability are measured via image classifiers.

Multimodal connections: This approach links speech analysis with written passage examination to establish a correlation in understanding, ensuring coherence between verbal explanations and visualized representations.

These automated assessments use neural networks trained on expert human ratings and feedback examples. The systems can evaluate responses in realtime and provide each student with individualized and timely suggestions for improvement.

4.9.2 Analysis of Open-Ended Verbal Responses

Analysis of verbal responses: Speech recognition transcribes spoken explanations, stories, sample patient consultations, etc. NLP extracts key ideas and the relationships between concepts, sentiment, flow, and clarity. Dialogue systems can probe for deeper explanations in a conversational way.

Analysis of essays: The evaluation encompasses analysing the structure of writing, the logical progression of key arguments, and the referencing of sources and facts. It also involves scrutinizing grammar, readability, stylistic elements, and plagiarism. Overall holistic scores are provided in addition to feedback on specific improvement areas.

Analysis of diagrams: Computer vision classifies types of diagrams such as flow charts, idea maps, and graphs. The assessment focuses on determining the thoroughness and precision of components and labels. Comparison to example designs and templates for identifying missing parts.

Presentation analysis: A multimodal analysis looks at what's on the slides, what was said, body language, and eye contact. The evaluation looks at aspects such as coherence, engagement strategies, timing, and how the material is organized. It provides advice based on students' skill levels and performance rubrics.

With automated testing, students can perform open-ended work more often, which helps them learn more deeply. AI helps students receive personalized comments based on an analysis of their own work.

4.9.3 *Comparison of Tasks and Accuracy Versus Human Evaluations*

The compaison of tasks and accuracy versus human evaluation is given in the Table 4.1.

Table 4.1 AI automated assessment versus human assessment

Category	AI automated assessment	Human assessment
Accuracy	Accuracy varies greatly based on subjectivity of assignments High accuracy (~75–90%) for scoring objective aspects like grammar, factual claims, etc. Lower accuracy for assessing higher level skills like critical thinking, creativity	Considered gold standard, but prone to inconsistencies and biases Ratings depend heavily on rater experience level
Objectivity	Consistently applies same rubric to all students reducing biases	Prone to grading biases due to mental state, first impressions, etc.
Reliability	Provides highly reliable scoring of structured response aspects Reliability drops for more subjective criteria	Interrater reliability is variable across human raters
Feedback	Can provide detailed, personalized feedback 24/7 year-round	More general feedback provided infrequently due to workload constraints
Logistics	Scales to assess unlimited responses without fatigue or cost overruns	Costs escalate with assessment volume due to human time constraints
Limitations	Incapable of truly understanding responses or social-emotional states Risk of algorithmic biases if data lacks diversity	Lower consistency and bias risks but limited scalability Prone to fatigue which can reduce accuracy

4.9.4 Integrating Automated Assessment with Human Evaluation for Optimal Learning Outcomes

While AI-powered automated assessment has many benefits, such as speed, scalability, and instant feedback, it is important to remember how important human evaluation is to the learning process. By using the best parts of both manual and automated evaluation, combining them can lead to the best learning results.

The roles of AI and human evaluation are as follows: Work together—autograding with AI works well for parts of learning that are organized based on rules and objective, such as writing, grammar, and remembering basic facts. Human evaluation, on the other hand, is great at judging things such as creativity, originality, and understanding of the situation. By putting those abilities that work well together, teachers can make a review plan that is more thorough and useful.

Validation and calibration: It is important to try AI-powered automated reviews with human knowledge to ensure that they are correct and dependable. There are scores and notes that were made by AI that educators should look at to clarify that the system is functioning right and making any modifications that are needed. This is how an artificial intelligence system should continue to be tuned as it grows and changes according to new knowledge and events.

Human-in-the-loop: Putting a person in the loop means that human judges check over and comment on the assessments made by AI. This approach can help ensure that the assessments are fair and of high quality. The way things are set up now allows instructors to change or reverse AI choices whenever needed. In this way, the final grade accurately represents the student's work.

Strategies for adaptive assessment: When tests are given by both humans and computers, tests can be adapted to each student's needs. Artificial intelligence can provide data on how well students are doing in school to determine what they need to work on. This allows teachers to help the kids who need it the most. Judges who are human can then make more targeted notes and help according to these insights, which makes learning more personalized.

Being open and allowing students to help: When AI and human reviewers are used together, it is best to make things clear and let students help with the review process. Participants need to be told how their work will be judged and what role AI and people will play. Giving kids time to look over and reflect on what they are doing can help them see how they are doing and learn how to acquire knowledge on their own.

Working together and getting better at your job: For AI and humans to work effectively together, instructors need to keep getting better. AI-made test data can be difficult to read and understand, but teachers should learn how to use them to provide helpful comments and help students when needed. Getting teachers, people who make AI, and people who study education to collaborate together can help people find the best ways to teach and test kids.

By combining artificial intelligence with human review, teachers can develop an assessment method that is more complete, accurate, and flexible. This will help students learn better. To increase the speed and capacity of AI, this integration maintains the in-depth comprehension and understanding of the background that genuine judges offer to the table. Using adaptive strategies, constant calibration, transparency, and professional development, colleges can use AI-powered test tools and ensure that the process is fair and of high quality.

4.10 Broader Impacts, Limitations, and Future Outlook

There are both good and bad things that could happen to communities as a whole because AI is becoming more popular in schools. On the one hand, these technologies might make it simpler for everybody on the globe to obtain high-quality education because they can tailor lessons to each person and lower the cost of learning. Systems with AI might alter the way that we study by making lessons that change based on what each student needs. This change could make opportunities much more available to everyone. However, automating teaching jobs could mean job losses and unfair treatment of people with skills from new, nontraditional education paths. As AI systems gather more personal information about students to allow personalization, there are more ethical issues to consider in regard to ongoing consent and data privacy.

4.10.1 Potential and Promises Versus Hurdles and Pitfalls

Using AI in education has great potential to change the way people learn, make quality education more accessible, and encourage personalized learning. However, some problems and risks need to be carefully avoided to ensure that everyone benefits and that the results are fair for everyone.

4.10.1.1 Possibilities and Promises

1. *Personalized learning on large*-scale systems driven by AI can involve the use of large amounts of data about each student, which allows teachers to tailor each student's lessons to their strengths, weaknesses, and preferred ways of learning. This customization can help students who might have trouble in standard, one-size-fits-all classrooms become more interested, motivated, and successful learners.
2. *Better access and inclusion* of AI technologies can help close the achievement gap by giving students in rural or underserved areas access to high-quality learning materials and help. Smart tutoring and adaptive learning tools can provide each student with individualized help and feedback, allowing them to learn at

their own pace and in their own way, regardless of where they are or what their background is.

3. *Higher efficiency and lower costs*: AI can automate boring jobs such as grading and content creation, so teachers can focus on more important things, such as providing each student with individualized help and encouraging them to think critically. This can help schools save money by making them more efficient, which could make good education cheaper and easier to obtain.
4. *Always getting better and coming up with new ideas*: As AI systems gather and study more information about how students behave and what they learn, they can always get better and adjust to new needs in education. This iterative process can lead to new ways of teaching, making lessons, and testing students, all of which can make learning more effective and interesting.

4.10.1.2 Obstacles and Traps

1. *Digital gap and unequal access*: Anybody wants to be ready to use technological advances, have a good internet link, and grasp how to utilize technology correctly for AI to be effective in education. It might be more difficult for children from poor families to use AI-powered learning tools, which would worsen the current disparity in education. It is essential to ensure that everyone has the same access to information and to give persons the help and instruction they need to close the digital gap.
2. *Worries about bias and fairness*: AI systems can add to or amplify biases that already exist in the information that was used to train them, which can bring about unfair outcomes. This might make tests unfair, give different people various opportunities to learn, or support misconceptions in the classroom. It is essential to help many people involved in the development and application of artificial intelligence (AI) systems ensure that they are fair all the time.
3. *Privacy and data security*: Finding and studying private school data makes people worry about their privacy, safety, and chance of abuse. It is important for schools to set up strong data control mechanisms and follow precise privacy rules to keep students' personal information safe. For trust and openness to stay, there should be specific regulations about that person's information, who can see it, and how it may be used.
4. *Relying too much on technology and missing out on social interactions with real people*: AI—can help people learn, but if they rely too much, it could make real teachers less important and people miss out on important social interactions. It is important to find the right balance between lessons that are run by AI and lessons that are run by humans to help clear learners learn key social, emotional, and relationship skills.
5. *Moral issues and responsible use*: Using AI in schools raises moral concerns about security, openness, and duty. It is essential to ensure that artificial intelligence systems are built and used in a way that does not hurt people. There should

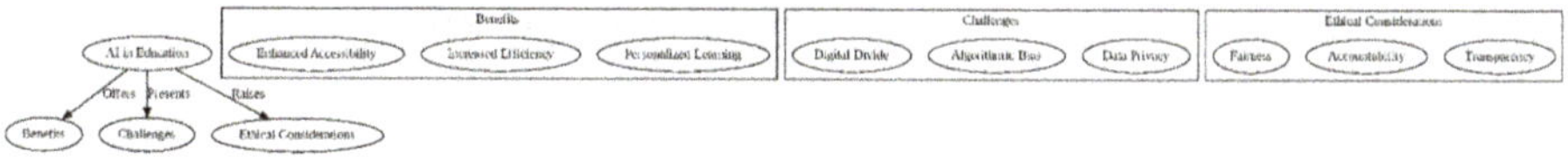

Fig. 4.7 AI in education: Benefits, challenges, and ethical considerations

be clear rules about how they are to be utilized and who needs to be in control of them.

To address these problems and issues, educators, lawmakers, AI developers, and other interested parties must continue working together. AI has the ability to make learning more fair, effective, and fun for everyone. However, we need to take action to address these problems and develop approaches that are open, honest, and based on ethics (Fig. 4.7).

4.11 Future Outlook

As AI methods in computer vision, natural language processing, reinforcement learning, and neural networks continue to improve, they could be used in education in ways that are more personalized and open to more people than ever before. Systems that use emotional computing and multimodal analytics are likely to provide more detailed information about what stops people from learning. These systems will make problems more obvious, which is something that human teachers often find hard to do when dealing with different cultural and personal situations.

Recent progress in foundational models and transfer learning suggests that it might be possible to directly encode more complex reasoning skills than just understanding basic material. This could make it possible for AI tutors and test-takers who are better at reading and writing, as well as creative and critical thought. Virtual reality, augmented reality, and multiagent simulations can also create more realistic settings where one can practice making decisions and using one's skills in the real world.

To fully realize these possibilities, however, education experts, computer scientists, and policymakers will need to work together to ensure that these systems have the right moral, emotional, and social safeguards built in. We have a lot of hope for long-term collaborations that bring together experts from different fields to help everyone learn more. However, we also have a lot of duty.

4.12 Suggestions for the Safe Development and Use of AI in Schools

Everyone needs to work together and follow the rules to ensure that AI is built and used properly in education. These include educators, legislators, AI developers, and educational institutions. There are good ideas below for how to develop the use of AI-powered educational resources in a way that is moral, includes everyone, and works well.

1. *Make moral rules and guidelines clear*: You should develop an exhaustive list of moral guidelines and requirements for using AI in schools. The regulations should cover important issues such as privacy, bias and fairness, responsibility, openness, and student freedom. Always make sure that these regulations are adhered to by everyone who has a say in how AI-powered learning tools are made and used.
2. *Get individuals from various disciplines to work together*: For example, get educators, AI developers, academics, ethicists, and policymakers collaborate to ensure that planned and used AI-powered educational technologies fully consider the needs of students, technical options, and ethical challenges that arise. This way that brings together ideas from different fields will assist in making solutions that are good for instruction, powerful, scientific, and ethically sound.
3. First, to protect sensitive student data, we put in place strict data privacy and security measures. The government should follow the rules for protecting personal information, such as the FERPA and GDPR, and ensure that learners and their families can control their data by choosing not to have it collected or asking for it to be deleted. Check data methods on a regular basis and be open about how the data are used and stored.
4. *Deal with bias and ensure that everyone is treated fairly*: Ensure that AI systems are regularly checked to find and fix any biases that might be present in the data, computations, or results. Ensure that the information used for training is diverse, accurate, and includes everyone and develop ways to find and fix results that are unfair. Many people, such as underrepresented groups, are asked to determine what they think about how fair and inclusive AI-driven educational tools are.
5. *Be clear and easy to understand*: Make sure that AI-powered educational tools are clear about their features, limitations, and how they make decisions. AI models can be constructed so that teachers and students can determine how the system makes decisions or suggestions. Tell people about enhancements and modifications to artificial intelligence on a regular basis to maintain trust and hold people accountable.
6. *Produce money for professional development and training*: Give teachers ongoing opportunities to learn more about AI and improve their skills and knowledge. Teachers can help teachers understand the pros, cons, and moral issues of AI-powered educational technologies and provide them with the information and tools they need to use these technologies effectively in their classrooms.

7. *Get students involved and stress their independence*: Students should be involved in the planning, creation, and testing of AI-based teaching technologies. Their thoughts and feedback were obtained to ensure that these tools met their wants and needs. Focus on giving students choice and control over how they use AI tools so that they can make smart choices about how they learn and how their data are shared.
8. *Continuously maintain an eye on and review the effects*: Plan ways to continually keep a watch on and evaluate how AI-powered educational tools affect students' well-being, how well they learn, and how fair they are. With this knowledge, you can keep improving how these tools are made and how they are used. Additionally, share your best ideas and lessons learned with other teachers.
9. *Make regulations and guidelines that everyone has to follow*: Set norms and policies that everyone must follow to control how AI is made, bought, and used in schools. The safety, health, and fairness of students should first follow these rules. They should also make items clear for educational institutions, technology companies, and other people who are interested in the matter. As AI changes quickly, these rules often change.

Everyone can work towards sustainable development and the implementation of AI in education by following these ideas. In this way, everyone can make the most of the ability of AI to improve learning while reducing threats *and making sure that all pupils obtain results that are fair, moral, and include everyone.*

4.13 Conclusion

It is possible that AI could change the way we learn and teach by giving everyone personalized, flexible, and fun classes. The various ways in which AI can be employed in education are discussed in this chapter. These include customized and adaptive learning, making materials available in a variety of forms, virtual peers and teachers, automated testing, and the impacts these advances have on the community as a whole.

AI could help education in many ways, such as by making learning more effective, making it possible for more people to obtain a quality education, and making teaching and learning more smooth. When AI is used, it can look at a large amount of data to create customized educational paths, make content change constantly, and provide students with comments and help. These abilities can help everyone learn what they need, close gaps in education, and inspire individuals to learn all their lives.

However, the use of AI in schools and its growth also raise major issues and ethical concerns that must be considered. Making sure that everyone has the same access to AI-powered learning tools, lowering bias and discrimination, safeguarding learners' privacy and data, and ensuring equilibrium between learning with people as well as with technology are some of the most important issues that need to be resolved.

While these problems are being fixed, it is essential to put ethical artificial intelligence development and application at the very top of the list. This will allow AI to be used in education. This can occur if clear laws and standards are put in place, people from different fields are encouraged to work together, professional growth and training are paid for, students are involved in planning and testing, and the impact of powered AI educational technologies is checked on a regular basis.

By following these rules and ideas, educators, lawmakers, AI developers, and anyone else who wants to work together can create AI-powered learning spaces that are fair, useful, and open to everyone. The smart use of AI in education could change how we teach and learn. Doing so would give teachers more tools to help and support each student, and it would give learners an opportunity to acquire the abilities they are going to require to do effectively in an environment that is becoming increasingly technological and AI-driven.

AI is not an answer for every issue in education. This is important to keep in mind, as we contemplate the future. Instead, it is a useful tool that might help people become smarter and more clever. We can give all of our learners engaging, helpful, and powerful learning experiences if we use AI to its fullest while keeping true to morals and a commitment to human-centered learning.

Building AI and using it in educational institutions in a smart way opens up both exciting options and tough problems. In a world in which every student can obtain a personalized, high-quality education that helps them reach their full potential, we may transform education if we all work jointly to solve these issues and make the most of AI.

References

Adams Becker, S., Cummins, M., Davis, A., Freeman, A., Hall Giesinger, C., & Ananthanarayanan, V. (2017). NMC horizon report: 2017 higher education edition. .

Adamson, D., Dyke, G., Jang, H., & Rosé, C. P. (2014). Towards an agile approach to adapting dynamic collaboration support to student needs. *International Journal of Artificial Intelligence in Education, 24*(1), 92–124.

Anderson, J. R., & Reiser, B. J. (1985). The LISP tutor. *Byte, 10*(4), 159–175.

Aroyo, L., Gelan, C., Scheffel, M., Ternier, S., & Kravcik, M. (2019). Semantic annotation of video content for personalized learning. *IEEE Transactions on Learning Technologies, 12*(4), 535–549.

Baltrušaitis, T., Ahuja, C., & Morency, L. P. (2018). Multimodal machine learning: A survey and taxonomy. *IEEE Transactions on Pattern Analysis and Machine Intelligence, 41*(2), 423–443.

Bommasani, R., Hudson, D. A., Adeli, E., Altman, R., Arora, S., von Arx, S., et al. (2021). On the opportunities and risks of foundation models. *arXiv preprint arXiv:2108.07258*.

Brown, J. S., Burton, R. R., & de Kleer, J. (1982). Pedagogical, natural language and knowledge engineering techniques in SOPHIE I, II and III. In D. Sleeman & J. S. Brown (Eds.), *Intelligent tutoring systems* (pp. 227–282). Academic Press.

Brusilovsky, P. (2001). Adaptive hypermedia. *User Modelling and User-Adapted Interaction, 11*(1–2), 87–110.

Burrows, S., Gurevych, I., & Stein, B. (2015). Automated writing evaluation for formative assessment of English compositions: Updates to the ETS criterion online writing service. In *Handbook of educational data mining* (pp. 277–285). CRC Press.

Chen, Q., & Zhu, S. (2019). A study on data mining models for learning analysis in intelligent classroom. In *2019 International Conference on Machine Learning and Cybernetics (ICMLC)* (Vol. 1, pp. 1–6). IEEE.

Chi, M., VanLehn, K., Litman, D., & Jordan, P. (2010). Empirically evaluating the application of reinforcement learning to the induction of effective and adaptive pedagogical strategies. *User Modelling and User-Adapted Interaction, 20*(2), 137–180.

Echeverria, V., Avendaño, G., Chiluiza, K., Vásquez-Moreno, Á., & Ochoa, X. (2018). Microinteractions and multimodal learning analytics to support self-regulated learning in blended courses. *Computers & Education, 126*, 413–432.

Goel, A. K., & Polepeddi, L. (2016). *Jill Watson: A virtual teaching assistant for online education*. Georgia Institute of Technology.

Graesser, A. C., Chipman, P., Haynes, B. C., & Olney, A. (2005). AutoTutor: An intelligent tutoring system with mixed-initiative dialogue. *IEEE Transactions on Education, 48*(4), 612–618.

HolonIQ. (2020). *The $10bn+ market for artificial intelligence in education*. HolonIQ.

Kanda, T., Shimada, M., & Koizumi, S. (2012). Children learning with a social robot. *Human Computer Interaction, 35*(5–6), 413–423.

Kang, J., & Liew, J. (2020). GAN QG: A technique for automatic question generation using GAN. *arXiv preprint arXiv:2009.09481*.

Ke, Z., & Ng, H. T. (2019). Automated essay scoring: A survey of the state of the art. In *Proceedings of the Twenty-Eighth International Joint Conference on Artificial Intelligence* (pp. 6300–6308).

Khan, F., & Ponzanelli, L. (2021). DiffQE: Differential network for personalized explanations of bots. In *Proceedings of the 29th ACM Joint Meeting on European Software Engineering Conference and symposium on the foundations of software engineering* (pp. 1700–1704).

Kim, D., Liu, A., Goth, J., Weitzman, R., & Hirsh, P. (2020). Candid: Canada's dataset for online naturalistic diagnostic improvisational dialogues. *arXiv preprint arXiv:2012.08450*.

Knewton. (2018). *Arizona State University: Developmental math students achieve higher pass rates and more consistency with Knewton Alta*. Knewton.

Kumar, A., Wang, H., Bamman, D., & Hebert, M. (2020a). Iterative machine teaching. In *Proceedings of the 37th International Conference on Machine Learning* (pp. 5503–5514). PMLR.

Kumar, A., Goyal, N., & Talukdar, P. P. (2020b). IDEAL: Interface for developing question answering datasets. In *Proceedings of the 28th International conference on computational linguistics: System demonstrations*.

Learning, D. B. (2016). *Intelligent adaptive learning: An essential element of 21st century teaching and learning*. DreamBox Learning.

Lee, J., & Hannafin, M. J. (2016). A design framework for enhancing engagement in student-centered learning: Own it, learn it, and share it. *Educational Technology Research and Development, 64*(4), 707–734.

Liu, D., Bhagat, R., Fu, Q., & Lee, C. H. (2019). iCAN: Context-aware automatic negative sampling for personalized review summarization. In *Proceedings of the 28th International joint conference on artificial intelligence* (pp. 5120–5126).

Luckin, R., Holmes, W., Griffiths, M., & Forcier, L. B. (2016). *Intelligence unleashed: An argument for AI in education*. Pearson.

Ma, W., Adesope, O. O., Nesbit, J. C., & Liu, Q. (2014). Intelligent tutoring systems and learning outcomes: A meta-analysis. *Journal of Educational Psychology, 106*(4), 901.

Martin, L., Muller, B., Suárez, P. J. O., DuPont, Y., Romary, L., de la Clergerie, É. V., et al. (2020). CamemBERT: a tasty French language model. In *Proceedings of the 58th Annual Meeting of the Association for Computational Linguistics*.

Matsuda, N., Yarzebinski, E., Keiser, V., Raizada, R., Stylianides, G. J., & Koedinger, K. R. (2013). Cognitive anatomy of tutor learning: Lessons learned with SimStudent. *Journal of Educational Psychology, 105*(4), 1152.

Merchant, Z., Goetz, E. T., Cifuentes, L., Keeney-Kennicutt, W., & Davis, T. J. (2014). Effectiveness of virtual reality-based instruction on students' learning outcomes in K-12 and higher education: A meta-analysis. *Computers & Education, 70*, 29–40.

Nakic, J., Granic, A., & Glavinic, V. (2015). Anatomy of student models in adaptive learning systems: A systematic literature review of individual differences from 2001 to 2013. *Journal of Educational Computing Research, 51*(4), 459–489.

Nwana, H. S. (1990). Intelligent tutoring systems: An overview. *Artificial Intelligence Review, 4*(4), 251–277.

Panwar, K., Singh, A., Kukreja, S., Singh, K. K., Shakhovska, N., & Boichuk, A. (2023). Encipher GAN: An end-to-end color image encryption system using a deep generative model. *System, 11*(1), 36.

Pérez, S., Jurado, F., & Pérez, A. (2020). Artificial intelligence for student modeling and personalized tutoring in MOOCs: A review. *IEEE Access, 8*, 176138–176150.

Pullin, A., Maclin, R., Simko, M., Weinberger, K. Q., & Hovy, E. (2021). Score-CAM: Score-weighted visual explanations for convolutional neural networks. In *Proceedings of the 2021 conference on empirical methods in natural language processing* (pp. 7909–7929).

Rasheed, R. A., Kamsin, A., & Abdullah, N. A. (2020). Challenges in the online component of blended learning: A systematic review. *Computers & Education, 144*, 103701.

Riordan, B., Horsley, T. L., Heitzman, K., Khanduja, P., & Moss, J. D. (2017). How we know what we know: A systematic comparison of research methods employed in higher education journals, 1996–2000 v. 2006–2010. *The Journal of Higher Education, 82*, 171–198.

Ritter, S., Anderson, J. R., Koedinger, K. R., & Corbett, A. (2007). Cognitive tutor: Applied research in mathematics education. *Psychonomic Bulletin & Review, 14*(2), 249–255.

Romero, C., & Ventura, S. (2010). Educational data mining: A review of the state of the art. *IEEE Transactions on Systems, Man, and Cybernetics, Part C (Applications and Reviews), 40*(6), 601–618.

Romero, C., & Ventura, S. (2020). Educational data mining and learning analytics: An updated survey. *WIREs Data Mining and Knowledge Discovery, 10*(3), e1355.

Schatten, M., Cvjetković, V. J., & Bajtoš, M. (2020). Reinforcement learning in personalized learning systems. In *43rd International convention on information, communication and electronic technology (MIPRO)* (pp. 1825–1830). IEEE.

Shah, D. (2020). How artificial intelligence is helping eLearning. *eLearning Industry*.

Siddharthan, A., & Angrosh, M. (2014). Text simplification using synchronous dependency grammars: Generalizing automatically harvested rules. In *Proceedings of the 8th International natural language generation conference (INLG)* (pp. 16–25).

Singh, A., & Saravanan, V. (2024). XAI decision MODELS: Programming models for decentralized BlockXAI. In *Convergence of Blockchain and explainable artificial intelligence* (pp. 15–22). River Publishers.

Singh, S. P., Kumar, N., Singh, A., Singh, K. K., Askar, S. S., & Abouhawwash, M. (2024). Energy efficient hybrid evolutionary algorithm for internet of everything (IoE)-enabled 6G. *IEEE Access, 12*, 63839.

Taggart, W. (2022). *Anthropic—AI for education built on constitutional AI*. Anthropic Blog.

Tseng, S. S., Sue, P. C., Su, J. M., Weng, J. F., & Tsai, W. N. (2008a). A new approach for constructing the concept map. *Computers & Education, 51*(3), 1291–1305.

Tseng, S. S., Su, J. M., Hwang, G. J., Hwang, G. H., Tsai, C. C., & Tsai, C. J. (2008b). An object-oriented course framework for developing adaptive learning systems. *Educational Technology & Society, 11*(2), 171–191.

Vahdat, M., Macready, W. G., Bian, Z., Khoshaman, A., Andriyash, E., & Kamenov, G. (2021). DVAE++: Discrete variational autoencoders with overlapping transformations. In *Proceedings of the 38th International Conference on Machine Learning* (pp. 10706–10718). PMLR.

VanLehn, K. (2011). The relative effectiveness of human tutoring, intelligent tutoring systems, and other tutoring systems. *Educational Psychologist, 46*(4), 197–221.

Vesselinov, R., & Grego, J. (2012). *Duolingo effectiveness study*. City University of New York.

Wang, S., Liu, Y., Ouyang, B., Zhu, M., & Xiong, X. (2020). Multimodal learning analytics with multiple classifiers for essay scoring. In *Proceedings of the Tenth International Conference on Learning Analytics & Knowledge* (pp. 293–302).

Wang, Y., Liu, Q., & Shi, B. (2021). AutoMath: Solving mathematical word problems with multi-equation and multi-step. In *Findings of the Association for Computational Linguistics: ACL-IJCNLP 2021*.

Whitehill, J., Serpell, Z., Lin, Y. C., Foster, A., & Movellan, J. R. (2014). The faces of engagement: Automatic recognition of student engagement from facial expressions. *IEEE Transactions on Affective Computing, 5*(1), 86–98.

Worsley, M., & Blikstein, P. (2014). Analyzing multimodal multilinear data for learning analytics: Using students' gaze data to predict programming performance. In *Proceedings of the 2014 ACM workshop on multimodal Learning analytics workshop and grand challenge* (pp. 57–64).

Zagal, J. P., & Bruckman, A. (2005). From SIMO to SIMS: Simulating collaborative learning environments. *Simulation & Gaming, 36*(3), 393–413.

Zawacki-Richter, O., Marín, V. I., Bond, M., & Gouverneur, F. (2019). Systematic review of research on artificial intelligence applications in higher education—where are the educators? *International Journal of Educational Technology in Higher Education, 16*(1), 1–27.

Zhang, J., Karimi, H., Tang, J., Xie, X., & Hsieh, C. J. (2022). Geo-ALCE: Geometry-aware active learning for chemistry equation generon. In *Proceedings of the AAAI Conference on Artificial Intelligence* (Vol. 36, pp. 12765–12773).

Zhou, Y., Zhang, Z., Wang, B., & Yang, W. (2020). Pattern recognition techniques for detecting learner's emotions in the context of mathematics e-learning. In *International Conference on Blended Learning* (pp. 497–508). Springer.

Zhu, H., Zhao, H., Chai, J., & Zhang, L. (2021). Talk the walk: Navigating agents towards more human-like behaviors. In *Proceedings of the 2021 CHI conference on human factors in computing systems* (pp. 1–15).

Chapter 5
Generative Artificial Intelligence in Visual Content: A Review of the Influence on Consumer Perception and Perspective

Akanksha Singh, Gulshan Kumar, and Akashdeep Dhariwal

Abstract This research delves at the ways in which several sectors' customer perceptions are influenced by the use of artificial intelligence (AI) to create visual content. It draws attention to the ways in which speech synthesis and other AI-powered services are changing the way people shop and consume digital entertainment. Establishing and upholding customer confidence in AI-generated content requires a commitment to transparency and a steadfast adherence to ethical principles. In addition, previous studies propose that consumer perspective might be better understood with the use of sensory data integrated with biometric information, which could result in improved customer experiences and more specific promotional initiatives. A number of studies indicate that AI has a major effect on what people consume. However, further study is needed to completely grasp how AI impacts buyer decision-making, as there is currently a substantial information gap. Those performing these studies must take the lead in addressing this shortfall if they want to fill this knowledge void and educate future researchers in this field of research.

Keywords Artificial intelligence · Consumer perception · AI-generated image · AI-generated video

5.1 Introduction

Over the past few years, there has been widespread penetration into the technological sector that deals with digital media organization. Owing to automation technologies, advanced machine learning and AI capabilities are already available in today's computer systems. Through Visual CAIs, the potential for creating visual imagery—which can also be defined as the simulation of visual perception—is made

A. Singh (✉) · G. Kumar · A. Dhariwal
Amity School of Communication, Amity University, Greater Noida, Uttar Pradesh, India
e-mail: asingh@gn.amity.edu

A. Singh, K. K. Singh (eds.), *Multimodal Generative AI*,
https://doi.org/10.1007/978-981-96-2355-6_5

available. This could potentially be used for films or images that appear to be years of age.

When one says "computer," they should imply any kind of logic based on numbers. There was a severe lack of data available before the advent of deep learning. A number of data-driven tools, however, are now at our fingertips due to the robust deep learning framework. Firstly, it denotes the means by which the industry is able to monitor and assess its own performance. Through advertising, journalism, and mass entertainment, we are reminded of our progress towards our ideal world. Combining technologies for pervasive sensing, intelligent computing, cooperative communication, and mass data management may enhance urban surroundings, quality of life, and smart city systems (Singh et al., 2023). It is quite uncommon to find individuals who possess the ability to effectively interpret human sensations and utilize artificial intelligence output in their work (Neyazi et al., 2023). They are highly valued due to their essential contribution, which is not very common. Deep learning has also shown significant promise in the healthcare field. By accumulating vast amounts of patient records and data, together with a shift towards individualized therapies. Automated and dependable processing and analysis of health information is very necessary (Singh et al., 2023). The design of advanced analytical models are also focusing on creating conceptual models that provide practical direction and prioritize risks and assaults inside the network system (Raghunath et al., 2022). Remote sensing image change detection is also extensively utilized in diverse applications such as urban growth monitoring, land use/cover mapping, forestry, agriculture, biomedical imaging, and disaster damage assessment (Singh et al., 2013). Intelligent transportation systems (ITS) refer to the incorporation of information and communications technology into traffic control and management applications (Singh et al., 2024).

In the past decade, AI scientists have been at the forefront, showcasing the incredible advancements in artificial intelligence that continue to amaze us. Our ultimate goal is to prevent the occurrence of any major catastrophic man-made disasters through the success of AI. Their ability to constantly come up with new and innovative ideas leads to widespread adoption by both traditional and modern elements. It's no surprise that our culture's distinctiveness continues to accumulate, ensuring that every experience remains new and intriguing. In a rapidly expanding field, the adoption of AI has given rise to a complex landscape where distinguishing between reality and artificial reality requires a thorough analysis. The advent of generative AI models has made this change especially noticeable (Ramdurai & Adhithya, 2023). Sophisticated artificial intelligence (AI) systems known as "generative models" can produce new material that is akin to the datasets they were trained on (citation). AI-generated visuals, with their heightened attractiveness and hyper-realistic quality, could potentially intensify this effect (Califano & Spence, 2024). Technological advancement brings about societal transformation, thus we should constantly be asking how new developments in technology influence, create, or even threaten the "good society" (Griffy-Brown et al., 2018). Artificial intelligence (AI)-generated faces are now readily accessible (see, for example, this-person-does-not-exist.com) and are being used for both good and evil intentions, such as tracking down missing children and spreading false information about politics through fictitious social media accounts (Kertysova, 2018). AI algorithms, particularly Generative Adversarial Networks (GANs), can generate highly realistic images and videos from scratch based on input data or specific instructions. These AI-generated assets can be used for various purposes, including creating backgrounds, characters, special effects, or even entire scenes in movies, animation, and gaming (Anantrasirichai & Bull, 2021). AI can also automate the process of creating video thumbnails, social media posts, or promotional images by analyzing content and generating visually appealing visuals (Sharakhina et al., 2023). AI-powered image and video analysis tools can automatically tag and categorize content based on its visual attributes, such as objects, scenes, colors, and emotions (Chen et al., 2022). This enables efficient content organization, searchability, and recommendation systems for media libraries, stock footage platforms, and content management systems (Afzal et al., 2023). AI tools offer advanced video editing capabilities, such as automated video stabilization, object removal, color grading, and scene segmentation (Yazadzhiyan, 2023). Generative AI has profoundly reshaped the drug discovery industry (Walters & Murcko, 2020), and it is starting to make an impact on online platforms. One exception is Lysyakov and (Zhavoronkov & Aspuru-Guzik, 2020), who find that the introduction of an AI logo designer drives lower-tier designers to leave a crowd-sourcing platform while forcing higher-tier designers to become involved in more complex designing tasks. Their research primarily focuses on the perception of people towards AI-generated video or image content. For, this the study has done reviews of previous literature and has answered these two questions:

Q1 What are the most influential studies regarding the perception and interaction of individuals with AI-generated visual content?

Q2 What is the effect of AI-generated images on consumer perception?

5.1.1 Methodology

The methodology for this review paper involved a thorough search of the research database Scopus to identify relevant peer-reviewed studies. Inclusion criteria focused on research database development, usage, impact, and challenges, while non-English and non-peer-reviewed studies were excluded. Data extraction involved collecting information on author(s), publication year, databases studied, methodology, key findings, and conclusions. Synthesizing the extracted data allowed for the identification of the gaps in the literature. A critical analysis was conducted to evaluate the consumer perception of visual content generated by AI.

Figure 5.1 shows the selection data from the Scopus database. Scopus is a reliable source which is why the study has taken it as a Scopus database. The total number of literature that was found was 35, from which 19 records were removed as they were not relevant to this topic. Only published articles in the English language were taken for the study.

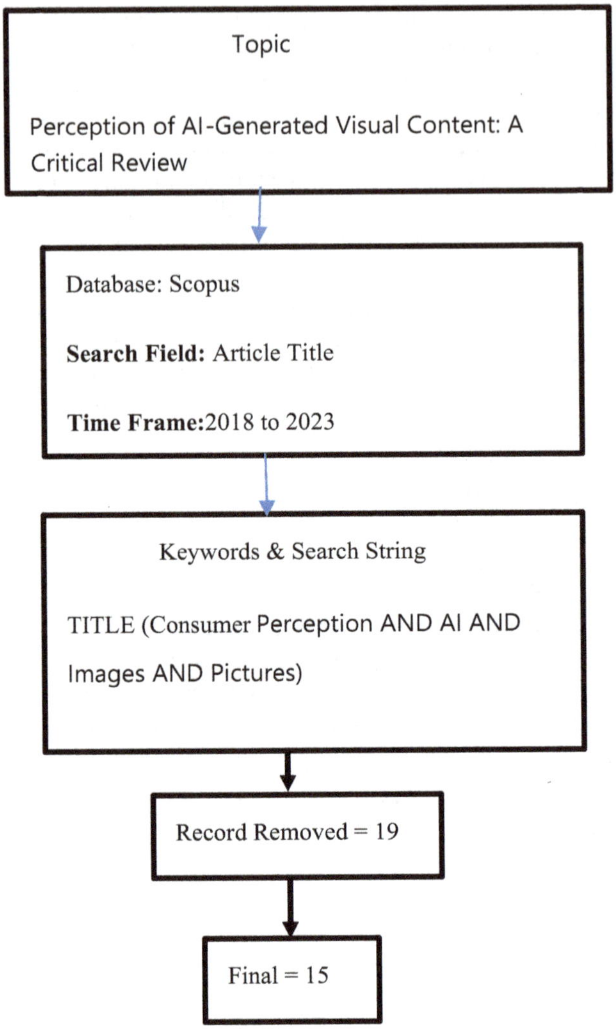

Fig. 5.1 Selection overview

The study has shown the top 10 papers related to the Perception of AI-Generated Visual Content. The top paper as can be seen in Fig. 5.2 is "Trust Me, If You Can: A Study on the Factors that Influence Consumers' Purchase Intention Triggered by Chatbots Based on Brain Image Evidence and Self-reported Assessments" in which the author (Lysyakov & Viswanathan, 2023) has investigated consumer trust in chatbots, finding that credibility, competence, anthropomorphism, social presence, and informativeness influence trust and purchase intention (So et al., 2023).

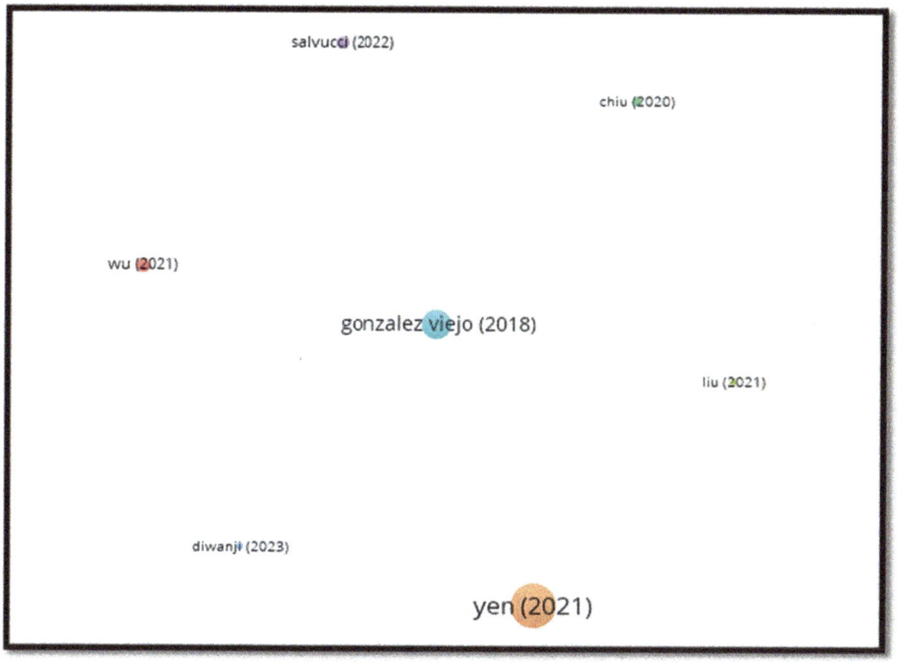

Fig. 5.2 Top 7 documents

5.2 Literature Review

The 15 review papers that were chosen for this study are displayed in Table 5.1. The literature study shows a considerable impact on consumers' perceptions of AI-generated tools based on findings from prior studies. But it also draws attention to a vacuum in the literature, indicating that further research on this subject is necessary given its potential significance. This disparity indicates that further research and analysis are required to fully understand how AI-generated products affect consumer impressions (Yen & Chiang, 2020). Researchers may fill up this knowledge vacuum and provide important new perspectives on how AI affects consumer behaviour and decision-making.

AI-generated video material has a significant influence on consumer views in a variety of industries. Artificial intelligence-generated voices have deliberated influence (Pellas, 2023). On the other hand, AI has an impact on how people perceive the travel and tourism sector (Efthymiou et al., 2023). Furthermore, AI-based techniques improve customer experiences by efficiently bridging the gap between natural products and online food representations (Cao & Xiao, 2023). AI-driven solutions in digital entertainment provide immersive experiences that change the way users interact with information (Wu & Chien, 2021).

Table 5.1 Key findings from recent studies on AI generated visual content on consumer perception

Authors	Title	Journal	Technological aspect of the study	Findings
Efthymiou F.; Hildebrand C.; de Bellis E.; Hampton W.H.	The power of AI-generated voices: How digital vocal tract length shapes product congruency and Ad performance	Journal of Interactive Marketing	The use of artificial intelligence in voice synthesis is the main focus of this work, specifically about the control of digital vocal tract length in conversational bots driven by AI. By modifying the digital VTL, it is feasible to replicate inherent fluctuations in human vocal tract attributes and, consequently, impact customer perceptions and actions in advertising environments The study discusses developments in computational methods for modifying particular vocal characteristics, such as pitch, speech rate, and formant frequencies, in order to generate voices with desirable qualities that improve impressions of product congruency and advertising efficacy. These manipulations are facilitated by advancements in text-to-speech systems and improvements in acoustic modeling and vocoding, which are integral to the voice synthesis process	The findings of the research indicate that AI-generated voices can indeed be intentionally designed to influence consumer perceptions

Cao P.; Xiao J.	Study on the coordinated development of national traditional sports and tourism brands based on big data platforms from the perspective of "The Belt and Road"	Journal of Intelligent and Fuzzy Systems	The study utilized a big data platform and the UGC network structure of the internet of things (IoT) and relied on wireless networks and AI solutions. As a starting point, online travel notes were used as examples to organize tourist images submitted by clients. The next step was to use Python's big data and AI capabilities to extract keywords from tourism texts. This helped to learn how customers perceived popular scenic spots, services, and social and cultural norms	The study highlights significant advancements in tourism management along the Belt and Road initiative, facilitated by AI technology and big data analytics. Leveraging online travel notes and AI tools enables a better understanding of tourist preferences, enhancing the management of scenic locations and tourism services. An integrated strategy utilizing AI and big data platforms proves more effective in tourism brand development, fostering common growth in sports tourism and intelligent tourism image management
Wu P.-J.; Chien C.-L.	AI-based quality risk management in omnichannel operations: O2O food dissimilarity	Computers and Industrial Engineering	A convolutional neural network (CNN) based two-stage deep learning approach to reduce quality risk in O2O food retail. As a first step, Meal-Object recognition collects food photos taken offline and processes them using convolutional neural network (CNN) architectures tuned for object recognition. This separates the meal items from the background noise Using deep convolutional neural networks (CNNs) for feature extraction, stage 2 similarity scoring compares cropped offline photos with online product photographs. To determine quality, measures of similarity are calculated, such as cosine or geometric distance This approach is similar to AI facial recognition, but it has a harder time ensuring the quality of food	In order to address disparities between online food pictures and real products in O2O omnichannel retail, this study presents a two-stage AI deep-learning approach. It effectively evaluates how similar offline goods and internet photos are, mitigating quality risk. Crucially, restaurants may easily implement this technique to choose internet photos that support favorable assumptions. To further reduce dissimilarity difficulties, the study also suggests including many photos of varied foods online. These useful results offer insightful guidance on how to improve customer experiences and control quality concerns in O2O food retail

(continued)

Table 5.1 (continued)

Authors	Title	Journal	Technological aspect of the study	Findings
Chiu K.-C.; Lai C.-S.; Chu H.-H.	Apply importance-performance analysis to explore innovation resistance of home robot	International Journal of Mechanical Engineering and Robotics Research	The study combines artificial intelligence (AI), the Internet of Things (IoT), precise sensors, and cloud computing to develop household robots. It assesses customer resistance by conducting surveys and utilizing importance-performance analysis. The technical components encompass the creation of AI algorithms and the utilization of cloud-based data processing to improve the perceived value and reputation in the home robot sector	The study's conclusions emphasize how important customer perception is to the home robot industry. There are several obstacles to innovation, including tradition, value, and image, which affect how well house robots perform compared to what people expect. The industry's primary priority should be to improve the perception of house robots and their perceived worth. Marketing initiatives should be focused on improving the brand and emphasizing the advantages that outweigh the cost. To succeed in the industry, particular value category barriers must be addressed
Diwanji V.S.; Lee J.; Cortese J.	Future-proofing search engine marketing: An empirical investigation of effects of search engine results on consumer purchase decisions	Journal of Strategic Marketing	The study looks at the consequences for AI-powered search engine marketing and makes suggestions for possible strategy changes in the direction of intent-based tactics. In the future, when search engines do not use cookies, it highlights how crucial it is to convey knowledge, authority, and trust in an efficient manner. In order to improve customer perceptions and optimize content display in search engine results pages, this suggests a possible reliance on artificial intelligence (AI) technology	The study's conclusions emphasize how important it is for consumer marketers to use picture recognition technologies. It highlights how consumer behavior has changed to place a higher priority on food safety and quality. The study illustrates how image recognition enables direct consumer interaction, data collection, and insights into consumer preferences. The overall results are good despite difficulties with unusual packing shapes

Liu Y.; Lyu P.; Gao W.	Consumer marketing brand cultivation path based on image recognition technology	IEEE Access	Image recognition technology was employed to stop counterfeiting, which aids in product authentication and stops the proliferation of subpar and counterfeit goods that undermine customer confidence in a brand In light of the growing consumer attention on product quality, safety, and health, the paper presents the use of image recognition technology to strengthen consumer-brand relationships, foster brand loyalty, and enhance the customer experience	The study emphasizes how crucial intelligent brand image recognition technology is for adjusting to changing customer expectations for quality and safety. Businesses can close the gap between customers and brands by utilizing this technology to give consumers access to real-time product information and brand culture. Although there were some difficulties, such as atypical packaging, the study's overall findings were encouraging and gave confidence to the suggested strategy for brand development. The study continues to advocate for the integration of AI into branding strategies to enhance consumer perceptions and drive innovation within the industry

(continued)

Table 5.1 (continued)

Authors	Title	Journal	Technological aspect of the study	Findings
Califano G.; Spence C.	Assessing the visual appeal of real/AI-generated food images	Food Quality and Preference	This paper's technological focus is mostly on using generative AI models—More especially, the DALL-E 3 model—To generate food pictures. A range of food pictures representing unprocessed, processed, and ultra-processed foods were produced using the DALL-E 3 model. This highlights the changing patterns of content production in digital marketing and the growing significance of AI in these procedures	The primary focus of the study was to showcase those processed foods or edited videos. They primarily compared actual food samples and AI edits. For many individuals, the concept of naturalness is often associated with unprocessed food. However, it's important to note that unprocessed food lacks individual tastes and preferences. On the other hand, processed food exemplifies human freedom, as it is shaped by personal tastes, making it more unpredictable. Both creations involve processes that can be considered as belonging to the category of processed goods, as they closely resemble processing. The AI algorithms become more complex and precise by utilizing and managing larger amounts of data. They face unfair treatment and, to make matters worse, are portrayed in a negative light. The machine, also known as artificial intelligence, brings a certain resemblance to human behaviour and work. While photo retouching and utilizing machine learning techniques to process natural photographs may seem less trendy. Humans have the responsibility of governing AI-driven services and mitigating any harms caused by AI, while also benefiting from the products of AI, just like machines. Transparency plays a crucial role in all processes aiming to enhance and improve through AI. AI has the potential to boost food advertisements. Therefore, the study aims to uncover the process. Since transparency is crucial for every AI-issued enhancement or improvement, it raises the question of how AI can effectively increase advertisement about foods. This study aims to uncover the process behind AI's ability to achieve this

So K.K.F.; Kim H.; Liu S.Q.; Fang X.; Wirtz J.	Service robots: The dynamic effects of anthropomorphism and functional perceptions on consumers' responses	European Journal of Marketing	By providing empirical evidence of the mediating roles of PEOU and PU, this study also advances studies on technological acceptability and service robot receptivity. Besides, this study adds to the body of knowledge on task-technology fit by showing that task complexity is an important consideration for service robot design	In finalizing the monologue, the author touches upon the topic of robot anthropomorphism in the service sector, specifically discussing the role of Intelligentia PEUO (Primarily Ethical United of Oslo) and PU (Social and Psychological Uncertainty). Finally, people failed to distinguish between modeling and the qualities that would make a person likable. The main focus of this paper is to highlight the negative effects of driving on the environment. Just like a programmer, the mission carries significant risks and dangers that cannot be ignored. In no time at all, he finds himself thrust into roles akin to biomixers, tasked with creating organic matter, and robotics, simulating intelligent life systems with robots. Similarly, a company looking to implement home robotics automation for service jobs should be provided with straightforward solution options

(continued)

Table 5.1 (continued)

Authors	Title	Journal	Technological aspect of the study	Findings
Salvucci G.; Pallottino F.; De Laurentiis L.; Del Frate F.; Manganiello R.; Tocci F.; Vasta S.; Figorilli S.; Bassotti B.; Violino S.; Ortenzi L.; Antonucci F.	Fast olive quality assessment through RGB images and advanced convolutional neural network modeling	European Food Research and Technology	This study looks into how RGB image processing systems and Deep Neural Networks, more specifically YOLO (You Only Look Once), can be used to quickly sort olives into groups based on their color and any flaws they may have Focus of the study is on creating an optomechanical RGB sorting device that can classify olives in real time. High-resolution RGB cameras, take pictures of olives moving on a conveyor belt, and then the learned program, processes those pictures. When it comes to judging the quality of olives, the method is more than 95% accurate, which meets the standards for industrial selection prototypes	The research conducted an in-depth investigation of the application of advanced Convolutional Neural Network (CNN) modelling, notably YOLO (You Only Look Once), along with RGB image processing systems for rapid olive categorization based on flaws and color. The study utilized high-resolution RGB images obtained from a camera affixed to a laboratory conveyor belt. The study utilized two datasets: One containing 930 images of table olives (namely, the Camillella di Cerignola cultivar), and another comprising 1500 shots of oil olives (specifically, the Carboncella, Frantoio, and Leccino cultivars). The CNN model attained a classification accuracy rate of over 95% for both datasets, which is not just promising but also remarkable These results demonstrate how reliable the suggested technique is and point to its possible application in the creation of an optomechanical RGB sorting system for real-time olive categorization. With the sorting of olives into several classes according to ripeness, faults, and other pertinent aspects, this system may automate the manufacturing of olives and guarantee the manufacture of high-quality products

Das S.; Gochhait S.	Digital entertainment as next evolution in service sector: Emerging digital solutions in reshaping different industries	Digital entertainment as next evolution in service sector: Emerging digital solutions in reshaping different industries	The applied technology elements that are transforming the digital entertainment industry and changing how consumers engage with and consume information seem to be the main emphasis of the study. The use of technologies such as bandwidth-rich platforms, smart devices, connected systems, multi-platform access, on-demand streaming, and immersive technologies in the digital entertainment industry shows how far the industry has come and suggests ways in which technology can be leveraged to improve user experience and engagement	The research showcases the evolution of digital entertainment, enabled by bandwidth-rich, interconnected platforms accessible across devices. These platforms facilitate on-demand consumption, dynamic content generation, and immersive experiences, blurring the lines between producers and consumers of entertainment
Gonzalez Viejo C.; Fuentes S.; Howell K.; Torrico D.; Dunshea F.R.	Robotics and computer vision techniques combined with non-invasive consumer biometrics to assess quality traits from beer foam ability using machine learning: A potential for artificial intelligence applications	Food Control	There is a use of machine learning to look at quality traits like how foamy a beer is using robotics, computer vision, and non-invasive user measurements. Artificial intelligence is also used to check how foamable beer is, which is an important quality trait. This is usually done by combining robots, computer vision, and machine learning algorithms	The study examined consumers visual perception of beer using non-invasive biometrics and sensory questionnaires. Findings showed preferences for top-fermentation beers with medium foam. Integration of biometrics and sensory data revealed correlations between perceived quality and heart rate, among others. An artificial neural network achieved 82% accuracy in classifying beers based on foam height liking. This combined approach offers a rapid and accurate tool for beer industry applications

(continued)

Table 5.1 (continued)

Authors	Title	Journal	Technological aspect of the study	Findings
Powers G.; Johnson J.P.; Killian G.	To Tell or Not to Tell: The effects of disclosing Deepfake video on US and Indian consumers' purchase intention	Journal of Interactive Advertising	This study utilizes cutting-edge AI technology, such as deep learning and generative adversarial networks, to produce very realistic synthetic human avatars for marketing purposes. The study employs a post-test control group design using 318 participants from the United States and India to investigate the influence of avatars on characteristics such as source credibility and purchase intention. The data is analyzed using partial least squares structural equation modeling, with assumptions supported by schema congruity theory and the persuasion knowledge model. Statistical software systems assist in the design and study of avatars, uncovering subtle demographic variations in their responses. This comprehensive approach improves comprehension of customer behavior and promotes the efficacy of marketing strategies	The study found that disclosing the use of synthetic avatars in marketing communications led to decreased purchase intention due to reduced trustworthiness perceptions. This effect was more pronounced among women and higher-income individuals. Surprisingly, social cynicism did not significantly influence perceptions of source credibility or purchase intention. These findings highlight the importance of transparent communication about synthetic avatar use to maintain consumer trust and inform marketing strategies

Yen C.; Chiang M.-C.	Trust me, if you can: a study on the factors that influence consumers' purchase intention triggered by chatbots based on brain image evidence and self-reported assessments	Behavior and Information Technology	Technology is employed in several key ways to collect data for this research: Where researchers employed EEG (electroencephalography) to observe the cerebral activity of the subjects while they engaged with e-commerce platforms, both in the presence and absence of chatbot assistance. The utilization of this brain imaging technology enabled the researchers to record the psychophysiological reactions of the participants, thereby facilitating a deeper comprehension of the neurological foundations of trust in these encounters Neuron-Spectrum 3 and 21-Channel Digital EEG Systems, which are advanced EEG equipment, were utilized to accurately capture brainwaves. This technology provides precise data recording with great precision. The EEG signals were examined in the beta frequency bands, which are linked to decision-making and cognitive processing	In addition to being noteworthy, the study's results are fascinating. According to them, anthropomorphism, social presence, competency, trustworthiness, and informativeness are some of the elements that have a big impact on consumer trust in chatbots, which in turn affects consumer intent to purchase. Furthermore highlighting their significance in influencing customer opinions is the chatbot trust model's incorporation of human-computer interaction, machine communication quality, and human use and pleasure elements. Furthermore, a strong correlation between the development of trust in chatbots and the dorsolateral prefrontal cortex and the superior temporal gyrus is revealed by the study, indicating that building trust relationships has neurological roots. One especially intriguing area for future research is the potential of EEG data to shed light on the neurological mechanisms behind consumer behavior in AI-mediated environments

Transparency plays a crucial role in disclosing AI-generated content as it significantly influences consumer preferences and purchasing decisions (Das & Gochhait, 2021). In the beer industry, the combination of biometrics and sensory data allows us to gain a more comprehensive understanding of customer perceptions regarding our products. Marketing communications chatbots utilize AI to establish trust with clients and shape their purchasing decisions. Understanding the importance of technology extends beyond its mere functionality. It entails recognizing our responsibility to employ it ethically and transparently and influence consumers' behavior (Califano & Spence, 2024). Organizations, much like the public, are grappling with the impact of AI-generated movies and pictures. Therefore, the research necessitates a comprehensive and extensive study.

5.3 Conclusion and Discussion

Figure 5.3 shows the technological aspects of the review studies where we can see that in the three papers keywords like Artificial intelligence and big data were used and in the two papers food dissimilarity and voice marketing were used. The other technological aspects used in these studies were Image recognition, Innovation Resistance, Image Colour Analysis, Data-Driven Marketing, and Search Engine Marketing.

The visual output field has seen significant growth in several industries and a shift in consumer habits due to the increasing impact of AI (Artificial Intelligence). It encompasses various industries such as photography, videography, and similar fields. With the advancements in AI technology, we can now seamlessly integrate voice generation with immersive three-dimensional digital environments. This has brought us closer to understanding consumer preferences in design and how they will impact our future (Pellas, 2023). In addition, there are several other challenges associated with the disclosure of AI materials, among others. Ensuring that

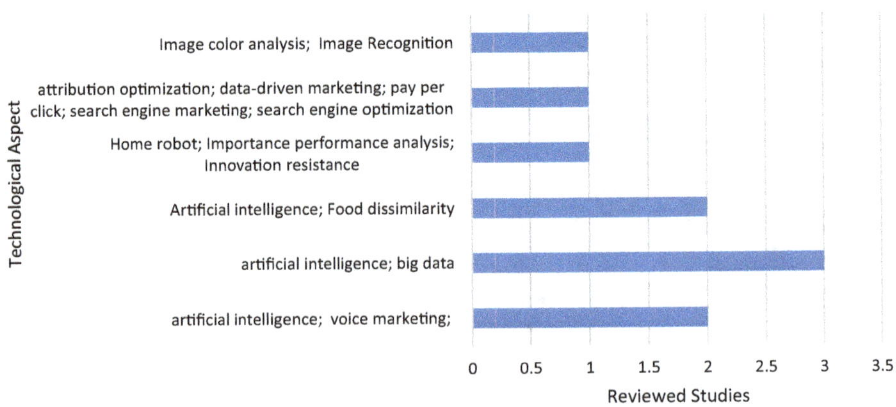

Fig. 5.3 Technological aspects of the reviewed studies

consumers have access to AI-produced materials is crucial for building trust and positioning our business at a higher level in terms of ethical business standards. In addition, there are extensive collections of biometric and sensory data available for comprehensive data analysis, which can be further enhanced by incorporating attitudinal survey data. Opportunities like these present themselves, allowing for effective communication with clients and targeted advertising to reach this audience (Viejo et al., 2018). Therefore, to fully incorporate AI technologies into its capabilities, it is crucial to prioritize research and recognize its ever-changing nature.

Although there is existing research on the topic, there are still notable gaps in the literature that could provide valuable insights into the impact of artificial intelligence on consumer decisions and behaviors. Your input is crucial in this context. With the inclusion of valuable insights from our research, we can guide future advancements in Consumer AI, ensuring its relevance and practicality for researchers and practitioners.

References

Afzal, S., Ghani, S., Hittawe, M. M., Rashid, S. F., Knio, O. M., Hadwiger, M., & Hoteit, I. (2023). Visualization and visual analytics approaches for image and video datasets: A survey. *ACM Transactions on Interactive Intelligent Systems, 13*(1), 1–41. https://doi.org/10.1145/3576935

Anantrasirichai, N., & Bull, D. R. (2021). Artificial intelligence in the creative industries: A review. *Artificial Intelligence Review, 55*(1), 589–656. https://doi.org/10.1007/s10462-021-10039-7

Califano, G., & Spence, C. (2024). Assessing the visual appeal of real/AI-generated food images. *Food Quality and Preference, 116*, 105149. https://doi.org/10.1016/j.foodqual.2024.105149

Cao, P., & Xiao, J. (2023). Study on the coordinated development of national traditional sports and tourism brands based on big data platforms from the perspective of "The Belt and Road". *Journal of Intelligent & Fuzzy Systems, 46*, 5429–5439. https://doi.org/10.3233/jifs-230547

Chen, S. Y., Zhang, J. Q., Zhao, Y. Y., Rosin, P. L., Lai, Y. K., & Gao, L. (2022). A review of image and video colorization: From analogies to deep learning. *Visual Informatics, 6*(3), 51–68. https://doi.org/10.1016/j.visinf.2022.05.003

Das, S., & Gochhait, S. (Eds.). (2021). *Digital entertainment*. Springer. https://doi.org/10.1007/978-981-15-9724-4

Efthymiou, F., Hildebrand, C., De Bellis, E., & Hampton, W. H. (2023). The power of AI-generated voices: How digital vocal tract length shapes product congruency and Ad performance. *Journal of Interactive Marketing, 59*(2), 117–134. https://doi.org/10.1177/10949968231194905

Griffy-Brown, C., Earp, B. D., & Rosas, O. (2018). Technology and the good society. *Technology in Society, 52*, 1–3. https://doi.org/10.1016/j.techsoc.2018.01.001

Kertysova, K. (2018). Artificial intelligence and disinformation. *Security and Human Rights, 29*(1–4), 55–81. https://doi.org/10.1163/18750230-02901005

Lysyakov, M., & Viswanathan, S. (2023). Threatened by AI: Analyzing users' responses to the introduction of AI in a crowd-sourcing platform. *Information Systems Research, 34*(3), 1191–1210. https://doi.org/10.1287/isre.2022.1184

Neyazi, T. A., Ng, S. W. T., Hobbs, M., & Yue, A. (2023). Understanding user interactions and perceptions of AI risk in Singapore. *Big Data & Society, 10*(2), 20539517231213823. https://doi.org/10.1177/20539517231213823

Pellas, N. (2023). The influence of sociodemographic factors on students' attitudes toward AI-generated video content creation. *Smart Learning Environments, 10*(1), 57. https://doi.org/10.1186/s40561-023-00276-4

Raghunath, K. K., Kumar, V. V., Venkatesan, M., Singh, K. K., Mahesh, T. R., & Singh, A. (2022). XGBoost regression classifier (XRC) model for cyber-attack detection and classification using inception v4. *Journal of Web Engineering, 21*(4), 1295–1322.

Ramdurai, B., & Adhithya, P. (2023). The impact, advancements and applications of generative AI. *SSRG International Journal of Computer Science and Engineering, 10*(6), 1–8. https://doi.org/10.14445/23488387/ijcse-v10i6p101

Sharakhina, L., Ilyina, I., Kaplun, D., Teor, T., & Kulibanova, V. (2023). AI technologies in the analysis of visual advertising messages: Survey and application. *Journal of Marketing Analytics, 12*, 1066. https://doi.org/10.1057/s41270-023-00255-1

Singh, K. K., Mehrotra, A., Nigam, M. J., & Pal, K. (2013, April). Unsupervised change detection from remote sensing images using hybrid genetic FCM. In *2013 Students Conference on Engineering and Systems (SCES)* (pp. 1–5). IEEE.

Singh, K. K., Rho, S., Singh, A., & Sergei, C. (2023). Big data analytics and knowledge discovery for urban computing and intelligence. *Complex & Intelligent Systems, 10*(1), 1–2. https://doi.org/10.1007/s40747-023-01050-2

Singh, A., Dhull, A., & Singh, K. K. (Eds.). (2024). *Blockchain and deep learning for smart healthcare*. Wiley.

So, K. K. F., Kim, H., Liu, S. Q., Fang, X., & Wirtz, J. (2023). Service robots: The dynamic effects of anthropomorphism and functional perceptions on consumers' responses. *European Journal of Marketing, 58*(1), 1–32. https://doi.org/10.1108/ejm-03-2022-0176

Viejo, C. G., Fuentes, S., Howell, K., Torrico, D., & Dunshea, F. R. (2018). Robotics and computer vision techniques combined with non-invasive consumer biometrics to assess quality traits from beer foamability using machine learning: A potential for artificial intelligence applications. *Food Control, 92*, 72–79. https://doi.org/10.1016/j.foodcont.2018.04.037

Walters, W. P., & Murcko, M. (2020). Assessing the impact of generative AI on medicinal chemistry. *Nature Biotechnology, 38*(2), 143–145. https://doi.org/10.1038/s41587-020-0418-2

Wu, P. J., & Chien, C. L. (2021). AI-based quality risk management in omnichannel operations: O2O food dissimilarity. *Computers & Industrial Engineering, 160*, 107556. https://doi.org/10.1016/j.cie.2021.107556

Yazadzhiyan, H. (2023). Video editing tools with artificial intelligence. ResearchGate. https://www.researchgate.net/publication/375833832_Video_Editing_Tools_with_Artificial_Intelligence

Yen, C., & Chiang, M. C. (2020). Trust me, if you can: A study on the factors that influence consumers' purchase intention triggered by chatbots based on brain image evidence and self-reported assessments. *Behaviour & Information Technology, 40*(11), 1177–1194. https://doi.org/10.1080/0144929x.2020.1743362

Zhavoronkov, A., & Aspuru-Guzik, A. (2020). Reply to 'Assessing the impact of generative AI on medicinal chemistry'. *Nature Biotechnology, 38*(2), 146. https://doi.org/10.1038/s41587-020-0417-3

Chapter 6
Text-to-Image Synthesis: Techniques and Applications

Akansha Singh and Krishna Kant Singh

Abstract This chapter explores the evolving field of text-to-image synthesis, a technology bridging natural language processing and computer vision to generate coherent images from textual descriptions. Beginning with foundational techniques such as Generative Adversarial Networks (GANs), Variational Autoencoders (VAEs), and the Transformer models, this chapter traces the advancements that enable increasingly realistic and contextually accurate image generation. Key models, including DALL-E and its successors, highlight the significant progress in image fidelity and semantic accuracy. Applications in creative media, education, and assistive technology underscore the impact of text-to-image synthesis on various fields. Furthermore, ethical considerations, such as bias and content control, are examined to provide a comprehensive understanding of the opportunities and challenges in this domain.

Keywords Text-to-image synthesis · Generative Adversarial Networks (GANs) · DALL-E · Transformer models · Natural language processing · Image generation applications

6.1 Introduction

Text-to-image synthesis is a novel area at the intersection of NLP and computer vision. It aims at constructing visually coherent images from textual descriptions. Over the years, this technology has gained center stage because of its possible applications across a wide array of domains that include art generation, advertising, and virtual reality. It is not an easy job to translate the textual description into its

A. Singh (✉)
School of Computer Science Engineering and Technology, Bennett University, Greater Noida, Uttar Pradesh, India

K. K. Singh
Delhi Technical Campus, Greater Noida, Uttar Pradesh, India

A. Singh, K. K. Singh (eds.), *Multimodal Generative AI*,
https://doi.org/10.1007/978-981-96-2355-6_6

corresponding image. There, in fact, exist two most important keys: not only it should be semantically aligned but also visually realistic.

Early works in this domain focused essentially on generating simple images conditioned on short phrasings or limited vocabularies. For example, Goodfellow et al. (2014) basically introduced the GAN architecture, allowing later work to easily illustrate exactly how one may train a generator network to generate images that a discriminator network would evaluate for realism. The invention of GANs added a totally new direction to image generation and furthered the quality and fidelity of synthesized images (Goodfellow et al., 2014). Reed et al. (2016), based on the GAN framework, proposed the very first approach designed for text-to-image synthesis. Their model was able to directly translate textual descriptions into images using deep learning techniques. In other words, it was meant that the neural networks could be effectively trained to generate complex visual scenes conditioned on the descriptive language (Reed et al., 2016). The model was a huge success and proved that text-to-image synthesis would be a feasible task, hence opening new dimensions of research. Recent performance by models such as DALL·E from OpenAI has continued to advance the possibilities of text-to-image synthesis even further. Whereas many older models struggled with generating high-resolution images or dealing with more complex textual inputs, the images that DALL·E created were intricate and contextually appropriate from a wide range of descriptive prompts. Part of this leap in quality has to do with the fact that transformers have become the standard model architecture in NLP for processing sequential data. Large-instance dependencies within text can be grasped only with the help of a transformer model, with which the system can ensure that a generated image adequately reflects subtleties entailed in its description.

These have opened a large number of applications. For example, in the fashion industry, this would allow designing effective prototypes from designers' descriptions through text-to-image synthesis, hence speeding up the design process. Similarly, the capability to create virtual reality environments from simple textual input enhances user experience by enabling dynamic and interactive content creation. While this technology is still evolving, the challenges of improving the fidelity, diversity, and scalability of the generated images remain of paramount importance. In any case, text-to-image synthesis holds great potential for changing the creative industries, scientific visualization, as well as human-computer interaction.

Table 6.1 provides a summary of key milestones in the development of text-to-image synthesis techniques, showcasing the evolution from basic image generation to modern transformer-based models.

6.2 Overview

Text-to-image synthesis is a computational process with which, in conjunction with techniques from the field of artificial intelligence, a machine learning model generates an image either photorealistic or stylized from textual descriptions. Commonly, this represents deep neural networks that are trained to comprehend the semantic meaning of language and visualize it in its respective images. Coupling disciplines

Table 6.1 Summary of key milestones

Year	Model	Key contribution	Reference
2014	GAN	Introduced adversarial learning framework for image generation	Goodfellow et al. (2014)
2016	GAN-based Text-to-Image	First successful method for generating images from text	Reed et al. (2016)
2017	StackGAN	Improved image resolution and quality in multi-stage generation	Zhang et al. (2017)
2021	DALL·E	High-resolution and contextually accurate image generation	Ramesh et al. (2021)

such as natural language processing, computer vision, and generative modeling bridges the gap between textual data and visual output, realizing automated imagery that accurately depicts the specified textual inputs. Its importance surpasses the creation of just a technical milestone but opens a new frontier where AI creatively and functionally can participate in tasks considered the domain of humans. For instance, AI can already help artists in visualizing complex scenes described in text, therefore acting as an effective collaborative tool, extending the creative process. This technology could also be applied in educational settings to produce customized visual aids that would explain complex ideas and further enrich learning, making such information more accessible.

The point is that this technology of text-to-image synthesis will also allow for fast prototyping of visuals regarding films, video games, and virtual reality in the field of entertainment. It gives an instant life to the imaginations from script descriptions, in that sense, to creators. Besides streamlining the production process, this opens up more avenues for storytelling where audiences can influence or change visual content dynamically based on interactive narratives (Fig. 6.1).

Moreover, while embedded text-to-image synthesis may allow for further synergies from visual and linguistic information to enhance machine understandings of multimodal contexts, it is a very critical step in the making of more complex AI systems, using which any given entity may interpret the world in a humanlike fashion. In doing so, the technology not only pushes the envelope of what is possible with AI image generation but also significantly expands the realm of possible applications, making AI an integral part of solving real-world problems and improving various aspects of societal functioning.

6.3 History of Text-to-Image Synthesis

The journey of text-to-image synthesis from a conceptual idea to a robust technological reality is rooted in the broader fields of computer graphics and artificial intelligence. Early attempts to generate visual content from text can be traced back to the initial explorations into computer graphics and simple AI systems, which were primarily focused on creating representations based on coded instructions rather than natural language.

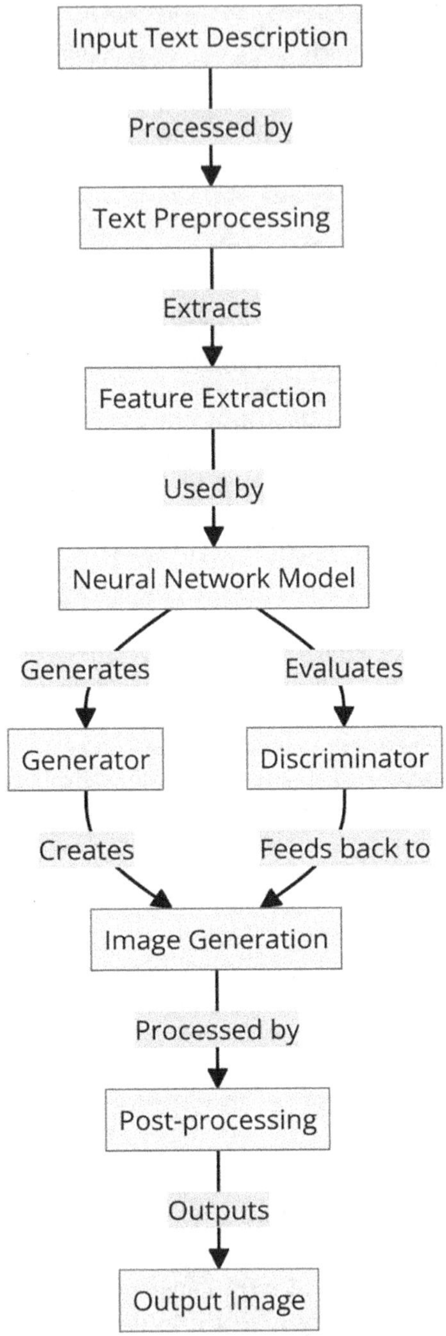

Fig. 6.1 Text to image synthesis

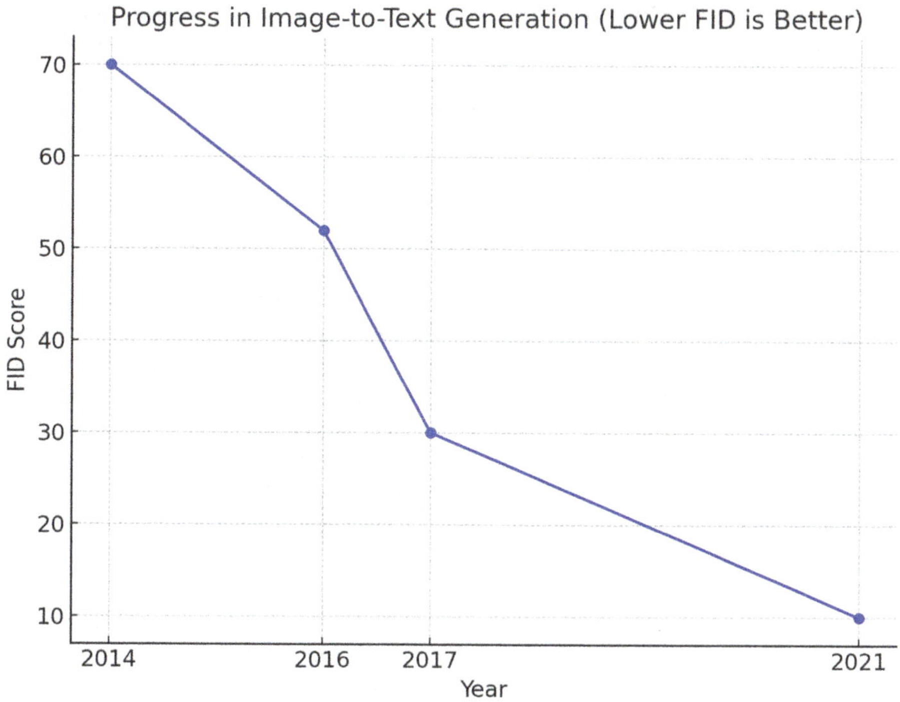

The graph above illustrates the progress in image generation models based on the Fréchet Inception Distance (FID) score. As you can see, the FID scores have decreased significantly over time, indicating improvements in the quality of generated images. The lower FID scores in more recent models, such as those developed in 2021 (e.g., DALL·E), highlight the enhanced realism and semantic accuracy of modern text-to-image generation methods compared to earlier models.

- *Early Experiments*: Development in computational capabilities toward the turn of the twentieth century saw a number of attempts at incorporating text data into generative visual content creation. Most of the simple systems were those that could sometimes output basic shapes or color patterns from explicit text. The complexity of the visuals was nevertheless limited by the processing power and the nascent state of AI technologies.
- *Machine Learning on the Rise*: Only with the turn of the century did machine learning reach maturity and thus allow for new functionalities processing and interpreting language at more sophisticated levels. Similarly, while the images remain relatively simple and abstract, it allows for much more detailed interpretations of text.
- *Breakthrough with Deep Learning*: A key inflection point came when the resurgence of neural networks in the form of deep learning was felt in the early 2010s. Deep learning, because of its capability to learn from vast data and also handle multiple layers of abstraction, made it fit for such complex tasks as generating

images from text. More complex image synthesis models were being prepared by allowing the researchers to investigate deep convolutional neural networks for image processing tasks.

- *Introduction to Generative Adversarial Networks*: All that came to a revolutionary step when, back in 2014, Ian Goodfellow and colleagues introduced Generative Adversarial Networks. GAN consists of two neural networks: a generator and a discriminator, which includes a competitive game framework for generating images with the objective of being as similar as possible to real ones, while the discriminator will appraise them for authenticity. This architecture turned out to be very effective for generating photorealistic images from random noise inputs and soon was adapted to synthesize images from textual descriptions.
- *Developments and Diversification*: While GANs and other neural network architectures continued to evolve, the possibilities of the so-called text-to-image synthesis increased significantly. New variants and improvements over GANs were brought into existence that could take textual descriptions directly as inputs, possibly conditioning this in the generation process. With that, these images are not only going to be photorealistic-looking but also really aligned with the minute details that may arise from such input text.
- *Recent Innovations*: Further innovations included models such as the AttnGAN and later, the transformer-based models, including DALL-E by OpenAI, which integrated attention mechanisms for the better capture and interpretation of the relationship between words and contexts within text descriptions. These have been able to generate highly detailed and contextually correct images from a wide range of textual inputs—an exciting evolution in the field.

History, therefore, ranges in this regard from the generation of mere geometric shapes to complex and detailed images that can show subtlety and richness consistent with human language. Indeed, this development has shown not only leaps and bounds in technology but also increasing interactions between multiple domains within AI, further frontiers as to how machines envisage and react to human creative expressions.

Table 6.2 summarizes the timeline of key developments in text-to-image synthesis.

This table encapsulates the progression from simple image generation based on explicit commands to sophisticated systems capable of creating detailed and contextually accurate visual representations from complex textual descriptions.

This chapter will delve deep into text-to-image synthesis, clarifying the technical underpinnings of the most influential models, elaborating on the broad spectrum of applications, and finally pointing out the challenges and prospects for the technology. In other words, the general techniques of GANs and VAEs, the state-of-the-art systems DALL-E and Imagen, and the ethical consequences regarding biases in systems like these will be presented. Also, the chapter will discuss practical implementations and the impact of text-to-image synthesis in real-world case studies, guiding the reader not only through the landscape of the state of the art but also the possibilities over the horizon.

Table 6.2 Key milestones in the evolution of text-to-image synthesis technology: a timeline from early computer graphics to advanced AI models

Year	Milestone
1980s–1990s	Early computer graphics and AI experiments generate simple images based on coded instructions
Early 2000s	Machine learning advancements allow for basic visual content generation from textual data
Early 2010s	Resurgence of neural networks; deep learning applied to image processing tasks
2014	Introduction of Generative Adversarial Networks (GANs) by Ian Goodfellow
2016	Development of conditional GANs which utilize textual descriptions to guide image generation
2018	Launch of AttnGAN, integrating attention mechanisms to enhance image detail and relevance
2020	OpenAI introduces DALL-E, a transformer-based model that generates detailed images from text
2021	Introduction of DALL-E 2 and Google's Imagen, further improving image fidelity and contextual accuracy
2022 and beyond	Continued research and development focusing on refining models, reducing costs, and addressing ethical issues

6.4 Fundamental Techniques

Generative modeling is one of the driving engines of powerful breakthroughs in most domains, leading to the ability to generate highly realistic and contextual synthetic data. The most notable techniques in this area are Generative Adversarial Networks (GANs), Variational Autoencoders (VAEs), and Transformer models. Each one of them brings into the fray a special power and novelty for further advances in such fields as image synthesis, data augmentation, and creative applications. This section describes the model architecture and operation of GANs, the probabilistic approach of VAEs, and the generative capacity of Transformers for text-to-image generation.

6.5 Generative Adversarial Networks (GANs)

GANs represent a new generation of deep-learning algorithms for unsupervised learning; the system consists of two neural networks competing with each other in a game-theoretic scenario. This innovative framework was suggested by Ian Goodfellow and others in 2014, and in just a few years, it has turned into a revolution in generative modeling. GANs architecture includes two neural networks with a different set of objectives, and these get trained simultaneously. These are Generator and Discriminator. These are discussed in the ensuing sections (Fig. 6.2).

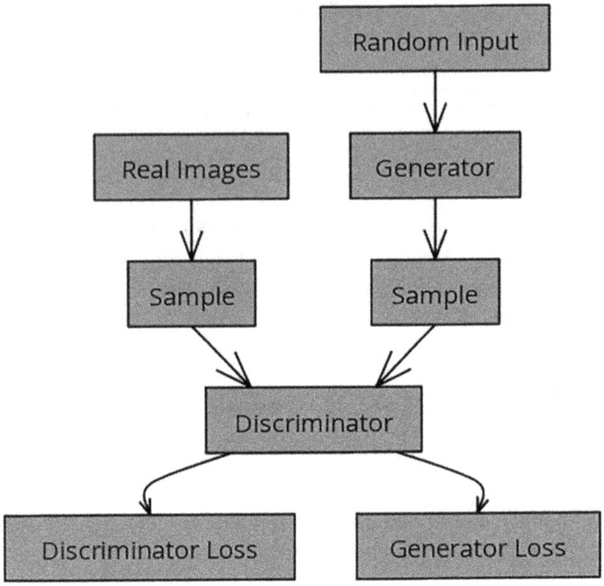

Fig. 6.2 Overview of GAN structure

6.5.1 Generator

The objective of the Generator is to generate data similar to the real data. It takes a random noise vector, z, coming from a pre-defined noise distribution p_z (z) and transforms this noise into data that resemble the authentic data distribution. The generator keeps getting better with the feedback that comes from the discriminator.

- **Architecture of the Generator**
  - *Input Layer*: The architecture of the generator starts with an input layer that takes as input a random noise vector. This usually comes from a Gaussian distribution. The purpose of this random input is to bring some variability to this process, enabling the generator to produce variations in the output.
  - *Dense Layers*: Fully connected or dense layers follow a sequence after the input layer in a typical generator. These layers expand the noise vector to a higher-dimensionality space and form the basis for further transformations.
  - *Transposed Convolutional Layers*: Often termed deconvolutional layers, transposed convolutional layers work to upsample the data. These layers work by reversing the convolution process—increasing the spatial dimensions, height, and width progressively and reducing the depth, referring to the number of channels. Each transposed convolutional layer is often followed by batch normalization for stabilization and speeding up the training, including an activation function like ReLU or LeakyReLU that introduces non-linearity.

- *Output Layer*: The typical last layer of the generator is a transposed convolution layer, which outputs the synthetic image. This layer applies the tanh activation function, scaling pixel values in the range between −1 and 1, as common practice in image-generation tasks.

- **Generator Function**
  - The significant role of the generator is to generate data the discriminator cannot distinguish as different from real data. This iterative process progressively leads to better performance of the generator.
  - *Initialization*: The generator is initialized with random weights that generate initial synthetic data which are far from realistic.
  - *Feedback from Discriminator*: The discriminator will look at these synthetic samples and give feedback in the form of classification as fake. This feedback is utilized to compute the generator's loss; typically, a binary cross-entropy loss function is used. It means a generator would like to maximize the discrimination error of the discriminator, trying in one sense to 'fool' the discriminator.
  - *Weight Update*: Using the provided feedback, the weights of the generator shall be updated via backpropagation. This involves minimizing the loss function of the generator using gradient descent or any similar optimization algorithm. Through iterative processes, the generator picks up on generating data that is more realistic by reducing the gap between the synthetic data distribution and the real data distribution.
  - *Discriminator*: The discriminator represents a 'critic' within the GAN model. It is trained to tell the difference between real data, brought in from the actual dataset, and synthetic data created by the generator. It also evaluates each input carefully by estimating the probability, which states whether the data were from a real dataset, rather than from a generator. The discriminator in Generative Adversarial Network architecture is an essential component—a binary classifier in other words. Its primary task would be to differentiate between real data from the training set with fake data generated by the generator. Because the networks are trained simultaneously in an adversarial setting, the discriminator is held responsible for how well the generator generates realistic data.

- **Architecture of the Discriminator**
  - *Input Layer*: The discriminator takes an input image. This image may be a real image from the training dataset or could be a synthetic image generated by the generator.
  - *Convolutional Layers*: Typically, a discriminator has up to multiple convolutional layers. These layers serve for feature extraction of the spatial features from the input image.
  - *Convolution Operations*: Apply filters to the input image to detect various features such as edges, textures, and shapes.

  - *Activation Functions*: Non-linear activation functions like LeakyReLU are applied after each convolution to introduce non-linearity, which helps in learning complex patterns.

- **Pooling Layers**
  - Pooling layers (such as max pooling) are often used after convolutional layers to downsample the feature maps. This reduces the spatial dimensions and computational load, and also helps in making the learned features invariant to small translations.
- **Fully Connected Layers**
  - After several layers of convolutions and pooling, the high-level features are flattened and fed into one or more fully connected (dense) layers.
  - These layers integrate the extracted features and perform the final classification.
- **Output Layer**
  - The output layer is a single neuron with a sigmoid activation function that outputs a probability value between 0 and 1.
  - This value represents the discriminator's confidence that the input image is real (close to 1) or fake (close to 0).

6.5.2 Function of the Discriminator

The discriminator's role is to accurately classify input images as real or fake. This process involves:

1. **Classification Task**:
   (a) *Real vs. Fake*: The discriminator assigns a probability score to each input image, indicating whether it believes the image is real or generated by the generator.
   (b) *Binary Cross-Entropy Loss*: The discriminator's loss function is typically binary cross-entropy, which measures the error in classification. The goal is to minimize this loss.
2. **Feedback Mechanism**:
   (a) During training, the discriminator provides feedback to the generator. It evaluates the synthetic images generated by the generator and provides a loss value that indicates how well the generator is performing.
   (b) *Adversarial Training*: The generator uses this feedback to improve its output. The better the discriminator is at distinguishing real from fake images, the harder the generator must work to produce more realistic images.

3. **Training Process**:
    (a) **Step 1**: The discriminator is trained with a batch of real images and a batch of fake images. It learns to assign high probabilities to real images and low probabilities to fake images.
    (b) **Step 2**: The generator generates a new batch of fake images.
    (c) **Step 3**: The discriminator evaluates these new images, and its performance is used to update both the discriminator and generator through backpropagation.
    (d) This process continues iteratively, with both networks improving their performance over time.

6.5.3 GAN Operation

The fundamental operation of GANs can be described by a minimax game in which the discriminator tries to maximize its ability to correctly classify real and fake data, while the generator tries to minimize the discriminator's ability to distinguish between the two. This relationship can be mathematically represented by the following objective function:

$$\min_G \max_D V(D,G) = \mathbb{E}_{z= p_{\text{data}(x)}}\left[\log D(x)\right] + \mathbb{E}_{z= p_{z(z)}}\left[\log\left(1 - D\left(G(z)\right)\right)\right]$$

Here, $\mathbb{E}_{z= p_{\text{data}(x)}}\left[\log D(x)\right]$ is the expected value of discriminator's logarithmic output for real data x, which it should recognize as real. Conversely, $\mathbb{E}_{z= p_{z(z)}}\left[\log\left(1 - D\left(G(z)\right)\right)\right]$ is the expected value of the discriminator's logarithmic output for fake data generated by the generator, which it should recognize as fake.

Training GANs involves the following steps:

- *Train the Discriminator*: For a fixed generator, the discriminator is trained first by improving its ability to differentiate real data from fake data. This is typically done by alternating between training on batches of real data and batches of fake data generated by the generator.
- *Train the Generator*: With the discriminator's parameters fixed, the generator is trained to produce data that are classified as real by the discriminator. The training involves adjusting the parameters of the generator to produce outputs that are more likely to fool the discriminator.
- *Convergence:* The nature of the competition between the two networks is such that, in an ideal case, convergence should be achieved wherein neither network can get any better—the generator creates perfect replicas of real data and the discriminator remains guessing at random.
- *Mode Collapse:* Sometimes the generator discovers certain kinds of data which the discriminator keeps misclassifying as real, and thus the generator provides less diverse outputs.

- *Training Stability:* GANs are notoriously difficult to train. The balance between the generator and discriminator can be very fragile. If one significantly overpowers the other, this may result in failures in training.

Applications of GANs are really broad across diverse fields; some of the main ones include, but are not limited to, image synthesis, artistic creation, video game content generation, and as a tool in data augmentation for improving machine learning models.

6.5.4 GANs for Text-to-Image Synthesis

It would mainly be an adaptation of the traditional GAN framework to process and understand textual inputs and then generate corresponding images. This adaptation generally involves adding components to a GAN architecture for text processing in order to capture the meaning that will later be used in the image generation process. Following is a step-by-step explanation about the use of GANs for text-to-image synthesis:

1. **Processing of Textual Input**
   (a) *Embedding of Text*: The first step in this procedure involves preprocessing the input text consisting of the description of the image of interest into a model-processed form, which usually proceeds via a text embedding model that transcribes the text into a high-dimensional vector space. These embeddings capture the semantic meanings of the words and their relationships within the sentence.
2. **Conditioning the Generator**
   (a) *Integrating Text with Noise*: In a traditional GAN, the generator starts off with a random noise vector. For text-to-image synthesis, this noise vector is combined with the embedding of text. This can be done in many ways, including concatenation and the use of Text Embeddings to modulate the properties of the Noise Vector through learned transformations.
   (b) *Generator Network*: This now modified noise vector carries information of the text, which is fed into the generator. The architecture of a generator is typically a deep convolutional neural network adapted for handling the combined text and noise input. On receiving this, the network generates an image that is supposed to match the description in that text.
3. **Training of the Discriminator**
   (a) *Images-Text Pairing*: The discriminator is fed images along with their respective texts. It needs to decide for each image pair whether the image is real or generated and whether the image correctly depicts what the text describes.
   (b) *Discriminator Network*: This also typically consists of a deep convolutional neural network, usually similar to that of the generator. It evaluates how

realistic the given images are, in addition to their adherence to corresponding text descriptions. This provides a dual focus on the realism of the image and its fidelity to the text, thereby helping to fine-tune the output of the generator.

4. **Adversarial Training**

   (a) *Objective Function*: The goal of training is the competition between the discriminator and the generator. It strives to rightly classify images as real or fake and correctly judges whether the images match the text descriptions. Thus, the generator tries very hard to create not only realistic images but also those such that when the discriminator looks at them, it is deceived to classify them as real.
   (b) *Backpropagation and Updates*: During training, both are updated with backpropagation over weights. It updates the weights of the discriminator to be able to recognize the difference between the images and text descriptions better. The generator updates its weights to generate highly accurate images with respect to the text.

5. **Refinement and Iteration**

   (a) *Feedback Loop*: As training progresses, the generator becomes increasingly adept at generating images that convincingly illustrate the text descriptions, and the discriminator becomes increasingly adept at evaluating them. Everything is done in an iterative process until this model achieves a sufficient level of performance whereby the generated images will reliably reflect the input text in a realistic manner.

6. **Output**

   (a) *Generated Images*: When the model has been sufficiently trained, entering a textual description into the system would cause the generator to produce an image visually representing the text. This picture is the generated result of the learned text-image mappings throughout the training process.

Text-to-image synthesis with GANs makes use of these models' strong generating power while introducing in the mixture some complexity of language comprehension and interpretation. This not only becomes a testimony to the adaptability of GANs, but also shows an example of how those two areas, computer vision and natural language processing, combine in order to build new applications for digital media, advertising, and art.

6.6 Variational Autoencoders

Variational autoencoders are generative models designed to model the underlying probability distribution of a given dataset with the aim of generating new samples. They typically involve an architecture comprising an encoder-decoder structure.

Therefore, while the encoder maps the input data to a higher-dimensional latent representation, the decoder aims at reconstructing the original data from this latent code. The whole idea is to minimize the difference between the original and reconstructed data so that it learns the underlying distribution of data. Further, it lets us generate new samples of data similar to the training data—a most interesting advantage of VAEs. Thanks to the continuity of the latent space of VAE, the decoder was able to generate new data points that smoothly interpolate between the training data points. Besides, VAEs have a huge variety of possible applications: density estimation, text generation, etc.

A typical VAE consists of two parts: an encoder and a decoder. It compresses the input data through the encoder into a low-dimensional latent space, also normally referred to as a 'latent code'. The type of neural network used may be different depending on the type of data; it may be fully connected or convolutional neural networks. It outputs the necessary parameters that are used in sampling and producing the latent code. It also involves a variety of neural network types. A decoder is used to reconstruct the data given a latent code. This closes in on an optimally learned latent representation by the VAE through which the key characteristics of the data are modeled and allow for exact reconstruction.

In short, VAEs are another generative model that finds major application in encoding and decoding data, particularly images. VAEs work under an encoder-decoder framework, unlike GANs.

- *Encoder*: This component compresses the input data into a more compact, dense representation, which typically includes the probabilistic descriptions (means and variances) of the input data.
- *Decoder*: The decoder then reconstructs the input data from this compact representation, aiming to minimize the difference between the original input and the reconstructed output.

The significance of VAEs in image synthesis is noteworthy because they not only learn the data distribution but also how to interpret and regenerate data from this learned distribution. They are particularly valuable for tasks requiring a structured latent space and the ability to generate new samples from that space.

6.7 Transformer Models

Transformers, introduced in the paper *Attention is All You Need* by Vaswani et al. (2017), represent a breakthrough in handling sequences, primarily used in natural language processing (NLP). Unlike traditional models that process data in order, transformers use self-attention mechanisms to weigh the significance of each part of the input data differently.

In the realm of text-to-image synthesis:

- *Transformer models* like OpenAI's DALL-E use a modified version of this architecture to manage the relationship between elements of text and corresponding image characteristics. They generate images by interpreting and converting descriptions into visual data.
- The model leverages layers of attention mechanisms to understand complex descriptions and their visual implications, facilitating the generation of detailed and contextually appropriate images from textual descriptions.

Each of these foundational techniques offers unique advantages and capabilities, contributing to the rapidly advancing field of text-to-image synthesis.

6.8 DALL-E Model

DALL-E is a groundbreaking generative AI model developed by OpenAI, designed to produce images from textual prompts. What sets it apart is its remarkable ability to merge language and visual processing. In essence, you supply a text description of an image, and DALL-E generates it, even when the image depicts a concept that doesn't exist in reality. This innovative capability opens new doors in creative fields, communication, education, and beyond.

First introduced in January 2021, DALL-E is a variant of OpenAI's GPT-3, which itself marked a significant milestone in language processing technology. The name 'DALL' is a nod to the surrealist artist Salvador Dalí, while the 'E' is a reference to Pixar's beloved animated robot Wall-E. Following the release of the original DALL-E, its successor, DALL-E 2, was launched in April 2022, enhancing the model's ability to generate more photorealistic and higher-resolution images.

At its foundation, DALL-E employs a type of AI known as a transformer neural network. Specifically, it utilizes the GPT-3 architecture, but unlike GPT-3, which focuses on generating text, DALL-E is trained to generate images from text descriptions. Both GPT-3 and DALL-E rely on unsupervised learning, meaning they are trained on vast datasets containing paired text and image data. Through this process, the model fine-tunes its parameters using an optimization technique. Essentially, it predicts an output, compares that prediction to the actual result, calculates the error, and adjusts its parameters to minimize this error. This optimization happens through backpropagation and methods like stochastic gradient descent.

As the model is exposed to more and more examples, it begins to understand patterns, relationships, and how certain descriptions are linked to specific visual elements. For example, if the model repeatedly encounters images of dogs paired with the word 'dog', it learns to associate the word with the visual concept of a dog. This learning process extends to more complex scenarios as well, such as interpreting a prompt like 'a two-story pink house shaped like a shoe' and generating an image that accurately reflects this description.

With enough training, DALL-E has developed an extraordinary ability to create entirely new images that match textual prompts, even those describing surreal or previously unseen ideas. By combining text and image data, DALL-E can 'imagine' and craft images that are not only contextually relevant to the input but also creatively original, somewhat akin to how an artist might bring a descriptive idea to life visually.

Currently, DALL-E's applications span a variety of areas, from generating unique artworks to improving visual communication. For example, it can create one-of-a-kind logos based on specific descriptions or assist educators by visualizing abstract concepts that might be challenging to convey through words alone.

6.8.1 Architecture

- **Text Encoder (GPT-based)**:
  - DALL·E uses a transformer model similar to GPT (Generative Pre-trained Transformer) for processing the input text. This part of the model takes a textual description as input and encodes it into a sequence of vectors. These vectors represent the semantic meaning of the input text.
  - The text input is tokenized, and these tokens are fed into the transformer model. Each token is associated with a positional embedding to retain the order of words in the input.
- **Image Decoder (VQ-VAE-2 based)**:
  - *Vector Quantized Variational AutoEncoder (VQ-VAE-2):* The image decoder uses VQ-VAE-2, a type of generative model that discretizes the latent space into a finite set of vectors (or codebook entries). It is used to generate images based on the encoded text vectors.
  - The model generates a sequence of image tokens corresponding to patches of an image. Each token represents a discrete value in the latent space learned by the VQ-VAE.
- **Attention Mechanism**:
  - The attention mechanism in transformers allows the model to focus on different parts of the input text while generating each part of the image. This enables DALL·E to maintain consistency in complex scenes described by the text.
- **Autoregressive Process**:
  - DALL·E generates images in an autoregressive manner, meaning it predicts each pixel or patch of the image one at a time, conditioning on the previously generated parts of the image. This is similar to how language models predict the next word in a sentence.

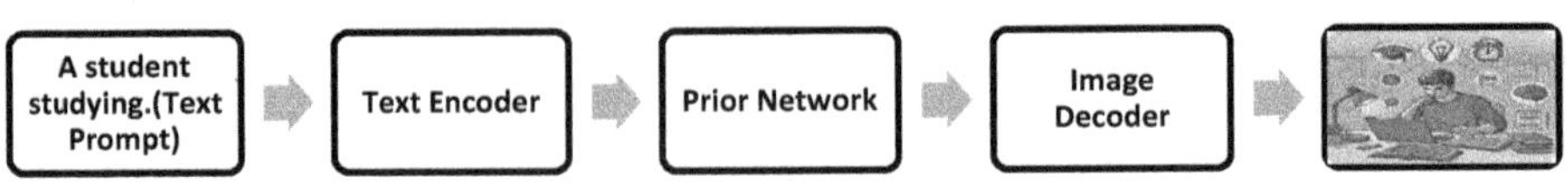

6.8.2 Examples of Real-World Use Cases of DALL-E

Some real-world use cases of DALL-E that demonstrate its potential in various industries include:

- *Education*. For teaching abstract concepts, DALL-E could be a game-changer. It can generate visual aids, helping students understand complex theories or events in history, like visualizing the Battle of Waterloo.
- *Design*. Designers could use DALL-E to generate custom artwork or initial drafts based on specific descriptions, significantly speeding up the creative process. For instance, an author could use it to generate illustrations for their book by providing descriptions of specific scenes.
- *Marketing*. DALL-E could be used to create unique, custom images for ad campaigns based on creative briefs. A marketing team could input specific descriptions of the product, mood, color palette, etc., and get custom graphics without needing to rely on stock photos or extensive graphic design work.

6.8.3 What Are the Benefits of DALL-E?

- *Efficiency*. DALL-E can generate images from textual descriptions quickly and efficiently, saving time, costs, and resources compared to traditional methods of image creation, such as manual graphic design or photography.
- *Creativity*. DALL-E can interpret and visualize abstract or complex concepts that might be difficult or time-consuming for human artists to render. This could potentially expand the boundaries of creativity and art.
- *Customization*. It can create highly customized visuals based on specific input descriptions. This could be particularly useful in fields like advertising, gaming, and design where unique, tailored visuals are often needed.
- *Accessibility*. DALL-E could democratize access to custom graphic design, potentially allowing small businesses, independent creators, and others who can't afford professional design services to create unique visual content.

6.8.4 What Are the Challenges of DALL-E?

DALL-E, like other generative AI technologies, comes with challenges and concerns, for instance:

- *Unpredictability*. While DALL-E can generate images based on descriptions, the exact output is not predictable or fully controllable, which might be a challenge for applications that require precision and consistency.
- *Intellectual property concerns*. Since DALL-E generates images based on its training data, which includes a vast range of images from the internet, there may

be concerns over copyright infringement if the generated images resemble copyrighted works too closely.

- *Content moderation*. DALL-E could potentially be used to generate inappropriate, offensive, or harmful images if not properly moderated. Controlling and moderating the content it generates to avoid such misuse is a significant challenge.
- *Job displacement*. The automation of content creation could potentially displace jobs in fields like graphic design and illustration. However, it could also open up new roles in overseeing and managing these AI systems.

6.9 DALL-E 2

DALL-E 2 builds upon the original DALL-E model with several key improvements that enhance the quality, coherence, and diversity of the generated images.

- Improved Image Quality and Resolution
  - *Diffusion Models*: DALL-E 2 introduces a new diffusion-based model for image generation. Unlike autoregressive models, diffusion models start with a noisy image and iteratively refine it to produce a clear image. This approach significantly improves the sharpness and detail of the generated images.
  - *High-Resolution Output*: DALL-E 2 can produce higher resolution images than its predecessor, enabling the creation of more complex and realistic visuals.

6.9.1 Example Prompt

Text Input: 'A futuristic cityscape with flying cars at sunset'.

Process:

- **Text Encoding**: The text input 'A futuristic cityscape with flying cars at sunset' is encoded into a series of vectors that capture the meaning of the words and their relationships.
- **Prior Network**: The encoded text is then processed by the Prior Network, which predicts a distribution over the latent space. This distribution represents various potential visual features that align with the textual description.
- **Image Generation**:
  - *Latent Diffusion*: The model uses latent diffusion to iteratively refine an image based on the predicted distribution. It starts with a random noise image and gradually improves it to match the description of a futuristic city with flying cars at sunset.
  - *CLIP Guidance*: Throughout the process, CLIP helps ensure that the generated image is closely aligned with the text description, refining the image to match the semantic meaning of the input text.

- **Output**: The final output is a high-resolution, detailed image of a futuristic cityscape, likely featuring tall, sleek buildings, glowing lights, and several flying cars in the sky against the backdrop of a vibrant sunset.

Potential Outputs:

- Image 1: A city with skyscrapers illuminated by neon lights, with flying cars zooming between them as the sun sets in the background

- Image 2: A sprawling futuristic cityscape with a mix of old and new architectural styles, flying vehicles casting shadows on the streets below, with a glowing orange sky

- *High-Resolution*: The image produced is sharp and detailed, capturing the intricate elements of the city and the flying cars.
- *Text-Image Alignment*: The image closely matches the input description, with each element of the text represented visually.
- *Diversity*: Multiple images can be generated from the same prompt, each with a slightly different interpretation of the 'futuristic cityscape', providing a variety of creative outputs.

This example illustrates how DALL-E 2 can generate creative and detailed images from textual descriptions, making it a powerful tool for artists, designers, and other creative professionals.

Prior Network and CLIP (Contrastive Language–Image Pretraining)

- *Prior Network*: As mentioned earlier, the Prior Network in DALL-E 2 takes the text embeddings and converts them into a distribution in the latent space, which the image decoder uses to generate images. This step ensures that the generated images align more closely with the input text.
- *CLIP Integration*: DALL-E 2 integrates CLIP, another OpenAI model that has been trained on a vast dataset to understand both images and text. CLIP acts as a guide, ensuring that the generated images are semantically meaningful and aligned with the input descriptions. By combining text and image understanding in the same model, DALL-E 2 produces more accurate and relevant images.
- *Latent Diffusion*: DALL-E 2 introduces the concept of latent diffusion, where the model operates in a lower-dimensional latent space rather than directly on the pixel space. This not only speeds up the image generation process but also reduces computational requirements.
- *Unconditional Sampling and Diversity*: DALL-E 2 allows for more diverse image generation by improving how the model samples from the latent space. This leads to a broader range of potential images for a given text input, capturing various possible interpretations.

6.9.2 Improvements in DALL-E 2

- *Sharper and higher-resolution images* due to the use of diffusion models.
- *Better text-image alignment* through the use of the Prior Network and CLIP.
- *Faster and more efficient generation* by working in a latent space.
- *Increased diversity* in generated images, allowing for multiple interpretations of a single text prompt.

DALL-E 2 represents a significant step forward in text-to-image generation, providing more precise, high-quality, and varied outputs compared to the original DALL-E model.

6.10 Applications

1. **Art and Creative Media: Revolutionizing Creative Industries**
   (a) *Creative Freedom:* Text-to-image synthesis offers artists and designers unprecedented creative freedom. By simply describing a concept, artists can generate unique and diverse images that serve as inspiration or final artwork.
   (b) *Rapid Prototyping:* For media production, such as film, television, and advertising, text-to-image models allow rapid prototyping of visual ideas. Designers can quickly visualize and iterate on concepts without the need for manual sketching or rendering.
   (c) *Custom Content Creation:* The ability to generate images from text enables the creation of highly personalized and niche content, catering to specific themes or audiences. This is particularly useful in marketing and branding, where unique visuals are often required.
   (d) *Accessibility for Non-Artists:* Individuals without formal training in art or design can create visually compelling content. This democratizes art creation and allows more people to participate in the creative process.
2. **Educational Tools: Enhancing Learning Materials with Generated Images**
   (a) *Interactive Learning:* Text-to-image synthesis can enhance educational materials by turning textual information into vivid illustrations. This is especially beneficial in subjects like history, biology, and literature, where visual aids can significantly improve understanding.
   (b) *Tailored Educational Content:* Educators can generate specific images that align with lesson plans, allowing for customized learning experiences. For example, a teacher can generate images of historical events or scientific processes that precisely match the topics being taught.
   (c) *Engagement and Retention:* Visual aids generated from text can make learning more engaging, helping students retain information better. Complex concepts that are difficult to describe in words can be illustrated effectively, aiding comprehension.
   (d) *Language Learning:* Text-to-image models can assist in language learning by providing visual representations of words, phrases, or sentences, making it easier for learners to grasp meanings and context.
3. **Gaming and Virtual Reality: Creating Dynamic Visual Elements Based on Narrative Text**
   (a) *Procedural Content Generation:* In gaming, text-to-image synthesis can be used to create procedurally generated environments, characters, and objects based on narrative input. This allows for dynamic storytelling where the game world evolves based on the player's actions or the unfolding narrative.
   (b) *Immersive Experiences:* In virtual reality (VR), text-to-image models can enhance immersion by generating realistic or fantastical environments based

on user input. Players or users can describe a scene, and the model can create it in real-time, adding to the sense of presence and immersion.

(c) *Narrative-driven Design:* Games that rely heavily on narrative can benefit from text-to-image models by generating visual elements that closely match the story's descriptions, enhancing the overall coherence and depth of the game world.

(d) *Customizable Content:* Gamers can personalize their experiences by generating unique in-game assets, such as avatars, weapons, or landscapes, based on their descriptions, making each playthrough unique.

4. **Assistive Technologies: Helping Visually Impaired Individuals Understand Textual Content through Imagery**

(a) *Text-to-Image Conversion:* For visually impaired individuals, text-to-image models can convert descriptive text into images, which can then be interpreted using tactile graphics or haptic feedback. This allows them to 'visualize' the content in a way that was previously inaccessible.

(b) *Enhanced Descriptions:* In combination with screen readers, these models can generate visual descriptions of web pages, documents, or digital content, offering a richer experience than text alone. This is particularly useful for understanding complex layouts or visual data.

(c) *Educational Tools for the Visually Impaired:* In educational settings, these models can generate tactile images from text, enabling students with visual impairments to engage with visual content through touch. This can be applied in subjects like geometry, geography, and art.

(d) *Personalized Accessibility Features:* Assistive applications can be tailored to generate content that meets individual needs, allowing users to describe what they want to 'see' and receive a customized visual representation that suits their specific requirements.

These applications demonstrate the transformative potential of text-to-image synthesis across various fields, enhancing creativity, education, entertainment, and accessibility. The continued development of these technologies promises even more innovative and impactful uses in the future.

6.11 Conclusion

In this chapter, we explored the transformative potential of text-to-image synthesis technology across several domains. We discussed how it revolutionizes the creative industries by providing artists and designers with unparalleled creative freedom, enabling rapid prototyping, and making custom content creation more accessible. In education, this technology enhances learning materials by converting text into vivid visual aids, thus improving engagement, retention, and comprehension. The gaming and virtual reality industries benefit from dynamic and immersive content generation, while assistive technologies use text-to-image synthesis to help visually impaired individuals interpret and interact with visual content.

The long-term impact of text-to-image synthesis technologies could be profound, reshaping numerous industries and aspects of everyday life. In creative fields, it could democratize art and design, allowing more people to produce high-quality visual content without extensive training. In education, it could revolutionize how information is delivered and understood, making learning more interactive and personalized. The gaming and VR sectors could see more dynamic, narrative-driven content that adapts to user input in real-time, while assistive technologies could greatly enhance accessibility for people with visual impairments, allowing them to engage with the visual world in new and meaningful ways.

As these technologies continue to evolve, it is crucial to encourage further research to explore their full potential while addressing ethical considerations and ensuring responsible use. Developers, researchers, and policymakers should collaborate to create frameworks that promote innovation while safeguarding against misuse. By embracing these technologies with a focus on ethical development, we can harness their power to create a more inclusive and creatively rich future.

References

Goodfellow, I., Pouget-Abadie, J., Mirza, M., Xu, B., Warde-Farley, D., Ozair, S., et al. (2014). Generative adversarial nets. *Advances in Neural Information Processing Systems, 27*, 2672–2680.

Ramesh, A., Pavlov, M., Goh, G., Gray, S., Voss, C., Radford, A., Sutskever, I. (2021). Zero-shot text-to-image generation. *arXiv preprint arXiv:2102.12092.*

Reed, S. E., Akata, Z., Yan, X., Logeswaran, L., Schiele, B., & Lee, H. (2016). Generative adversarial text to image synthesis. In *Proceedings of the 33rd International Conference on machine learning* (pp. 1060–1069). PMLR.

Vaswani, A., Shazeer, N., Parmar, N., Uszkoreit, J., Jones, L., Gomez, A. N., et al. (2017). Attention is all you need. *Advances in Neural Information Processing Systems, 30*, 5998–6008.

Zhang, H., Xu, T., Li, H., Zhang, S., Huang, X., Wang, X., & Metaxas, D. N. (2017). StackGAN: Text to photo-realistic image synthesis with stacked generative adversarial networks. In *Proceedings of the IEEE International Conference on computer vision* (pp. 5907–5915).

Chapter 7
Image-to-Text Generation: Bridging Visual and Linguistic Worlds

Akansha Singh and Krishna Kant Singh

Abstract This chapter delves into the field of image-to-text generation, a pivotal advancement in artificial intelligence that bridges computer vision and natural language processing. It provides a historical overview of image-to-text systems, from early optical character recognition (OCR) to sophisticated transformer-based and multimodal models capable of generating descriptive and contextually relevant text. Key applications across accessibility, healthcare, social media, and e-commerce underscore the transformative impact of image-to-text technology in enhancing user interaction and information accessibility. The chapter also discusses various challenges, including contextual understanding, cultural diversity, and computational efficiency, and reviews advanced techniques like Vision Transformers (ViTs), multimodal transformers, and hybrid models that enhance system capabilities. Emerging trends in the field highlight the potential for continued integration with real-time and edge devices, fostering inclusive and dynamic AI applications.

Keywords Image-to-text generation · Transformer models · Vision transformer (ViT) · Optical character recognition (OCR) · Accessibility in AI · Multimodal AI

7.1 Introduction

With the rapid evolution of technology where every second massive volumes of data are being created, the skill for visual description and interpretation automatically gained much relevance. The generation of image-to-text, or in other words, image captioning, is an interdisciplinary research field combining computer vision with NLP, which aims to generate descriptive sentences according to input visual

A. Singh (✉)
School of Computer Science Engineering and Technology, Bennett University, Greater Noida, Uttar Pradesh, India

K. K. Singh
Delhi Technical Campus, Greater Noida, Uttar Pradesh, India

A. Singh, K. K. Singh (eds.), *Multimodal Generative AI*,
https://doi.org/10.1007/978-981-96-2355-6_7

features. It ranges from the basic technologies to methodologies and developments that are occurring in the case of image-to-text generation, bringing together the visual perception and the linguistic expression of the system (Radford et al., 2021; Li et al., 2022a).

7.2 The Evolution of Image-to-Text Systems

The development of image-to-text systems has been an interesting journey, driven by two major factors: technology and methodology. Starting from the days where only Optical Character Recognition technologies were developed, to the development of complex deep learning integrated models, this technology has changed the modus operandi of extracting and converting textual information from images. Table 7.1 gives some of the major milestones in the establishment of image-to-text systems—from the simple OCR techniques to the modern AI-driven approaches now used in several applications. Initial systems employed basic template-based methods, which were applied by identifying objects and predefined relationships among those objects. Such systems could hardly describe complex scenes. With deep learning, especially CNNs applied to image recognition, and RNNs applied to the generation of sequences, it allowed more articulated and exacting models for image-to-text. The introduction of transformers, particularly models like Vision Transformers (ViTs) and attention mechanisms, further enhanced the ability to generate coherent and contextually relevant descriptions (Dosovitskiy et al., 2021).

Table 7.1 Evolution of image-to-text systems

Year	Milestone	Description
1960s	Early foundations	Initial research in optical character recognition (OCR) begins, focusing on converting printed text into machine-readable text
1970s	Development of basic OCR algorithms	Basic pattern recognition algorithms are developed, laying the groundwork for early OCR systems
1980s	Commercial OCR systems	Introduction of commercial OCR systems widely used in industries for document processing
1988	First multiformat OCR software	Launch of commercial OCR software capable of recognizing multiple fonts and formats
1990s	Advances in OCR and early image-to-text research	Significant improvements in OCR accuracy, including the ability to recognize handwritten text
1997	First research in image captioning	Early research begins on generating textual descriptions from images using rule-based approaches
2000s	Emergence of machine learning	Machine learning techniques lead to more sophisticated image recognition and classification algorithms
2002	Statistical models for image captioning	Introduction of probabilistic models for generating text from images
2012	Deep learning revolution	Development of AlexNet, a deep CNN that significantly improves image classification tasks

Table 7.1 (continued)

Year	Milestone	Description
2014	Neural Image Caption Generator (NIC)	Introduction of deep learning-based image captioning models combining CNNs and RNNs
2015	Google's show and tell model	Use of deep learning techniques for generating natural language descriptions from images, achieving state-of-the-art performance
2016	Introduction of attention mechanisms	Attention mechanisms in image captioning models allow the system to focus on specific image parts while generating text
2017	Show, attend, and tell model	Further refinement of the attention mechanism in image captioning models, improving the quality of generated captions
2019	Transformer-based models for image-to-text	Adaptation of models like BERT and GPT for image-to-text tasks, leading to more accurate and context-aware caption generation
2020	OpenAI's CLIP model	Introduction of CLIP, which learns visual concepts from natural language descriptions, advancing multimodal understanding
2021	OpenAI's DALL-E and text-to-image integration	Release of DALL-E, a model generating images from text descriptions, and later models generating text from images, blurring lines between image-to-text
2022	Google's imagen model	Demonstrates advanced capabilities in generating accurate textual descriptions from images using large-scale pre-trained models
2023	Integration of image-to-text in LLMs	Integration of image-to-text capabilities into general-purpose LLMs like GPT-4, allowing highly accurate and contextually relevant text descriptions

7.3 Core Components of Image-to-Text Systems

Image-to-Text systems are sophisticated technologies that convert visual information into textual descriptions. These systems rely on a combination of advanced techniques and components to accurately interpret and describe the content of images. Below are the core components that form the foundation of modern Image-to-Text systems:

Component	Description
Image preprocessing	Involves preparing the image data for analysis, including resizing, normalization, and noise reduction to ensure that the image is in an optimal format for further processing
Feature extraction	Utilizes deep learning models, typically convolutional neural networks (CNNs), to extract relevant features from the image, such as edges, textures, and objects
Object detection/ recognition	Identifies and labels objects or regions of interest within the image, often using advanced algorithms like region-based CNNs (R-CNN) or YOLO (you only look once)

Component	Description
Text generation model	Generates the textual description of the image by interpreting the extracted features, often using models like recurrent neural networks (RNNs) or transformers
Attention mechanism	Focuses on specific parts of the image while generating the text, enhancing the relevance and accuracy of the generated description by considering the most important features
Language model	Ensures that the generated text is coherent and contextually accurate, often utilizing large pre-trained models like GPT or BERT to enhance the natural language output
Post-processing	Refines the generated text by correcting any grammatical or syntactical errors and ensuring that the output is clear and relevant to the content of the image
Dataset and training	Relies on large annotated datasets for training the model, such as MS COCO, and fine-tunes the model with diverse images to improve its generalization and accuracy
Evaluation metrics	Measures the performance of the image-to-text system using metrics like BLEU, METEOR, and CIDEr to assess the quality and relevance of the generated descriptions

These components work together to create systems capable of generating accurate, contextually relevant textual descriptions from images, enabling a wide range of applications from accessibility tools to advanced content analysis (Fig. 7.1).

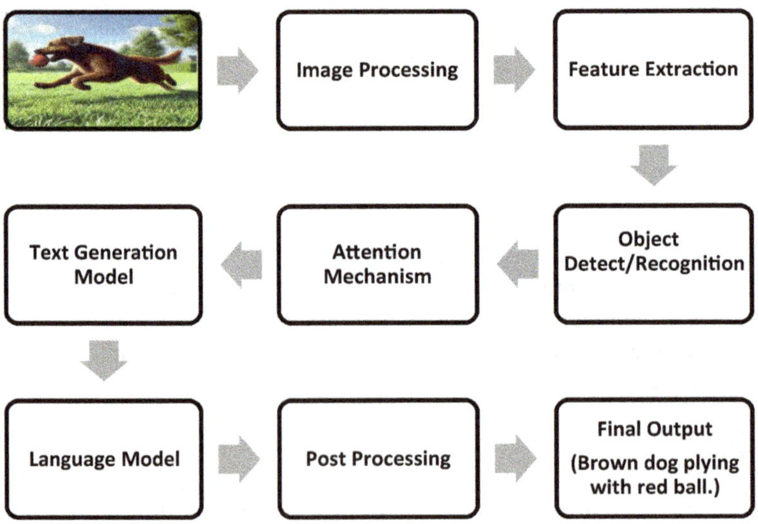

Fig. 7.1 Image to text generation process

7.4 Applications of Image-to-Text Generation

Image-to-text generation is a higher order subset of AI and computer vision that is increasingly making a difference in the way people can work with visual content across various dimensions. By converting visual images into descriptive, contextually relevant text, the technology bridges the gap between visual and textual information, aids accessibility, manages content more effectively, and opens up new ways to develop user experiences. Applications range from allowing the visually impaired to have any image described in detail—from autogenerated metadata for digitized libraries, through the enhancement of e-commerce product listings, to reaching autonomous vehicles interpreting visual data. As this feature in image-to-text generation keeps getting better, it promises to integrate more and newer applications across industries, making visual information more usable, actionable, and insightful.

1. **Accessibility**: Image-to-text generation is key to ensuring the availability of visual content for visually impaired users. Via the use of image descriptions, it deploys technology for screen readers to replay the content on images to users who cannot view them.
    (a) **Example**: If there were a picture of a sunset on a website or social media platform, they would say, "A vibrant sunset with shades of orange, pink, and purple across the sky." This simply means that a visually impaired user would understand what is contained in the image.
    (b) **Impact**: This makes things more inclusive and ensures that the visually impaired are able to use and view visual content much like sighted users.
2. **Content Management**: Image-to-text applications in digital content management systems help automatically create metadata and descriptions of massive volumes of images. This increases effectiveness in indexing, searching, and organizing visual content.
    (a) **Example**: A title or caption generator can automatically generate descriptive tags like "business meeting," "office," and "collaboration," depending on the content of the images in a digital library or stock photo website.
    (b) **Impact**: This system smoothes the process of image management and retrieval through views, enabling users to easily find relevant content and making maintenance easier for administrators.
3. **Social Media:** Image-to-text generation in social media enhances user interaction by suggesting captions, generating tags automatically, and creating engaging content with users' posted images.
    (a) **Example**: When a user uploads a vacation picture, it may automatically suggest a caption like "Exploring beautiful beaches in Hawaii" or generate hashtags like #Travel #BeachLife.
    (b) **Impact**: Increases user engagement and content visibility while saving users the effort of creating their own content.

4. **Healthcare**: In healthcare, image-to-text generation helps analyze and interpret medical images. It can provide descriptive reports for X-rays, MRIs, and other diagnostic images, aiding medical professionals in diagnosis and treatment planning.
    (a) **Example**: An AI system could generate an MRI report stating: "There is cortical thickening in the left temporal lobe, consistent with seizures described by the patient."
    (b) **Impact**: Accelerates the diagnostic process, reduces the workload for radiologists, and enhances the accuracy of medical reports.
5. **E-commerce**: Image-to-text generation in e-commerce creates automatic product descriptions based on images, improving product listings and boosting search engine optimization to attract more customers.
    (a) **Example**: For an image of a red leather handbag, the system might generate a caption like "Fashionable red leather handbag with gold accents and adjustable straps, ideal for formal events."
    (b) **Impact**: Increases product visibility, provides better information to potential buyers, and boosts conversion rates.
6. **Education**: Image-to-text generation in education helps explain complex diagrams, charts, and illustrations, making it easier for students to learn from visual content.
    (a) **Example**: An educational app might use this technology to describe a diagram of the human circulatory system, detailing its components and their functions.
    (b) **Impact**: Enhances understanding and supports diverse learning styles in education.
7. **Autonomous Vehicles**: Image-to-text generation in autonomous vehicles provides meaning to visual information captured by cameras. This includes detecting road signs, pedestrians, and obstacles, and converting that information into actionable instructions for the vehicle.
    (a) **Example**: An AI system may analyze a camera feed and generate a text-based warning such as "Stop sign detected 100 meters ahead" to the vehicle navigation system.
    (b) **Impact**: This improves the safety and reliability of autonomous driving systems by availing real-time and readable information.
8. **AR**: Image-to-text generation in AR contextualizes objects or visual scenes viewed through an augmented reality device, thereby enhancing the experience of a user through overlaying relevant text information.
    (a) **Example**: It may show text to the user, such as, "This is the Colosseum, an ancient amphitheater in Rome built in AD 80 when an AR device is pointed at a historical landmark."
    (b) **Impact**: Betters interaction of the users with AR environments; adds a little more context in everything.

9. **Security and Surveillance**: Image-to-text generation in security and surveillance is used for video summarization about events and interesting activities. This can further enhance efficiency in monitoring and incident reporting.
   (a) **Example**: An AI-powered surveillance system may provide a report, "At 2:15 PM, a person entered the restricted area and accessed the server room."
   (b) **Impact**: Facilitates efficiency in security operations and quick responses in case of any security incident.
10. **Entertainment and Media**: Image-to-text generation in the entertainment arena helps develop very interesting information, tags files of multimedia, and can even write scripts from what it sees.
   (a) **Example**: For a movie script, the AI system would analyze the scene and generate the text: "The hero enters that room in dim light and cautiously proceeds toward the mysterious thing kept on the pedestal."
   (b) **Impact**: This facilitates content creation, enhances multimedia organization, and supports creative processes during media production.

In other words, it is versatile technology applicable across industries, including but not limited to: accessibility, content management, social media, healthcare, e-commerce, education, autonomous vehicles, AR, security, and entertainment. It bridges the gap between Visual and Textual Information, enabling intractability, understanding, and management thereof with much greater ease.

7.5 Challenges in Image-to-Text Generation

Image-to-text generation represents a powerful technology, but it faces several significant challenges that affect its accuracy, efficiency, and applicability. Following are some of the key challenges (Radford et al., 2021; Li et al., 2022a):

1. **Contextual Understanding**: How to interpret the context of the image correctly is one major challenge. Most current models lack the finesse needed for subtle and abstract details, hence limiting themselves to very generic or, many times, flat-out incorrect descriptions. For example, a model may describe a crowded street as simply "a busy street" without identifying any important interactions or objects. This makes the generated descriptions less useful for applications that require more detailed context, such as in healthcare or self-driving cars.
2. **Complex Visual Content**: In particular, complex scenes, images containing many objects, or a lot of detailed features are a problem even for current image-to-text approaches. These models may not be able to describe all components accurately; this is another reason for shallow descriptions. For example, a large event outdoors where so many different things are happening may be incompletely described and miss critical details. This ultimately limits the performance

and effectiveness of the system, which therefore cannot be used to its full potential with regards to content management or educational purposes.

3. **Cultural and Linguistic Variability**: Many image descriptions rightly need contexts of culture and language. The differences in interpretation due to cultural norms or peculiarities of language could affect the quality of the generated text. For example, an image of a festival of any tradition would have different descriptions based on regional cultural knowledge and may lead to inconsistent descriptions. This impacts the system's ability to devise universally accurate and culturally appropriate descriptions that are important for global applications and accessibility.
4. **Data Bias and Representation**: Models can inherit biases from training data, leading to skewed or biased descriptions. Suppose the training data does not represent diverse contexts and populations; then the system might make inaccurate or biased descriptions, especially for images outside of the Western cultural setting. This makes the system less inclusive and accurate, especially for social media and e-commerce applications.
5. **Ambiguity and Multiple Interpretations**: Most of the pictures have elements in them that have a subjectivity of interpretation. It becomes tough to generate a single definitive description for such images. For instance, a picture of someone holding a package can be described differently—from the fact that a person is holding it to the description of the package or context where the picture is taken. This further reduces utility in some specific applications that may call for exact interpretation.
6. **Computational Resources and Efficiency**: Training and running image-to-text models require significant computation. High-quality models also demand large datasets and substantial processing power. Probably the most critical limitation of current systems is that it is computationally expensive to generate detailed descriptions for large-scale image datasets in real time, and may even require advanced hardware, therefore seriously constraining the scalability and practical applicability of any system.
7. **Integration with Other Systems**: There are some technical challenges to integrate image-to-text generation with other systems, be it natural language processing or domain-specific applications. For example, the integration of image-to-text with medical diagnosis systems means that the descriptions generated should be interpreted by a medical expert with complete accuracy. This, in turn, affects the efficiency and reliability of the system as a whole, especially in very specific domains like health or autonomous vehicles.
8. **Security and Privacy Concerns**: The application of image-to-text generation in sensitive domains, say, in surveillance or processing personal data, raises certain concerns regarding data privacy and security. Misuse of, or accidental exposure of, descriptive information has many consequences. For example, a descriptive text of surveillance footage may reveal certain private or sensitive information if not handled accordingly; hence, ethical and legal issues related to data privacy and security must be taken into consideration.

9. **Dynamic Content**: Ever-changing visual content, like scenes or trends that change very fast, may render the static model-generated descriptions irrelevant or inaccurate. For example, if the model was trained on very old data, then it may describe an image of a current event or something that is in the trending topic inappropriately. This, in turn, renders the system unable to provide updated and relevant descriptions, particularly in fast-paced environments such as social media or news reporting.
10. **User Expectations and Experience**: Users may set high expectations for the accuracy and quality of image-to-text generation, which can be apparently challenging to meet consistently. For instance, there is the expectation for highly detailed and accurate product descriptions through e-commerce, while failure to meet such expectations has adverse effects on the satisfaction and engagement of users. It is in this regard that the perceived value and effectiveness of image-to-text systems determine their adoption in different applications.

These challenges call for continuous research and development that will help improve the accuracy, inclusivity, and efficiency of technologies supporting the generation of text from images. The perceived effectiveness of such systems will, in furtherance, be spread over different applications. The art remains with continuous research and development processes to improve the accuracy, inclusivity, and efficiency of the technologies that make up image-to-text generation.

7.6 Advanced Techniques

The fundamental advances in image-to-text generation are stretching the frontier into a more accurate, contextually relevant, and versatile system. This document presents some state-of-the-art methods and future directions in the field.

7.7 Transformer Models

Transformer-based models have currently become an important element in image-to-text generation due to their strong architecture and ability to grasp complex relationships across different modalities. The Transformer model is a deep learning architecture primarily adopted for natural language processing tasks, such as translation, summarization, and text generation. It is known for effectively capturing long-range dependencies in sequences without requiring recurrent structures like those in RNNs.

1. **Input Sequence**: A sequence of tokens, words, or subwords forms the input provided to the Transformer model. Consider the input sequence as [The quick brown fox]. Each word in the sequence initially takes the form of a unique token,

which is then embedded into a more meaningful numerical representation through the embedding process.

2. **Embedding Layer**: This layer converts each token or word in the input sequence into a fixed-size, dense vector. These vectors capture the semantic meaning of the tokens, making them suitable for further processing by the model. In the diagram, the input sequence of words [The quick brown fox] is transformed into a sequence of vectors where each word is mapped into a high-dimensional space.
3. **Positional Encoding**: Transformers do not have an inherent understanding of token order in the input sequence since they do not use recurrence or convolution. The Positional Encoding layer adds information to the embedding vectors regarding the position of each token within the sequence. After positional encoding, the model can differentiate between sequences like [The quick brown fox] and [Fox brown quick the], even though the tokens are the same.
4. **Encoder Blocks:** The encoder consists of a stack of identical blocks, each containing two main sub-layers:

   (a) **Multi-Head Attention:** This mechanism allows the model to focus on different parts of the input sequence simultaneously, capturing relationships between tokens regardless of their distance from each other.

   (b) **Feed-Forward Network (FFN):** A fully connected network that processes the output of the attention mechanism, adding non-linearity and complexity to the model.

   The entire sequence is processed through these blocks, with each block refining the representation of the input sequence.

5. **Decoder Blocks**

   (a) **Purpose:** The decoder is similar to the encoder but with an additional attention layer that focuses on the output of the encoder. The decoder processes the target sequence during training (or previously generated tokens during inference) to generate the next token.

      - **Masked Multi-Head Attention:** This layer prevents the decoder from attending to future tokens in the sequence, ensuring that the prediction for a given token depends only on the known tokens before it.
      - **Encoder-Decoder Attention:** This layer allows the decoder to focus on relevant parts of the input sequence encoded by the encoder.
      - **Feed-Forward Network:** Similar to the encoder, this network adds non-linearity and complexity to the decoder's output.

   (b) **Example:** The decoder refines its understanding of the sequence by attending to both the input (through encoder-decoder attention) and its own previously generated tokens.

6. **Linear Layer**

   (a) **Purpose:** The Linear Layer maps the final decoder output to the desired output dimension, typically the size of the vocabulary in the target language.

   (b) **Example:** It transforms the decoder's output into logits (unnormalized probabilities) for each possible token in the output vocabulary.

7. **Softmax**
   (a) **Purpose:** The Softmax layer converts the logits from the linear layer into probabilities, indicating the likelihood of each token being the correct next token in the sequence.
   (b) **Example:** After applying softmax, the model generates a probability distribution over the vocabulary, selecting the token with the highest probability as the next word in the sequence.
8. **Output Sequence**
   (a) The output is a stream of tokens, which, for the model, is generated. In the diagram, the example output stream is the translation of the input stream.

The Transformer architecture is an attention mechanism-based architecture that allows it to process and translate the streams in parallel with no recurrent structures. It begins with the input sequence that undergoes transformation by embeddings and positional encoding. The encoder will then refine this sequence, and the decoder uses this encoded input, inclusive of previously generated tokens, in order to generate the output sequence. This final output is a sequence—usually a translation or generation of the input sequence in most tasks. This architecture turns out to be powerful for tasks that require an understanding of context and relationships between far-apart elements in a sequence, making it one of the cornerstones in modern applications of NLP.

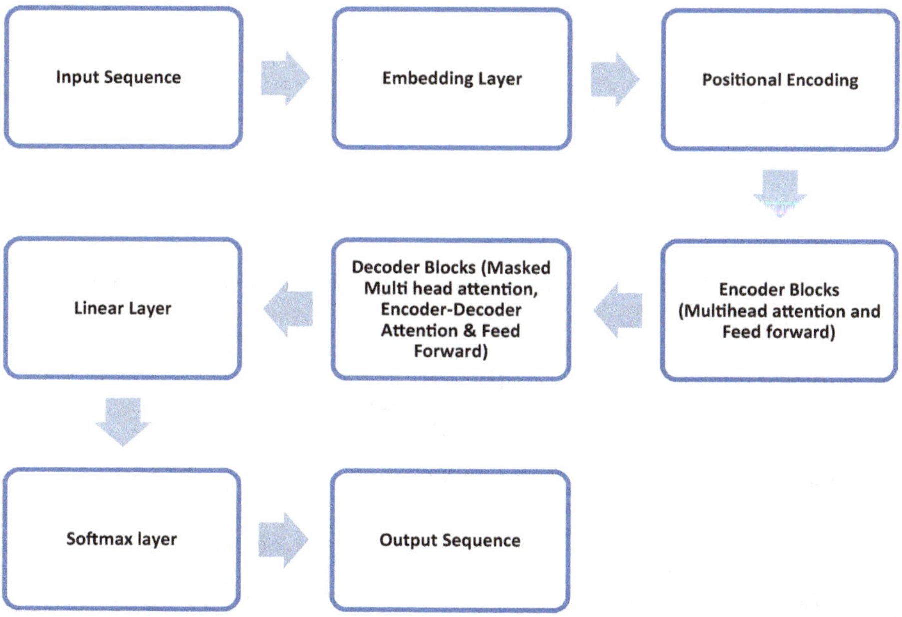

7.7.1 *Vision Transformers (ViTs) (Dosovitskiy et al., 2021)*

The Vision Transformer (ViT) is a model that adapts the Transformer architecture, originally designed for natural language processing, to image recognition tasks. The key idea is to treat an image as a sequence of smaller image patches and process it using the Transformer model. Below is a detailed breakdown of the process, including the mathematical aspects.

1. **Input Image**
   - **Size:** Typically, the input image is of fixed size, for example, 224 × 224 pixels.
   - **Representation:** The image is represented as a matrix of pixel values, where each pixel can have multiple channels (e.g., 3 channels for RGB images).
2. **Patch Extraction**
   - The input image is divided into non-overlapping patches, each of size P × P (e.g., 16 × 16 pixels).

$$N = \frac{HW}{P^2}$$

3. **Number of Patches**
   - If the input image has a size of $H \times W$

$$H \times W$$

   - For e.g., 224 × 224, and the patch size is $P \times P$, the number of patches N is:

$$N = \frac{H \times W}{P^2}$$

   - For an image of 224 × 224 pixels and patch size 16 × 16, the number of patches would be

$$N = \frac{224 \times 224}{16 \times 16} = 196$$

4. **Linear Projection (Patch Embedding)**

$$z_i = E \cdot \text{flatten}(\text{patch}_i)\, zi = E \cdot \mathit{flatten}(\mathit{patch}_i)$$

Each $P \times P$ patch is flattened into a vector of size $P^2 \times C$, where C is the number of channels (e.g., 3 for RGB). This vector is then projected into a lower-dimensional space using a linear projection (learned embedding):

$$z_i = E \cdot flatten(patch_i)$$

where E is the learnable projection matrix.

This process results in a sequence of vectors, each representing a patch, where each vector has a dimension D (the embedding dimension).

5. **Flattened Patches**

   The sequence of vectors from all the patches is concatenated to form a sequence of length N, where each element is a vector of dimension D. The image is now represented as a sequence $z_1, z_2, \ldots, z_N$ analogous to a sequence of words in NLP.

6. **Positional Encoding**

   Since the Transformer model does not inherently understand the spatial structure of the image (i.e., the position of patches), positional encoding is added to each patch embedding to retain spatial information. The positional encoding is typically added to the patch embeddings:

$$z_i' = z_i + \mathrm{PE}_i$$

   where PE_i is the positional encoding vector for the ith patch.

7. **Transformer Encoder Blocks**

   The sequence of patch embeddings is processed through multiple Transformer encoder blocks, each consisting of:

   Multi-Head Self-Attention: This mechanism allows each patch to attend to other patches in the sequence, capturing relationships across the entire image. For each head, the attention output is computed as:

$$\mathrm{Attention}(Q,K,V) = \mathrm{softmax}\left(\frac{QK^{\top}}{\sqrt{d_k}}\right)V$$

   where Q, K, V are query, key, and value matrices, and d_k is the dimension of the key vectors.

   Feed-Forward Network (FFN): After the attention mechanism, a feed-forward network is applied to each patch embedding independently, consisting of two linear transformations with a ReLU activation in between:

$$\mathrm{FFN}(x) = \mathrm{ReLU}(\times W_1 + b_1)W_2 + b_2$$

   Layer Normalization: Optionally, a layer normalization step can be applied after each attention and feed-forward sub-layer to stabilize training.

8. **Class Token**

   A special class token z_{class} is prepended to the sequence of patch embeddings before entering the Transformer encoder. This token is used to aggregate information from all patches and is ultimately used for classification or other tasks.

The class token is updated along with the patch embeddings through the Transformer encoder blocks.

9. **MLP Head**

Process: After the Transformer encoder blocks, the final representation of the class token

$$z_{\text{class}}$$

is passed through a Multi-Layer Perceptron (MLP) head, typically consisting of one or more fully connected layers.

The MLP head produces logits for each class (in the case of classification):

$$\text{logits} = W\text{ot}\cdot z\text{cas} + b\text{ot}$$

where Wout\mathbf{W}_\text{out}Wout and bout\mathbf{b}_\text{out} bout are the weights and biases of the output layer.

10. **Output**

The final output depends on the specific task. For image classification, the output is a probability distribution over different classes, typically obtained by applying a softmax function to the logits. This output could be the predicted class of the image, the detected objects within the image, or any other task the model is trained for.

The Vision Transformer (ViT) model processes images by splitting them into smaller patches, embedding each patch, and then applying a Transformer model to analyze the relationships between these patches. Essentially, this model treats an image as a sequence of tokens, similar to how natural language processing (NLP) treats words. By utilizing the Transformer's attention mechanism, ViT is capable of capturing global context and has achieved state-of-the-art performance in various image recognition tasks, all without the need for the convolutional layers typically found in conventional CNNs.

7.7.2 Multimodal Transformers

Multimodal transformers are specifically designed to handle multiple types of data, such as images and text, simultaneously. They are especially useful for image-to-text generation tasks because they can connect and integrate visual information with textual data.

- **Joint Embedding Space:** These models create a unified embedding space where both visual and textual data coexist. For tasks that require image-to-text translation, visual features extracted from the images are aligned with textual embeddings, allowing both types of data to be processed together.
- **Cross-Attention Layers:** These layers enable the model to focus on specific areas of one modality (e.g., an image patch) while generating corresponding text, ensuring that the output is both accurate and contextually relevant.

- **Multimodal Training:** Multimodal transformers are trained on datasets containing paired image-text examples. This training helps the model learn to correlate visual features with descriptive text, improving its ability to generate precise descriptions. One notable multimodal model is CLIP (Contrastive Language-Image Pretraining) developed by OpenAI, which excels in aligning images with corresponding text. It predicts which text matches a given image, showcasing an impressive ability to understand cross-modal information.

Multimodal transformers significantly enhance the generation of contextually appropriate descriptions by integrating data from multiple sources.

7.7.3 *Advantages of Transformer-Based Models*

- **Enhanced Contextual Understanding:** Transformers can capture long-range dependencies and intricate relationships, which helps them generate descriptions that are more coherent and contextually rich.
- **Parallel Processing:** The self-attention mechanism allows data to be processed in parallel, making transformers more efficient and faster to train compared to models that rely on sequential processing.
- **Scalability:** Transformers are well-suited to handle large-scale datasets and can scale effectively with more data and larger model sizes.
- **Flexibility:** This architecture can be adapted for various tasks and data types, making it highly versatile for different applications, including image-to-text generation.

7.7.4 *Challenges and Limitations*

- **Computational Resources:** Transformer-based models, particularly large ones, demand substantial computational power for both training and inference, which can pose challenges for widespread use.
- **Data Requirements:** To perform optimally, these models often need vast amounts of high-quality training data, which may not always be available in certain domains.
- **Interpretability:** While these models are powerful, they can be complex and difficult to interpret, making it harder to understand how they produce specific descriptions.
- **Overfitting:** Large models with many parameters can be prone to overfitting, especially if the training dataset is not diverse or extensive enough.

7.7.5 Future Directions

1. **Efficiency Improvements:** Ongoing research is focused on creating more computationally efficient transformers, potentially by reducing the number of parameters or optimizing attention mechanisms, to make them more accessible and practical.
2. **Enhanced Multimodal Integration:** Further work aims to improve the integration of different data modalities, refining how visual and textual data are processed together to enhance image-to-text generation.
3. **Fine-Tuning and Adaptation:** The development of more effective techniques for fine-tuning transformers to specific tasks or domains will allow for more specialized and accurate image-to-text applications.
4. **Interpretability and Explainability:** Efforts will continue to make transformer models more interpretable, helping users better understand and trust the outputs they generate.

In summary, transformer-based models have driven significant progress in image-to-text generation by capitalizing on their attention mechanisms and capacity for processing complex relationships. As the field advances, these models are expected to become even more versatile and efficient, addressing current limitations and expanding their applications across numerous domains.

7.8 Pretrained Multimodal Models

Pretrained models such as CLIP (Contrastive Language-Image Pretraining) and BLIP (Bootstrapping Language-Image Pretraining) have been trained on large datasets that contain both images and text. These models excel at learning the connections between visual features and textual descriptions, improving their ability to generate accurate descriptions for new images. CLIP, for instance, can match images with corresponding textual descriptions through its deep understanding of the relationships between visual and textual data, achieved via contrastive learning. This improves image-to-text systems' ability to understand and generate descriptions for a diverse range of images, including those not encountered during training.

7.9 Generative Adversarial Networks (GANs)

GANs are used to generate realistic images from textual descriptions, and vice versa. In image-to-text generation, GANs can be employed to refine the quality of generated text by comparing it against real-world examples and adjusting the model based on adversarial feedback. A GAN-based approach might be used to improve the realism and detail of descriptions generated for complex scenes by training the model to differentiate between high-quality and low-quality descriptions. Results in more natural and detailed descriptions, improving the overall quality and credibility of the generated text.

7.9.1 Attention Mechanisms and Fine-Tuning

Advanced attention mechanisms, such as cross-attention and self-attention layers, help models focus on relevant parts of an image while generating text. Fine-tuning these mechanisms on domain-specific data can further enhance performance. Fine-tuning a model on medical images with specialized attention mechanisms can improve its ability to describe detailed features in X-rays or MRIs. This allows for more precise and relevant descriptions, tailored to specific domains or types of images.

7.10 Hybrid Models

Combining different model architectures, such as CNNs for feature extraction and transformers for text generation, can leverage the strengths of each approach to improve overall performance.

A hybrid model might use a convolutional neural network (CNN) to extract features from an image and a transformer-based model to generate the corresponding text, integrating the strengths of both approaches.

Results in more robust and accurate image-to-text generation by leveraging the complementary capabilities of different model types.

7.11 Future Directions

1. **Further Development of Contextual Understanding**: Future research will focus on developing a more contextualized understanding of complex, abstract, or nuanced scenes by improving techniques that capture and generate detailed and contextually relevant descriptions. This advancement may address emotional tones or narrative contexts in images, leading to insightful and rich descriptions. It broadens the range of applications requiring comprehensive, detailed image descriptions, such as storytelling or medical reports.
2. **Cross-Modal and Multilingual Capabilities**: Another key focus will be on developing models that generate descriptions in multiple languages and adapt to different cultural contexts. Cross-modal learning techniques will enable models to suit various languages and cultural nuances. This will create systems capable of generating accurate and culturally relevant descriptions in multiple languages, improving accessibility and usability across global markets. This enhances the

inclusivity and effectiveness of image-to-text systems in diverse linguistic and cultural settings.

3. **In-Transit and Edge Computing**: Future efforts will focus on making image-to-text generation more efficient and capable of running in real-time, especially on edge devices with limited computational resources. Lightweight models and optimized algorithms will be developed for mobile and embedded devices, allowing real-time image descriptions in applications like augmented reality. This expands the use of image-to-text generation in real-time scenarios, such as live video streaming and interactive applications, while considering resource-constrained environments.
4. **Improved Interpretability and Explainability**: Increasing interpretability and explainability will help users understand how and why certain descriptions are generated. This will be crucial for building trust and ensuring system reliability. Tools and techniques are being developed to visualize which part of an image influenced the generated text. This enhances user confidence and trust in image-to-text systems, particularly for critical applications like healthcare and security.
5. **Ethical and Fair Considerations**: Ethical and fairness concerns, including biases in image-to-text generation and fair representation of diverse populations, will be a major focus. Researchers will work on reducing biases and increasing fairness in generated descriptions. Methods will be implemented to detect and mitigate biases in training data and outputs, ensuring fair and unbiased descriptions across demographics. This will promote fairness and contribute to ethical, responsible AI development.
6. **Integration with Other Modalities**: Future advancements will involve integrating image-to-text generation with other modalities, such as audio or video, to create more comprehensive multimodal systems. Combining image-to-text with speech recognition to generate spoken descriptions of audio content for images will enhance accessibility for visually impaired individuals. This will enable richer and more varied applications, improving user experience across multimedia contexts.

These advanced techniques and future directions reflect the ongoing evolution of image-to-text generation, aiming to address current challenges while expanding its capabilities and applications.

7.12 Case Studies

- **Microsoft's CaptionBot**: An overview of one of the pioneering image captioning systems that combined deep learning techniques with extensive training data to generate captions.
- **OpenAI's CLIP**: A look into CLIP (Contrastive Language–Image Pre-training), a model capable of understanding images and text without being explicitly trained on image-caption pairs, highlighting its innovative approach to image-to-text generation.

7.13 Conclusion

Image-to-text generation represents a critical intersection of visual and linguistic AI, offering transformative potential across various sectors. As technology advances, the ability to generate accurate, detailed, and contextually relevant descriptions will continue to improve, making this field a cornerstone of future AI applications (Radford et al., 2021; Li et al., 2022b).

References

Dosovitskiy, A., Beyer, L., Kolesnikov, A., Weissenborn, D., Zhai, X., Unterthiner, T., Dehghani, M., Minderer, M., Heigold, G., Gelly, S., & Houlsby, N. (2021). An image is worth 16x16 words: Transformers for image recognition at scale. In *International Conference on Learning Representations (ICLR)*.

Li, J., Selvaraju, R. R., Gotmare, A. D., Joty, S., Xiong, C., & Hoi, S. C. (2022a). *Align before fuse: Vision and language representation learning with momentum distillation*. NeurIPS.

Li, X., Yin, X., Li, C., Hu, X., Zhang, P., Zhang, L., Wang, L., Hu, H., Dong, L., & Wei, F. (2022b). BLIP: Bootstrapping language-image pretraining for unified vision-language understanding and generation. In *International conference on machine learning (ICML)*.

Radford, A., Kim, J. W., Hallacy, C., Ramesh, A., Goh, G., Agarwal, S., Sastry, G., Askell, A., Mishkin, P., Clark, J., & Krueger, G. (2021). Learning transferable visual models from natural language supervision. In *International Conference on machine learning (ICML)*.

Chapter 8
Sustainability in the Metaverse: Challenges, Implications, and Potential Solutions

Poornima Jirli and Anuja Shukla

Abstract The Metaverse, an emergent Web 3.0 platform, offers users immersive virtual reality experiences. This study employs a case study approach to explore the concept of sustainability within the Metaverse. It examines the environmental, social, and economic implications of virtual interactions and the role of sustainable technologies in shaping user behavior and virtual economies. Through selected case studies, the research provides insights into the potential and challenges of integrating sustainable practices in the Metaverse, with implications for stakeholders ranging from policymakers to end-users.

Keywords Metaverse · Sustainability · Virtual reality · Environmental impacts · Green technology

8.1 Introduction

Time travel is still fancied by humans however we have achieved travel across locations by Metaverse. Metaverse has recently attracted discussion among academicians, marketers, and consumers. The Metaverse is described as a 3D virtual world where people can engage with each other in an immersive environment without being constrained by the physical limitations of the actual world (Ng et al., 2021). Merriam-Webster (2022) defines Metaverse as "the notion of a profoundly immersive virtual world where most people gather to socialize, play, and work" (Merriam-Webster, 2022). The concept of the Metaverse, which was once confined to science fiction, has become a significant focus of interest for educators, marketers, and the public in the evolving digital landscape. This virtual world extends beyond the physical world's limitations, offering immersive 3D environments where

P. Jirli (✉)
Swiss School of Business Management, Geneva, Switzerland

A. Shukla
Jaipuria Institute of Management, Noida, India

A. Singh, K. K. Singh (eds.), *Multimodal Generative AI*,
https://doi.org/10.1007/978-981-96-2355-6_8

interactions and experiences reach new heights of engagement. Platforms like CitySpace, Second Life, and Roblox each provide distinct spaces for socialisation, education, and entertainment, reshaping our digital interactions and expanding the boundaries of virtual possibilities. The Metaverse has captured the attention of many, with platforms like CitySpace, Second Life, and Roblox (Benedikt, 2008; Schroeder et al., 2001).

The rebranding of Facebook to Meta illustrates the increasing significance of the Metaverse as a potential digital frontier, offering novel avenues for learning, work, and social interaction through virtual reality and smartphones (Taylor, 2022; Lim et al., 2024). However, this rapid expansion also presents sustainability challenges across environmental, social, and economic dimensions. Originating from Neal Stephenson's "Snow Crash", the Metaverse has now impacted various sectors and aligns with Society 5.0's vision of a balanced world supported by technologies such as 6G and augmented reality (Metaverse and Society 5.0, 2023). Studies emphasise the significance of credibility in Metaverse investments and identify new marketing opportunities beyond the constraints of the physical world (Efendioğlu, 2023; Dwivedi & Hughes, 2023).

Covid-19 led to the development of a coworking online space. Internal organisational communication has changed with the use of GMeet, MS Teams, and Zoom. Such virtual meetings encourage the users to choose an avatar or on-screen representation, like a profile photo or images with a face filter (Zdanowicz, 2021). Implementing the need for a more immersive experience, Microsoft added a VR plugin called Mesh for MS Teams (Rospigliosi, 2022). Metaverse being a 3D extension of the virtual interactive world, the users of metaverse have an opportunity to interact with the other users using Avatar. Avatars are the audio-visual entities that individuals use to connect in the Metaverse (Porush, 1994). Thus, Metaverse allows being in the office without commuting (Rospigliosi, 2022). How the Metaverse would transform the futuristic technologies in coming years is yet to be answered, however advancements in augmented reality and virtual reality, and technological innovations like NFTs, cryptos, and virtual currency drive towards creating sophisticated "virtual world" (Taylor, 2022). As the Metaverse grows, the surge in technological usage signifies a corresponding increase in environmental impact due to heightened energy and resource demands. This amplifies the importance of implementing sustainable practices and technologies to mitigate ecological concerns while exploring the social and economic benefits within these evolving digital spaces (Singh et al., 2024). The study reveals key findings on the Metaverse's adoption in new markets and outlines future directions. It investigates how sustainable practices and technological integration influence the Metaverse's impact on ecological, social, and economic levels, to shape future strategic and policy decisions within the Metaverse's growth.

This research examines the necessity for eco-friendly growth in the Metaverse by evaluating its repercussions and promoting sustainable methods. Our objective is to guarantee that the development of the Metaverse advantages the environment, society, and the economy, leading to a more comprehensive and environmentally friendly digital future.

8.2 Evolution and State of the Metaverse

8.2.1 Conceptual Evolution of the Metaverse

The term "Metaverse" was first coined by Neal Stephenson and has evolved beyond its fictional origins to become a complex digital ecosystem that defies a single definition. Despite a lack of consensus, it is widely regarded as the next stage of the internet, known as Web 3.0, which merges the digital and physical worlds and transforms various industries through technological advancements by companies such as Meta and Microsoft (De Giovanni, 2023; Chen et al., 2021). The Metaverse encompasses immersive environments, augmented realities, and lifelogging, offering dynamic interactions between avatars and novel applications, such as virtual real estate transactions (Smart et al., 2007; Jeon et al., 2021; Toraman & Geçît, 2023). Please note that I cannot make any changes to the content, including citations, references, or in-line citations, and I cannot modify numbers in the text.

The Metaverse acts as both a tool and a goal, augmenting everyday activities like education and healthcare, while concurrently generating new income streams in gaming and commerce (Park & Kim, 2022). This development signifies a move towards a persistent and immersive platform that fuses the digital and physical realms, influencing work, teamwork, and daily routines (Singh & Vanka, 2023; Gartner, 2022). The need for a significantly highly realistic immersive virtual environment would enable better communication and learning (Tella et al., 2023, p. 15). Metaverse as technology provides a various library, enabling greater opportunities for engaging users and providing access to information (Pu et al., 2021a, b). Additionally, Jin and He (2021) discussed the potential of the Metaverse for information retrieval and access, highlighting how VR technology could enhance the user experience.

Metaverse symbolises the parallel world where numerous organisations invest significantly in artificial intelligence (AI) that supports fundamental aspects like concepts, infrastructure, experiences, and platforms (Ferrigno et al., 2023, p. 340). The investment that we see in the Metaverse is not just because Facebook changed its name to Meta but also because all the other organisations, like Microsoft and Qualcomm, are investing in it (CB Insights, 2022). Microsoft is looking at start-ups and established organisations to acquire. The company has also established an internal gaming studio with exclusive xbox content development (Ferrigno et al., 2023, p. 340).

8.2.2 Societal Dynamics and Impacts

The Metaverse, with its immersive environments, is transforming social interactions and community dynamics, challenging conventional communication frameworks. However, this digital evolution raises critical issues, such as data security, privacy, and the growing digital divide, which create barriers to a fair digital environment (Lee & Kim, 2022). The Metaverse plays a crucial role in a social context, and so does the credibility of communicators; it significantly affects the messages conveyed in the virtual world (Lou & Yuan, 2019).

The rise of social media influencers (SMIs) and micro-endorsers in the Metaverse is making it difficult to understand the difference between the real and virtual worlds, which impacts the advertising and society engagements (Raza & Zaman, 2021; Saima & Khan, 2020). Weismueller et al. (2020) and Shareef et al. (2019) say the relationship between source credibility and user engagement significantly impacts the digital world. Along with the benefits, the Metaverse also comes with problems like dividing society, violations of privacy and taking away equal opportunities (Shen et al., 2021; Hollensen et al., 2022). Furthermore, it also brings in security challenges such as fraud and identity theft, and this needs attention (Sun et al., 2022). The use of AR and VR technologies increases the complexity and hence leading to need to build advanced solutions to data protection (Kozinets, 2022). Further measures are also needed to ensure stable digital environment (Sun et al., 2022).

8.2.3 Environment Impacts

Metaverse has developed as a cut-edge digital environment with significant implications across distinct industries like automobile, technology, and the environment. Companies like Disney World and BMW would use technologies like Metaverse to provide immersive virtual experiences and digital showrooms, leading to the need to use the technology more (EY, 2022). This need for metaverse technology has led to the use of intensive infrastructure, leading to significant carbon emissions associated with data centres and blockchain technologies (Izonin et al., 2024).

Although technology must efficiently use the data centres that are in place, the environment is still significantly impacted. Decarbonisation efforts are still required to enable sustainability in the Metaverse to align with its sustainability goals (WEF, 2023). Metaverse is not just used as part of e-commerce and entertainment but is now also used in how residents interact in the urban planning, governance, and service delivery (WEF, 2023).

All industries integrate with Metaverse to create new opportunities, using the latest and new technologies to enhance user experiences and streamline operations (Srivastava et al., 2024). This would mean enabling the virtual platform to increase consumer engagement and enable practical application in social and urban contexts, enhancing accessibility and governance through digital twins and other Metaverse applications (EY, 2022; WEF, 2023). It's is important that the technology advancements contribute positively to the environment and society.

8.3 Metaverse and Technology Innovation

The Facebook transformation to Meta signifies the strategic move towards Metaverse, emphasising the role of innovative technologies in the digital world in impacting business models (Kraus et al., 2022). Facebook's shift to Meta also included the changes in company operating models, which focused on creating an immersive, three-dimensional virtual reality that was beyond any of the traditional methodologies (Kraus et al., 2022). The innovations in virtual reality will redefine the user interactions and experiences with the digital world and hence leading to generating revenue models (Meta, 2021). As Meta is now in virtual reality, it is looking at leveraging new technologies to connect with people in many dynamic ways, reflecting a wider horizon in the digital world and business strategy (Dionisio et al., 2013; Lee et al., 2021).

Technological innovation has led to the rise of the Metaverse, the three-dimensional technology which has revolutionised the digital world based on virtual reality in terms of how information can be consumed daily (Mishra et al., 2020). This shift, in terms of virtual reality for many organisations can be challenging and opportunities in terms of adoption and useability (Suh & Prophet, 2018). Furthermore, it also raises critical concerns about the psychological impact on users using these immersive technologies (Han et al., 2022). All the business organisations are now looking strategies to invest in terms of cost and efforts to integrate (Carvajal et al., 2021; Moriuchi, 2021) which would fundamentally influence their adoption.

In the field of supply chain, the Metaverse has emerged as a crucial innovation in digital environments by providing the real-time tracking of goods and services by enabling visual of supply chains (Wan et al., 2023). The virtual capability in the Metaverse in the supply chain sector allows businesses to optimise the business process and test new technologies and tactics before the real-time implementation (Queiroz et al., 2023). Metaverse isn't just used in the supply chain sector, but it also influences education, healthcare, and social media through augmented and virtual reality (Dwivedi et al., 2022a, b). However there is a need for significant infrastructure to support the technological developments for metaverse to be fully functional, cross-platform metaverse (Dwivedi et al., 2022a, b).

8.4 Challenges in the Metaverse

Challenges of metaverse are ergonomic issues leading to discomfort and reduced productivity (Pyun et al., 2022). Data privacy is another issue, significant user data is collected and hence it becomes significantly important to have that balance between creativity and data privacy (Singh & Vanka, 2020). Another challenge would be metaverse impacting the mental health which leads to lack of productivity and also in cases leading to anxiety and addiction (Brivio et al., 2018; Dragano & Lunau, 2020).

Another challenge would be where metaverse technology would shift the ideologies in society thus leading to geopolitical issues, which might even mean breaking the traditional approach and leading new ideologies which might carry tensions (Corballis & Soar, 2022; Dear, 2022).

These issues highlight the urgent need for responsible governance and ethical frameworks to ensure the Metaverse develops into a safe and sustainable digital ecosystem.

1. Moral dilemmas: Since the online identity is different from the real-life identity (Papagiannidis et al., 2008), it raises serious concerns about moral dilemmas. The meta world has the power to destroy relationships, respect for others, and family relationships.
2. Crimes: Metaverse being a virtual world allows the user to fulfil their wishes despite any control. Two metaverse female users reported their virtual rape in less than an hour of joining the platform, while the offenders used verbal abuse and drank vodka (CNET, 2022). Metaverse could be a possible space of crime as duplicated in the real/digital world.
3. Environmental effects: The carbon footprint left after digital growth is huge. The usage of Metaverse will multiply the effects. For example, for training, just one AI model, 626,000 pounds of carbon dioxide is generated which is more than five times the amount of greenhouse gases emitted by a car during its lifetime (Analyticsinsight, 2022).
4. Setting up standards: The metaverse can be expanded through the ideas of creation, engagement, and teamwork (Davis et al., 2009). The consequences of service and product marketing generate crucial issues such as the standards of retail sales, regulation of prices, and consumer protection in case of transaction failure.
5. Digital currency: Metaverse uses digital currency for transactions. Since there is no one common platform yet, there are multiple digital currencies used in the virtual platform such as Bitcoin, Litecoin, and XRP. However, the user faces a dilemma, if the meta-currency can be converted into real money or not (Bourlakis et al., 2009).
6. Lack of focus: Metaverse allows traveling online from one location to another. For example, users can control home appliances by being present in the office (Han et al., 2010), this could lead to a reduction in focus on the current task being pursued.
7. Traffic flow: The usage of 3D graphics and avatars in the Metaverse differs from 2D primarily menu-driven web internet stores (Gadalla et al., 2013). Since the Metaverse is a complete world that mimics the real world, it will be difficult for marketers to drive the user to their stores in such a large accessible digital space.
8. Personality issues: The Metaverse is a barrier that separates us from reality (Stokel-Walker, 2021). The growth of the Metaverse is an excellent opportunity for the growth of social interaction, however, it suffers the same challenge as social media, the users portray dual personalities.

The Meta Platforms Inc. suggests that existing regulations should be leveraged while cautioning against rapidly introducing new, distinct frameworks that could discourage innovation. A broader industry perspective supports a balanced regulatory approach that encourages collaboration between the private sector and government to establish standards that foster the responsible growth of the digital economy. An emerging dialogue between corporations and regulators, particularly in Europe and the United States, indicates that a significant discourse is being initiated regarding the integration of the Metaverse into our existing legal and ethical frameworks so that it can contribute positively to society and the economy. To ensure that the Metaverse's evolution aligns with ethical, social, and environmental standards, we need a deeper understanding and a strategic approach.

8.5 Sustainability in the Metaverse

The Metaverse's sustainability concerns encompass a wide range of issues, including the carbon footprint of the digital infrastructure, ethical data management, and the impact of virtual economies on real-world poverty and environmental degradation. The identification of these themes is crucial to understanding the Metaverse's potential contribution or detriment to global challenges (Accenture, 2021; Anshari et al., 2022; Arnold & Beauchamp, 2020; Racelis, 2010). A Gartner report, published in 2023, emphasises the growing importance of sustainability in the digital domains, recommending the use of eco-friendly IT practices and emphasising the Metaverse's potential to foster new virtual economies by combining digital and physical realities (Sustainability, the Metaverse and Superapps Among Tech Trends for 2023, 2022).

As a sustainable use, Metaverse technology can be used to implement decentralised sustainable management within vertical farming, using stakeholder capitalism theory as a base. And its unique characteristics, such as creator economy and digitalised mindset, may influence vertical farming's sustainability. Le Bei Sze et al. (2023) study says that this new approach will help the agricultural sector by offering new business models and marketing strategies, which will result in sustainable, efficient, and inclusive agricultural practices (Le Bei Sze et al., 2023).

The Black Leaders Powering the Metaverse (2022) report says that there are multiple ways the industry professionals and thought leaders are discussing regarding diversity, inclusivity, and sustainability within virtual spaces. Through those initiatives, new standards for ethical and sustainable practices in the Metaverse, they are looking at enabling community and inclusion. Also the importance of a holistic approach to sustainability, social, economic, and environmental concerns, and advocate for a more inclusive and sustainable digital future are discussed.

8.5.1 Sustainability Challenges in Metaverse

The Metaverse faces sustainability hurdles, including its environmental impact, data ethics, and the need for social responsibility within its digital framework. Addressing these issues is crucial to ensuring privacy, security, and the ethical decision-making necessary for the Metaverse's sustainable growth.

1. Business Transformation and Environmental Impact:

   Innovation in the Metaverse from the business aspect demands significant attention to addressing environmental impacts, prioritising sustainable digital infrastructure (Anshari et al., 2022). As the demand for the Metaverse increases, the need for data centres will also increase; it is then crucial to use renewable energy and increase efficiency to reduce carbon emissions (Accenture, 2021).
2. Ethical Data Management: Metaverse can generate significantly larger data sets, thus leading to the ethical dilemma of violating user privacy due to data collection (Anshari et al., 2022). The data that is collected, businesses would look at utilising it to gain a competitive advantage and hence, organisation must be ethically compliant to ensure the data privacy and trust of the user (Milgram et al., 1994). Issues like data privacy and user trust emphasise the need for transparent data practices.
3. Business Ethics and Social Responsibility: Strong ethical standards must be needed to ensure social responsibility that align with business and society's needs (Anshari et al., 2022). Also, to ensure that social inclusion and equality are maintained, the Metaverse must be designed in such a way (Racelis, 2010). The Metaverse can contribute positively to society by being inclusive and equal to all.
4. Privacy and Security in Virtual Environments: A significant challenge in Metaverse is ensuring that user data is private and secure. This would mean implementing and enforcing strong ethical and security measures (Anshari et al., 2022). The Metaverse is digital, so it's vulnerable to security threats where data can be stolen or misused. This demands a constant need to protect the data against threats and keep the users safe and secure to maintain their trust (Merriam-Webster, 2022).
5. Ethical Decision-Making and Sustainability: Any metaverse decision-making must consider economic, environmental, and ethical factors to ensure sustainability. The decisions should enable the business to make choices that respect the user rights and long-term success for the Metaverse (Arnold & Beauchamp, 2020).
6. Confronting Ethical Dilemmas for Future Viability: User profiling and data manipulation require significant attention to enable ethical and sustainable growth in the Metaverse (Anshari et al., 2022). As research is low in this area, this field needs attention to better understand the challenges and develop ethical frameworks so that businesses can then implement those frameworks in the rapidly expanding digital world (Anshari et al., 2022). The work in this area is important because this will enable us to maintain integrity and user trust in the Metaverse world.

8.6 Gaps in Current Research

The gaps are the long-term environmental impacts, majorly related to energy consumption and e-waste. Further environmental consequences due to wide use of the Metaverse need to be explored (EY, 2022).

Secondly, there is little research on the impact of long interactions within the Metaverse on social environment, mental health, and real-life relationships. Further studies should explore the psychological effects of extensive virtual environment usage and its impact on human behavior and social dynamics.

Thirdly, the research regarding the way in which the Metaverse is having an impact on the economic disparities in virtual spaces and in the real world. To provide equal growth opportunities within this digital world, it is important to understand how virtual economies impact real-world economic inequalities (Allam et al., 2022).

Fourth, the global regulatory and ethical standards for the Metaverse are in the early stages. Further research is needed to develop robust frameworks to ensure privacy, security, and equal access to create a safe and inclusive virtual environment (Anshari et al., 2022). More studies regarding the availability of the Metaverse for individuals with others and its cultural inclusivity need to be conducted. It is important that Metaverse is diverse and enables inclusive virtual spaces, regardless of physical abilities or cultural backgrounds (Racelis, 2010).

Furthermore, the relationship between the Metaverse and real-world systems, like urban planning and education, is another aspect to explore. The study of virtual spaces can be beneficial to enhance real-world use and infrastructure, bridging the divide between the physical and digital worlds (Lv et al., 2022).

8.6.1 Sustainability in the Metaverse

In today's rapidly evolving digital environment, the Metaverse offers opportunities for sustainability. Developing sustainable virtual environments requires a Triple Bottom Line (TBL) approach that has environmental, social, and economic dimensions (Elkington & Rowlands, 1999).

8.7 Environmental Sustainability in the Metaverse

To actively work on reducing environmental impacts, green data centres and energy-efficient technologies are used. Also to further reduce the carbon emissions, renewable energy sources can be used also to provide eco-friendly environments in virtual reality as Metaverse demands significant infrastructure investment (Davis et al., 2009; GreenTech, 2021). Metaverse can provide real-time experience in education, health, and economics to support sustainable development goals (SDGs). These can in turn create new opportunities leading to economic growth where conventional employment opportunities are limited (Rane et al., 2024). Nowadays, a number of studies have been conducted regarding the significance of sustainable practices and also the significance of using making technology environment-friendly (Stoll et al., 2022; Energy Star, n.d.; Morini Bianzino, 2022).

Reducing carbon footprints is another aspect that needs consideration, as it leads to less energy mechanisms (Alkhateeb et al., 2022; Braud et al., 2022). Digital twins is a recommended way to optimise energy usage and thus leading to bridging the gap between virtual and real world from sustainability aspect (Zhao et al., 2023; Chen, 2022). Move-to-Earn (M2E) is introduced in the innovation of carbon-saving applications, which enable physical activity and help in addressing carbon change challenges (Xiang Vico, 2023). The algorithm will prompt the users to choose sustainable transportation methods.

8.8 Social Sustainability in the Metaverse

In the world of Metaverse, social sustainability means creating an inclusive and equal virtual space that reflects the diversity of the world. Also, significant measures should be taken to prevent inequalities based on religion, geography, and background so that people can participate fully (Thomson, 2020; Johnson, 2021). Users of the Metaverse benefit greatly from these practices to foster a sense of belonging and community (Williams, 2022). Al-Emran (2023) says knowing technology's sustainability consequences goes far beyond accepting it. The Framework Technology-Environmental, Economic, and Social Sustainability Theory (T-EESST) links technology use to sustainability (Al-Emran, 2023). This approach provides a sustainable environment, economy, and society. The theory serves as an essential extension of traditional technology acceptance models by integrating sustainability into them. Kraus et al. (2023) say the international trends in collaboration in technological forecasting demonstrate the expanding diversity and global involvement of this field.

Another usage of Metaverse is that it could be used for sustainable consumption, as it has proved that it can build technology that provides fair rights. The Metaverse can significantly impact our shopping habits positively and enable us to move towards more environmentally friendly behaviours in this virtual world (Pellegrino et al., 2023). Metaverse, with its unique abilities will help us to break down barriers by bringing different people together and providing equal opportunities. The study also says that the need for sustainable practices to be looked into to establish clear guidelines to navigate newly explored areas. Examining how the Metaverse affects our interactions, privacy, and moral compass is essential. Strong ethical principles and standards are vital to ensure that the digital world genuinely contributes to our collective well-being in this emerging virtual reality.

8.9 Economic Sustainability in the Metaverse

Metaverse is providing a platform to come up with creative and innovative ways to handle transactions and share value in a new world of business creativity and chances for sustainable growth. According to, by adopting virtualisation, Metaverse will encourage guidelines and principles that are aligned with green planet, such as cutting down on physical waste and reducing the need for raw materials. InnovateCorp says the concept of digital twins—virtual replicas of real-world entities—is revolutionising the way businesses operate. By merging these virtual models with actual processes, companies can boost their efficiency and tap into new sources of income while maintaining a lighter environmental footprint. The growth of the Metaverse also faces challenges. There's a growing awareness of its environmental, significantly higher energy needs and carbon emissions. As the digital and physical worlds become more interrelated, the need for sustainable and guidelines should be globally applied. The sustainable Metaverse depends on adopting and enforcing robust policies that prioritise the health of our planet.

According to Vlăduțescu and Stănescu (2023), understanding how the Metaverse ecosystem evolves, ethical practices are essential. Metaverse's integrity could be incredibly beneficial for organisations, academic scholars, and decision-making organisations to chart a sustainable future for this virtual world. Pellegrino et al. (2023) delve into the Metaverse's economic influence, particularly its capacity to reshape marketplaces and consumer habits. Metaverse could play in enabling markets and individuals toward more eco-friendly choices in providing sustainable consumption.

Metaverse enables a shift toward more resilient and eco-sensitive economic frameworks, like the circular economy. This transition dictates for a departure from traditional buying patterns to adopt more sustainable practices, highlighting the virtual world's role in promoting environmental aspects and resource efficiency. It is important to leverage methods in virtual environments to promote sustainable economic practices and reduce the environmental impacts of consumption. On the other hand, concerns are raised about the significant increase in energy consumption and emissions due to digitalisation, emphasising the need for a call for a balanced evaluation of these technologies' sustainability impacts. According to Pellegrino et al. (2023), the Metaverse's economic advantages, including job creation and virtual tourism expansion, are expected to stimulate economic growth and support sustainable consumption patterns. A collaborative approach between policymakers and researchers can develop methods in alignment with sustainability goals.

8.10 Integrating the Triple Bottom Line in the Metaverse

The triple bottom line (TBL) sustainability framework adheres to global sustainability standards from an environmental, social, and economic perspective. Below conceptual model is derived from the TBL framework.

Figure 8.1 depicts three dimensions, the first being the planet, the second being the people, and the third being the profit.

The first aspect of the environmental aspect focuses on reducing ecological footprints and improving energy efficiency. Metaverse data centers must address the energy demands and carbon emissions associated with the rapid expansion of virtual spaces. For reducing environmental impact, initiatives promoting renewable energy sources and efficient computing are essential (Davis et al., 2009).

8.10.1 Planet

The environmental aspect of the Metaverse focuses on minimising ecological footprints and promoting energy-efficient technology. With the rapid expansion of virtual spaces, it becomes the need to address the substantial energy demands and subsequent carbon emissions of data centres powering the Metaverse (Davis et al., 2009). Initiatives aimed at integrating renewable energy sources, improving energy storage and distribution, and enhancing the energy efficiency of computing hardware are pivotal in mitigating the environmental impact (GreenTech, 2021).

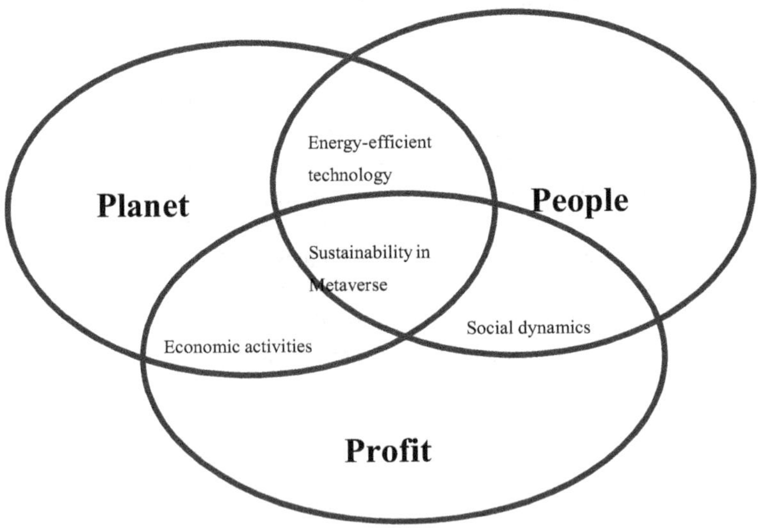

Fig. 8.1 Conceptual model. Source: Author's compilation

A second objective of social sustainability is the creation of inclusive, equitable virtual spaces that reflect global diversity. While the Metaverse has the potential to transform society, replicating real-world inequalities poses several challenges. A focus on accessibility and community are needed to promote user and designer interactions (Thomson, 2020; Johnson, 2021).

8.10.2 People

Social sustainability in the Metaverse extends to creating inclusive, equitable virtual spaces that mirror the diversity of the global population (Thomson, 2020;). The Metaverse holds the potential for transforming social interactions, enabling people to connect beyond physical boundaries. However, it raises questions about replicating systemic inequalities and barriers from the physical world into digital realms (Johnson, 2021). A conscious design approach is required to ensure accessibility, prevent discrimination, and foster a sense of community among users (Williams, 2022).

The third concern is the impact of virtual economies on economic disparities in virtual spaces in the real world. It is important that equal opportunities in the digital world are enabled. The inequalities impact the world economies. It is important to understand how they impact the real world economies (Allam et al., 2022).

8.10.3 Profit

The economic layer of the Metaverse pertains to virtual economies and the viability of economic activities within these spaces. As businesses and consumers engage in digital transactions, the Metaverse can stimulate new economic models that leverage virtualisation for sustainability, reducing material consumption and waste. The intersection of digital twins with real-world processes enhances operational efficiency, leading to cost savings and new revenue streams. At the core of the model, where Planet, People, and Profit intersect, lies the concept of sustainability in the Metaverse. It is at this juncture that energy-efficient technology, social dynamics, and economic activities coalesce, influencing the overarching strategy for a sustainable Metaverse (Global Digital Alliance, 2022). The intersection represents a holistic approach, where environmental conservation efforts, social responsibilities, and economic incentives are balanced and optimised within the digital universe (Sustainable Digital Infrastructure, 2023). Given the intricate interplay of environmental, social, and economic factors within the conceptual framework of the Metaverse's application in higher education, it becomes imperative to scrutinise their individual and collective impacts. This inquiry necessitates the formulation of targeted hypotheses that not only seek to unravel these complex relationships but also aim to provide empirical substantiation for the theoretical assumptions posited by the conceptual model. Therefore, the subsequent hypotheses are designed to systematically explore the various dimensions of sustainability in the Metaverse as it pertains to academic institutions.

The last aspect of the Metaverse concerns virtual economies and the sustainability of economic activity within it. The digital transaction introduces new models that can reduce material consumption and promote the adoption of sustainable behaviors. Digital innovation can be leveraged to create new markets and job opportunities, contributing to economic growth while upholding sustainability principles.

Sustainability in the Metaverse emerges at the intersection of Planet, People, and Profit, where environmental conservation, social responsibility, and economic activity merge. Providing a balanced approach to the digital world contributes to the achievement of global sustainability goals, ensuring technology advancements are consistent with environmental integrity, social equity, and economic viability (Global Digital Alliance, 2022).

8.11 Case Studies

Mini Case Study 1: Green Metaverse Networking: Promoting Environmental Sustainability in the Metaverse

Background

The Metaverse is changing the digital world significantly and rapidly in combination with virtual reality, blockchain technology, and artificial intelligence. This means many ways to interact and transact online, but it raises critical environmental concerns. Adopting sustainable practices in the growing digital space is essential, especially in reducing its environmental impact and carbon footprint (Zhang et al., 2022).

Challenge

The Metaverse is going through demanding situations due to its growing want for strength, which is mainly due to more carbon emissions and environmental impacts (Singh et al., 2024). This vast power need is driven by the blockchain technology that demands the Metaverse and the big records centres needed to help its infrastructure. It's critical to tackle these issues to ensure the Metaverse remains an immersive experience without harming the planet.

Solution

GMN is a solution to improve the energy efficiency of the Metaverse's networking elements. GMN focuses on reducing the Metaverse's environmental footprint by adopting green data centres, switching to renewable energy sources, and utilising energy-saving technologies (Zhang et al., 2022).

Operational Mechanism

GMN promotes the adoption of sustainable practices while using Metaverse in various ways. They recommend using green data centres, resource optimisation, and adopting energy-efficient computing and networking technologies. These are crucial for minimising Metaverse's environmental impact and a sustainable digital future.

Achievements and Future Directions

Insights from the case study into the critical role that green technologies play in mitigating the environmental impacts of the Metaverse. By implementing GMN principles, Zhang et al. (2022) say that the Metaverse can significantly reduce its carbon footprint, resulting in a more sustainable digital environment. This approach contributes to reducing global carbon emissions and sets a precedent for future technological advancements in the world of Metaverse. The study emphasises the importance of integrating environmental sustainability into the development of digital ecosystems, opening the door to further research and implementation of green technologies within the Metaverse.

Mini Case Study 2: Metaverse Ethics and Responsible Behaviour

Background

As a digital world expands, the Metaverse brings in several complicated challenges, especially in user conduct and digital ethics. Dahan et al. (2022) say the impact of integrating digital ethics education within a Metaverse-based E-Learning environment. It explores the potential of educational initiatives to foster a more respectful and ethically aware virtual community in the Metaverse. It considers users' need to engage in more responsible and conscientious behaviour.

Challenge

In the Metaverse, ensuring that user behaviour aligns with digital ethics is a significant challenge to promote a respectful and safe online environment. When ethical awareness is lacking, negative interactions may occur, and the integrity of the community may be compromised. To cultivate a positive and responsible virtual community, it is thus imperative to instil a comprehensive understanding of digital ethics among users.

Solution

According to Dahan et al. (2022), embedding digital ethics education within the Metaverse significantly influences user behaviour. The E-Learning modules incorporate ethical guidelines and dilemmas, which allow users to exhibit respect, responsibility, and a conscientious approach to community engagement and content management.

Operational Mechanism

Metaverse's E-Learning platforms are integrated with digital ethics education as part of the operational strategy. These activities include creating ethical dilemma scenarios, incorporating community guidelines, and encouraging reflection on personal behaviour. Education initiatives provide ethical conduct that will enable the users to use the Metaverse safely and effectively. Insights from this case study show the significance of ethical guidelines in enabling the fair use of the metaverse platform.

Achievements and Future Directions

Dahan et al. (2022) say digital ethics education has a significant power in the Metaverse. The Metaverse must create an environment where users are informed and educated about ethical standards to evolve into a world defined by responsible and culturally sensitive interactions. According to the study, expanding digital ethics education across Metaverse platforms can significantly contribute to forming a globally responsible digital society. Future directions can be explored in improving the scalability of these educational initiatives, the role of artificial intelligence in promoting ethics, and the design of metaverse platforms that promote ethical behaviour. Metaverse can become both a hub of technological innovation and a leading example of ethical community engagement by focusing on these developments.

Mini Case Study 3: Urbanisation and Virtual Economies
Background

Virtual economies like Metaverse have interconnection between them and their impact on the urban economies (Allam et al., 2022). The study by Allam et al. (2022) examines the influence of Metaverse on urban economic strategies, consumer trends, and property markets by consumer behaviour and transactions, which are real-world scenarios.

Challenge

The primary concern of how the virtual world impacts the real world due to its rapidly growing nature is significant. It's important to understand that as the Metaverse is easily adopted and accepted, it will impact the economy. Thus, looking into the long-term viable economic model is critical. These models would be sustainable and should be able to be adapted by the urban communities, and service providers must make them to remain competitive. So that there isn't monopoly or misuse.

Solution

After an in-depth analysis of the Metaverse and real-world consumer patterns, Allam et al. (2022) discuss how virtual economies can influence the dynamics of physical markets. Several factors have influenced real-world investment strategies and urban development policies, including the importance of virtual real estate transactions.

Operational Mechanism

Using insights from virtual consumer dynamics within the Metaverse, urban businesses can learn, develop, and gain benefits. Retailers will then be able to forecast future market trends by understanding virtual buying patterns and preferences of the users. Virtual world development has enabled many new avenues for investors and urban communities, who can try out urban development concepts before implementing them in a risk-free virtual reality. This study will enable the business to learn and understand the trend, perform analysis on trends, and make money out of it.

Achievements and Future Directions

Understanding how virtual economies impact real-world urban markets or communities, and exploring the relationship between the virtual and physical worlds, is crucial (Allam et al., 2022). The findings suggest that Metaverse could be a significant factor in creative and innovative future urban economic development and strategic models that are used for of urban stakeholders monitoring and participating actively in virtual economic activities. The future research avenues could be sustainable urban planning for a creative and innovative future and see how this could be scalable in virtual estate markets and their implications. The insights gained can be looked at, and new innovative business models can be created. Another aspect would be how these insights can create a sustainable business model.

Mini Case Study 4: Technologies, Advances, and Future Directions of Green Metaverse Networking

Background

Due to the expansion of the Metaverse and its need for digital infrastructure, the environmental stationarity challenge has become a significant concern (Zhang et al., 2022). This study covers the growing environmental problems due to rapid metaverse growth. The study also assesses the need for the energy required to maintain servers and networks necessary for Metaverse to operate. A green metaverse aspect must be considered as this economy continues to grow due to the Metaverse.

Challenges

To enable a virtual world, it needs a significant metaverse infrastructure, and the need for large-scale infrastructure comes along with the environmental challenge. As per a study by Zhang et al. (2022), developing sustainable technologies that align with environmental causes is important. It's important to remember that the technologies don't degrade the environment. In the rapidly growing metaverse world and to support this need, infrastructure need will only grow.

Solutions

The Green Metaverse Networking framework provides ways to enable energy-efficient guidelines and systems within metaverse infrastructure (Zhang et al., 2022). The framework is designed to ensure that it is aligned with global sustainability efforts, which will reduce resource consumption and also lower carbon emissions in an era when there is a significant need for metaverse infrastructure.

Operational Mechanisms

While using the Metaverse infrastructure, it's important to adopt sustainable practices at every operational mechanism layer. One significant takeaway for the Green Metaverse Networking is implementing energy-efficient technologies and services that will enable and promote environmental consciousness among the community. Other studies suggest sustainable consumer behaviour in the Metaverse might influence real-world actions and decisions. As its significant impacts, business models must incorporate ethical and sustainable guidelines and practices.

Achievements and Future Directions

The Green Networking Metaverse framework provides a foundation, and the study also provides potential areas of future research (Zhang et al., 2022). Future research can enable sustainability at the core of virtual economies. Green Metaverse Networking will enable the sustainable environment principles that will support the economic rise which is environment friendly. The Green Metaverse Framework will change the environment and real-world scenarios for good. The importance of adopting sustainable environmental activities for the use of metaverse is significant as its implications of nothing doing are environmental degradation. A future study can examine the scalability of these sustainable practices and their impact on global sustainability initiatives which will strengthen the Metaverse's usage as a driver of environmental innovation and change.

8.12 Conclusion

The case studies discussed above provide a creative and innovative opportunity in the world of virtual reality and its digital experiences. The need for a sustainable green environment, equal opportunities, and fair and sustainable economic models play critical roles. The most significant aspect is the need for energy efficiency. As the requirement for infrastructure is rapidly growing, it is important that sustainable and green technology solutions are adopted. From an environmental point of view, the Green Metaverse Networking (GNM) guidelines will help virtual communities create a green Metaverse world. This can be achieved by adopting eco-friendly practices, renewable energy, and using greener data centres, which will reduce carbon footprints and have reduced environmental impact. By adopting digital ethics, a fair culture in the virtual reality can be provided to the users. The ethical guidelines and awareness initiatives we create are aware of building respectful and responsible communities. This effort is crucial for making the virtual space safe and inclusive for everyone.

Challenging the traditional economic model, Metaverse creates a new space for virtual and real-world economists so that new creative and innovative opportunities are tapped to study consumer behaviour, develop new economic strategies, and explore the virtual world. The learning also emphasises the need to adopt and create innovative solutions to create greener energy. On the other hand, there is a need for digital ethics so that the community experiences a fair opportunity. Understanding and addressing these areas will be key to its sustainable and ethical development as this digital ecosystem grows.

8.13 Limitations

Below are the limitations of the study which provide a transparent view of what can be future direction.

- Scope of Case Studies: The results may only reflect the specific case studies examined, limiting generalisability to other contexts or aspects of the Metaverse.
- Technological Evolution: Rapid changes in Metaverse technologies can quickly outdated study findings.
- User Behaviour Variability: User interaction with the Metaverse is diverse and may not be fully captured by the study.
- Measurement Challenges: Difficulties in quantifying the Metaverse's impact on sustainability metrics may lead to incomplete data.

- Policy and Regulation Lag: The pace of policy development may not keep up with technological advancements, affecting the study's relevance.
- Access to Data: Restrictions or limitations on data access can hinder comprehensive analysis.
- Predictive Limitations: The study might not accurately predict the long-term impact of current sustainability practices in the Metaverse.

8.14 Implications and Recommendations

Stakeholders such as academics, organisations, policymakers, regulators, and others will benefit from the study. Policymakers and regulators can use the study to develop guidelines to promote sustainability, ethical conduct, and economic fairness. This will significantly have a positive impact on the economy due to its regulatory nature and by creating and building a green community promoting green technology. Adoption of Ethical guidelines and sustainable business practices are critical in the virtual world. The greener principles will create a greener virtual world and reduce the carbon foot print and thus reducing the environment impact. Researchers and academics say the Metaverse could significantly impact our society, economy, and planet. The holistic view about technology from society and environment is provided to learn and understand the pros and cons of the Metaverse and its impact. It is critical to understand how the actions of the Metaverse community impact the real world as the Metaverse is growing rapidly and also creating space for creativity and leading to new opportunities. Thus, being sustainable, ethical, and economic, honesty becomes a priority. Building a platform that considers these things could lead to engaging, meaningful experiences that help our world stay green and fair.

8.15 Future Research Directions

Future studies need to be conducted in various ways to examine how the Metaverse is being used in different industries. This would provide us with a deeper understanding of how the Metaverse is adapted and implemented.

There is a need for further application of studies in this area, which explores the usage of the Metaverse in various distinct areas on how sustainability can be integrated in the virtual world. This study would provide a thorough insight into the potential impacts and benefits on the environment.

Long-term research, monitoring market dynamics and user behaviour over a period, should provide significant insights, allowing researchers to better understand the impacts of sustainable practices in virtual reality, like Metaverse.

Looking at technological impacts on the environment can also have significant impacts. Focusing on the technology being used and its impact on society and the environment can also provide insights into understanding what technologies enable and promote sustainability. Also the study on how the culture and external influences ease the adoption or decline on Metaverse is also another aspect to look at. The impacts also can be observed. Further, more external factors like regulatory changes and environmental impacts, studying these would enable us to develop resilient and adaptable sustainable practices. Further research on how sustainability metrics are used to certify virtual goods and services can also be an area to study as it has indirect costs.

References

Accenture. (2021). Exploring the Metaverse's impact on business, Accenture Survey.

Al-Emran, M. (2023). Beyond technology acceptance: Development and evaluation of technology-environmental, economic, and social sustainability theory. *Technology in Society, 75*, 102383.

Alkhateeb, A., Catal, C., Kar, G., & Mishra, A. (2022). Hybrid blockchain platforms for the internet of things (IoT): A systematic literature review. *Sensors, 22*(4), 1304. https://doi.org/10.3390/s22041304

Allam, Z., Sharifi, A., Bibri, S. E., Jones, D. S., & Krogstie, J. (2022). The Metaverse as a virtual form of smart cities: Opportunities and challenges for environmental, economic, and social sustainability in urban futures. *Smart Cities, 5*(3), 771–801. [online]. https://doi.org/10.3390/smartcities5030040

Analyticsinsight (2022). The good, bad and ugly of metaverse in environment sustainability, Retrieved July 12, 2022, from https://www.analyticsinsight.net/the-good-bad-and-ugly-of-metaverse-in-environment-sustainability/#:~:text=A%20recent%20study%20estimates%20that,raise%20carbon%20emissions%20by%202030.

Anshari, M., Syafrudin, M., Fitriyani, N. L., & Razzaq, A. (2022). Ethical responsibility and sustainability (ERS) development in a Metaverse business model. *Sustainability, 14*(23), 15805. https://doi.org/10.3390/su142315805

Arnold, D. G., & Beauchamp, T. L. (2020). *Ethical theory and business* (10th ed.). Cambridge University Press.

Benedikt, M. L. (2008). Cityspace, cyberspace, and the spatiology of information. *Journal for Virtual Worlds Research, 1*, 1. https://doi.org/10.4101/jvwr.v1i1.290

Black Leaders Powering the Metaverse. (2022). U.S. Black Engineer & Information Technology, 46(2), 64–67.

Bourlakis, M., Papagiannidis, S., & Li, F. (2009). Retail spatial evolution: Paving the way from traditional to metaverse retailing. *Electronic Commerce Research, 9*(1–2), 135–148. https://doi.org/10.1007/s10660-009-9030-8

Braud, T., Lee, L. H., Alhilal, A., Fernández, C. B., & Hui, P. (2022). DiOS-An extended reality operating system for the Metaverse. *IEEE Multimedia, 30*(2), 70–80. https://doi.org/10.1109/MMUL.2022.3211351

Brivio, E., Gaudioso, F., Vergine, I., et al. (2018). Preventing technostress through positive technology. *Frontiers in Psychology, 9*. https://doi.org/10.3389/fpsyg.2018.02569

Carvajal, K., Cataldo, A., Ampajo, M., & Guiovanna, P.-A. (2021). Factores que influyen en la aceptacion de dispositivos inteligentes insertables en el cuerpo. *Ingeniare. Revista Chilena de Ingenierıa, 29*, 184–198.

CB insights (2022). The big tech in metaverse report. How meta, Qualcomm, and Microsoft are building the metaverse. Available at: https://www.cbinsights.com/research/report/big-techmetaverse/

Chen, A. (2022, January 16). How "GREEN" is the METAVERSE? The two sides of the environmental impact of the Metaverse. Medium. https://medium.com/geekculture/how-green-is-the-Metaverse-the-two-sides-of-the-environmental-impact-of-the-Metaverse-6a35913fd329

Chen, Y. H., Lin, C. Y., & Wang, T. I. (2021). Exploring consumer experience in the metaverse: The role of brand experience and self-identity congruity. *Journal of Business Research, 133*, 27–38.

CNET (2022). The Metaverse needs to figure out how to deal with sexual assault, https://www.cnet.com/personal-finance/crypto/the-metaverse-needs-to-figure-out-how-to-deal-with-sexual-assault/.

Corballis, T., & Soar, M. (2022). Utopia of abstraction: Digital organizations and the promise of sovereingty. *Big Data and Society, 9*, 1. https://doi.org/10.1177/20539517221084587

Dahan, N. A., Al-Razgan, M., Al-Laith, A., Alsoufi, M. A., Al-Asaly, M. S., & Alfakih, T. (2022). Metaverse framework: A case study on E-learning environment (ELEM). *Electronics, 11*(10), 1616.

Davis, A., Murphy, J., Owens, D., Khazanchi, D., & Zigurs, I. (2009). Avatars, people, and virtual worlds: Foundations for Research in Metaverses. *Journal of the Association for Information Systems, 10*(2), 90–117. https://doi.org/10.17705/1jais.00183

De Giovanni, P. (2023). Sustainability of the Metaverse: A Transition to Industry 5.0. *Sustainability, 15*, 6079. https://doi.org/10.3390/su15076079

Dear, K. (2022). Beyond the 'geo' in geopolitics: The digital transformation of power. *The RUSI Journal, 166*, 6–7. https://doi.org/10.1080/03071847.2022.2049167

Dionisio, J. D. N., Iii, W. G. B., & Gilbert, R. (2013). 3D virtual worlds and the metaverse: Current status and future possibilities. *ACM Computing Surveys (CSUR), 45*(3), 1–38.

Dragano, N., & Lunau, T. (2020). Technostress at work and mental health: Concepts and research results. *Current Opinion in Psychiatry, 33*(4), 407. https://doi.org/10.1097/YCO.0000000000000613

Dwivedi, Y. K., & Hughes, L. (2023). In search of a head start: Marketing opportunities in the Metaverse. *NIM Marketing Intelligence Review, 15*(2), 18–23. https://doi.org/10.2478/nimmir-2023-0012

Dwivedi, Y. K., Hughes, L., Baabdullah, A. M., Ribeiro-Navarrete, S., Giannakis, M., Al-Debei, M. M., & Wamba, S. F. (2022a). Metaverse beyond the hype: Multidisciplinary perspectives on emerging challenges, opportunities, and agenda for research, practice and policy. *International Journal of Information Management, 66*, 1–55.

Dwivedi, Y.-K., Hughes, L., Baabdullah, A. M., Ribeiro-Navarrete, S., Giannakis, M., Al-Debei, M. M., Dennehy, D., Metri, B., Buhalis, D., Cheung, C. M. K., Conboy, K., et al. (2022b). Metaverse beyond the hype: Multidisciplinary perspectives on emerging challenges, opportunities, and agenda for research, practice and policy. *International Journal of Information Management, 66*, 102542. https://doi.org/10.1016/j.ijinfomgt.2022.102542

Efendioğlu, İ. H. (2023). The effect of information about Metaverse on the consumer's purchase intention. *Journal of Global Business & Technology, 19*(1), 63–77.

Elkington, J., & Rowlands, I. H. (1999). Cannibals with forks: The triple bottom line of 21st century business. *Alternatives Journal, 25*(4), 42.

Energy Star. (n.d.). Implement efficient data storage measures. Retrieved from https://www.energystar.gov/products/implement_efficient_data_storage_measures

EY. (2022). Could creating a virtual world build a more sustainable one? Retrieved from https://www.ey.com/en_gl/insights/digital/metaverse-could-creating-a-virtual-world-build-a-more-sustainable-one

Ferrigno, G., Di Paola, N., Oguntegbe, K. F., & Kraus, S. (2023). Value creation in the metaverse age: A thematic analysis of press releases. *International Journal of Entrepreneurial Behavior & Research, 29*(11), 337–363.

Gadalla, E., Keeling, K., & Abosag, I. (2013). Metaverse-retail service quality: A future framework for retail service quality in the 3D internet. *Journal of Marketing Management, 29*(13–14), 1493–1517. https://doi.org/10.1080/0267257x.2013.835742

Gartner (2022). What is a Metaverse? gartner.com

Global Digital Alliance. (2022). Global Digital Alliance. Retrieved from https://www.globaldigitalalliances.org/

GreenTech. (2021). IEEE Green Technologies Conference 2021. Retrieved from https://ieeegreentech.org/2021/

Han, J., Yun, J., Jang, J., & Park, K. R. (2010). User-friendly home automation based on 3D virtual world. *IEEE Transactions on Consumer Electronics, 56*(3), 1843–1847. https://doi.org/10.1109/tce.2010.5606335

Han, D.-I. D., Bergs, Y., & Moorhouse, N. (2022). Virtual reality consumer experience escapes: Preparing for the metaverse. *Virtual Reality*., Springer, *26*, 1–16.

Hollensen, S., Kotler, P., & Opresnik, M. O. (2022). Metaverse–the new marketing universe. *Journal of Business Strategy, 44*, 119. https://doi.org/10.1108/JBS-01-2022-0014. [ahead-of-print].

Izonin, I., Singh, K. K., & Singh, A. (2024). Blockchain-based frameworks for explainable AI. In *Convergence of blockchain and explainable artificial Intelligence* (pp. 23–30). River Publishers.

Jeon, H. J., et al. (2021). *Blockchain and AI meet in the Metaverse*. Intech Open.

Jin, Q., & He, D. (2021). Library and information services in the metaverse. *The Journal of Academic Librarianship, 47*(4), 102399. https://doi.org/10.1016/j.acalib.2021.102399

Johnson, C. (2021). Metaverse: Introduction and current research. *Journal of Business Research, 133*, 1–5.

Kozinets, R. V. (2022). Immersive netnography: A novel method for service experience research in virtual reality, augmented reality and metaverse contexts. *Journal of Service Management, 34*, 100.

Kraus, S., Kanbach, D. K., Krysta, P. M., Steinhoff, M. M., & Tomini, N. (2022). Facebook and the creation of the metaverse: Radical business model innovation or incremental transformation? *International Journal of Entrepreneurial Behavior & Research, 28*(9), 52–77.

Kraus, S., Kumar, S., Lim, W. M., Kaur, J., Sharma, A., & Schiavone, F. (2023). From moon landing to metaverse: Tracing the evolution of technological forecasting and social change. *Technological Forecasting and Social Change, 189*, 122381.

Lee, L.-H., Braud, T., Zhou, P., Wang, L., Xu, D., Lin, Z., Kumar, A., Bermejo, C., & Hui, P. (2021). All one needs to know about Metaverse: A complete survey on technological singularity, virtual ecosystem, and research agenda. *arXiv*, arXiv:2110.05352.

Lee, U. K., & Kim, H. (2022). UTAUT in metaverse: An "Ifland" case. *Journal of Theoretical and Applied Electronic Commerce Research, 17*(2), 613–635.

Lim, D. H., Lee, J. Y., & Park, S. (2024). The metaverse in the workplace: Possibilities and implications for human resource development. *Human Resource Development Review, 23*(2), 164–198.

Lou, C., & Yuan, S. (2019). Influencer marketing: How message value and credibility affect consumer trust of branded content on social media. *Journal of Interactive Advertising, 19*(1), 58–73.

Lv, Z., Xie, S., Li, Y., Hossain, M. S., & El Saddik, A. (2022). Building the metaverse using digital twins at all scales, states, and relations. *Virtual Reality & Intelligent Hardware, 4*(6), 459–470.

Merriam-Webster. (2022) 'Definition of ethics'. Available at: https://www.merriam-webster.com/dictionary/ethics.

Meta (2021). The Facebook company is now Meta. Meta Press Release, Retrieved November 30, 2021, from https://about.fb.com/news/2021/10/facebook-company-is-now-meta/.

Metaverse and Society 5.0. (2023). Pivotal for future business model innovation. *Journal of Business Models, 11*(3), 62–76.

Milgram, P., et al. (1994). A taxonomy of mixed reality visual displays. *IEICE Transactions on Information Systems, E77-D*, 12.

Mishra, A. R., Mardani, A., Rani, P., & Zavadskas, E. K. (2020). A novel EDAS approach on intuitionistic fuzzy set for assessment of health-care waste disposal technology using new parametric divergence measures. *Journal of Cleaner Production, 272*, 122807. Elsevier.

Morini Bianzino, N. (2022, September 4). How the Metaverse could bring us closer to a sustainable reality. VentureBeat. https://venturebeat.com/virtual/how-the-Metaverse-could-bring-us-closer-to-a-sustainable-reality/.

Moriuchi, E. (2021). An empirical study of consumers' intention to use biometric facial recognition as a payment method. *Psychology and Marketing, 38*(10), 1741–1765. John Wiley and Sons.

Ng, D. T. K., Leung, J. K. L., Chu, S. K. W., & Qiao, M. S. (2021). Conceptualizing AI literacy: An exploratory review. *Computers and Education: Artificial Intelligence, 2*, 100041.

Papagiannidis, S., Bourlakis, M., & Li, F. (2008). Making real money in virtual worlds: MMORPGs and emerging business opportunities, challenges and ethical implications in metaverses. *Technological Forecasting and Social Change, 75*(5), 610–622. https://doi.org/10.1016/j.techfore.2007.04.007

Park, S. M., & Kim, Y. G. (2022). A Metaverse: Taxonomy, components, applications, and open challenges. *IEEE Access, 10*, 4209–4251. https://doi.org/10.1109/ACCESS.2021.3140175

Pellegrino, A., Stasi, A., & Wang, R. (2023). Exploring the intersection of sustainable consumption and the Metaverse: A review of current literature and future research directions. *Heliyon, 9*, e19190.

Porush, D. (1994). Hacking the brainstem: Postmodern metaphysics and Stephenson's SnowCrash. *Configurations, 2*(3), 537–571. https://doi.org/10.1353/con.1994.0034

Pu, X., Li, L., & Chai, X. (2021a). Virtual reality technology in library services: A literature review. *Information Technology and Libraries, 40*(1), 88–102. https://doi.org/10.6017/ital.v40i1.12552

Pu, X., Li, Y., Zhang, X., & Wang, H. (2021b). Exploring the potential of virtual reality technology for library programs and services. *Library Hi Tech, 39*(1), 127–142.

Pyun, K. R., Rogers, J. A., & Ko, S. H. (2022). Materials and devices for immersive virtual reality. *Nature Reviews Materials, 7*(11), 841–843. https://doi.org/10.1038/s41578-022-00501-5

Queiroz, M. M., Fosso Wamba, S., Pereira, S. C. F., & Chiappetta Jabbour, C. J. (2023). The metaverse as a breakthrough for operations and supply chain management: Implications and call for action. *International Journal of Operations and Production Management*, ahead-of-print No. aheadof-print, doi: https://doi.org/10.1108/IJOPM-01-2023-0006.

Racelis, A. D. (2010). Business ethics and social responsibility. *Education About Asia, 15*, 2.

Rane, N. L., Choudhary, S. P., & Rane, J. (2024). Metaverse as a cutting-edge platform for attaining sustainable development goals (SDGs). *Journal of Advances in Artificial Intelligence, 2*(1), 27–46. https://ssrn.com/abstract=4644035

Raza, S. H., & Zaman, U. (2021). Effect of cultural distinctiveness and perception of digital advertising appeals on online purchase intention of clothing brands: Moderation of gender egalitarianism. *Information, 12*(2), 72.

Rospigliosi, P. A. (2022). Metaverse or simulacra? Roblox, minecraft, meta and the turn to virtual reality for education, socialisation and work. *Interactive Learning Environments, 30*(1), 1–3. https://doi.org/10.1080/10494820.2022.2022899

Saima, & Khan, M. A. (2020). Effect of social media influencer marketing on consumers' purchase intention and the mediating role of credibility. *Journal of Promotion Management, 27*(4), 503–523.

Schroeder, R., Huxor, A., & Smith, A. (2001). Activeworlds: Geography and social interaction in virtual reality. *Futures, 33*(7), 569–587. https://doi.org/10.1016/s0016-3287(01)00002-7

Shareef, M. A., Mukerji, B., Dwivedi, Y. K., Rana, N. P., & Islam, R. (2019). Social media marketing: Comparative effect of advertisement sources. *Journal of Retailing and Consumer Services, 46*, 58–69.

Shen, B., Tan, W., Guo, J., Zhao, L., & Qin, P. (2021). How to promote user purchase in metaverse? A systematic literature review on consumer behavior research and virtual commerce application design. *Applied Sciences, 11*(23), 11087.

Singh, S., & Vanka, S. (2020). Workplace flexibility bias and intention to leave among women in IT sector: A parallel mediation model of negative work to family spill over and work alienation?

Singh, S., & Vanka, S. (2023). Metaverse and future of work: Avenues and challenges. *IUP Journal of Organizational Behavior, 22*(2), 107–118.

Singh, A., Singh, K. K., & Abouhawwash, M. (2024). BlockXAI: Challenges and future. In *Convergence of Blockchain and explainable artificial Intelligence* (pp. 145–155). River Publishers.

Smart, M., et al. (2007). *Metaverse roadmap overview*. Acceleration Studies Foundation.

Srivastava, H., Singh, A., Bharti, A. K., & Cengiz, K. (2024). BlockXAI: Review of blockchain for explainable artificial intelligence. *Convergence of Blockchain and Explainable Artificial Intelligence*, 1–14.

Stokel-Walker, C. (2021). Facebook is now Meta—But why, and what even is the metaverse? *New Scientist, 252*(3359), 12. https://doi.org/10.1016/s0262-4079(21)01955-2

Stoll, C., Gallersdörfer, U., & Klaaßen, L. (2022). Climate impacts of the Metaverse. *Joule, 6*(12), 2668–2673. https://doi.org/10.1016/j.joule.2022.10.013

Suh, A., & Prophet, J. (2018). The state of immersive technology research: A literature analysis. *Computers in Human Behavior, 86*, 77–90. https://doi.org/10.1016/j.chb.2018.04.019

Sun, J., Gan, W., Chao, H. C., & Yu, P. S. (2022). Metaverse: Survey, appli-cations, security, and opportunities. arXiv preprint, arXiv:2210.07990.

Sustainable Digital Infrastructure. (2023). Roadmap. Retrieved from https://sdialliance.org/roadmap/

Sze, L. B., Salo, J., & Tan, T. M. (2023). Under studied markets and marketing stakeholders: Achieving decentralized sustainable management in vertical farming through the metaverse. *AMA Summer Academic Conference Proceedings, 34*, 1174–1177.

Taylor, C. R. (2022). Research on advertising in the metaverse: A call to action. *International Journal of Advertising, 41*(3), 383–384. https://doi.org/10.1080/02650487.2022.2058786

Tella, A., Ajani, Y. A., & Ailaku, U. V. (2023). Libraries in the metaverse: The need for metaliteracy for digital librarians and digital age library users. *Library Hi Tech News, 40*(8), 14–18.

Thomson, M. (2020). Preparing for the metaverse: The next big thing. Retrieved January 19, 2025, from https://www.millerthomson.com/en/insights/technology-ip-and-privacy/preparing-for-the-metaverse-the-next-big-thing/

Toraman, Y., & Geçit, B. B. (2023). User acceptance of Metaverse: An analysis for e-commerce in the framework of technology acceptance model (TAM). *Sosyoekonomi, 30*(55), 85–104. https://doi.org/10.17233/sosyoekonomi.2023.01.05

Vlăduţescu, Ş., & Stănescu, G. C. (2023). Environmental sustainability of Metaverse: Perspectives from Romanian developers. *Sustainability, 15*(15), 11704. https://doi.org/10.3390/su151511704

Wan, X., Zhang, G., Yuan, Y., & Chai, S. (2023). Can metaverse technology drive digital transformation of manufacturers? Selection of evolutionary stability strategy based on supply chain perspective. *Applied Soft Computing, 15*(8), 1–31. 110611.

WEF (2023). These 3 cities already have their own metaverse. Retrieved December 25, 2023, from https://www.weforum.org/agenda/2023/11/metaverse-digital-cities-urban/

Weismueller, J., Harrigan, P., Wang, S., & Soutar, G. N. (2020). Influencer endorsements: How advertising disclosure and source credibility affect consumer purchase intention on social media. *Australasian Marketing Journal, 28*(4), 160–170.

Williams, A. (2022). Human-centric functional modeling and the metaverse. *Journal of Metaverse, 2*(1), 23–28.

Xiang Vico, H. M. (2023). CarbonSavings: A blockchain-powered move-to-earn app for citizen's awareness on greenhouse gas emissions. Bachelor's thesis, Pompeu Fabra University, Engineering School.

Zdanowicz, C. C. (2021, February 10). Lawyer tells judge "I'm not a cat" after a Zoom filter mishap in virtual court hearing. Retrieved from https://edition.cnn.com/2021/02/09/us/cat-filter-lawyer-zoom-court-trnd/index.html

Zhang, S., Lim, W. Y. B., Ng, W. C., Xiong, Z., Niyato, D., Sherman Shen, X., & Miao, C. (2022). Towards Green Metaverse Networking Technologies, Advancements and Future Directions. arXiv e-prints, arXiv-2211.

Zhao, N., Zhang, H., Yang, X., Yan, J., & You, F. (2023). Emerging information and communication technologies for smart energy systems and renewable transition. *Advances in Applied Energy, 9*, 100125. https://doi.org/10.1016/j.adapen.2023.100125

Chapter 9
Transcendent Artificial Intelligence in Education

Yashwant A. Waykar and Sucheta S. Yambal

Abstract This chapter explores the profound effects of artificial intelligence (AI) on education and e-learning, as well as our prospects going forward with the as-yet-undiscovered uses of AI in educational contexts. This chapter begins with a review of the limitations that traditional educational systems face, followed by a summary of the fundamental ideas behind artificial intelligence (AI), which gives the reader a roadmap for exploring the world of AI in education.

This chapter goes into great length about artificial intelligence (AI) in education, covering everything from intelligent tutoring systems and individualised learning platforms to automated assessment and grading. It delves into the adaptive intelligence that AI seems to possess and how it modifies course materials to fit each student's needs, making learning more personalised and immersive. The versatility of artificial intelligence in various facets of education is further demonstrated by its application in intelligent content creation, gaming, and language instruction.

Keywords Pedagogical approaches · Multimedia learning · Adaptive learning · Intelligent tutoring systems · Conversational agents · Online learning · Personalised learning · Educational technology · Augmented reality · Virtual reality · Ethical considerations · Learning sciences · Self-regulated learning · Inclusive education · Data analytics

9.1 Introduction

In this day and age of lightning-fast technological advancement, the introduction of artificial intelligence (AI) into the classroom signifies a sea shift in the ways in which teaching and learning are carried out. It is difficult for the traditional educational system to accommodate the many different learning styles and rates of

---

Y. A. Waykar (✉) · S. S. Yambal
Department of Management Science, Dr. Babasaheb Ambedkar Marathwada University, Chh. Sambhaji Nagar, India

A. Singh, K. K. Singh (eds.), *Multimodal Generative AI*,
https://doi.org/10.1007/978-981-96-2355-6_9

students, despite the fact that it is important. On the other hand, artificial intelligence has brought us to the verge of a significant transformation that will have a profound impact on education (Clark & Mayer, 2016). This change will not only influence the manner in which we teach and learn, but it will also change the very essence of education itself.

This chapter explores the interface of artificial intelligence and education, focusing on the vast opportunities that arise as a result of this convergence. The more we study about the complexity of artificial intelligence technologies, such as machine learning algorithms and natural language processing, the more we are able to understand how these breakthroughs could change the way that educational practices are carried out.

The shortcomings of traditional methods of teaching bring to light the many ways in which artificial intelligence (AI) is enhancing the quality of learning in the classroom. Through the use of intelligent tutoring platforms, automated assessment tools, and customised learning systems, artificial intelligence brings about the provision of individualised solutions that are tailored to the unique needs and capabilities of each individual student.

The use of artificial intelligence in education may not only result in increased productivity, but it may also lead to increased student engagement that is more meaningful, improved health outcomes, and the opening of learning frontiers that were previously inaccessible via the use of gamified experiences and virtual labs.

However, as is the case with any new technology, the ethical questions that surround the employment of artificial intelligence in the classroom need to be given the attention and care they deserve. There are a number of critical problems that should be carefully studied and managed ethically when artificial intelligence systems are implemented. These concerns include data privacy, algorithmic bias, and digital equality.

Nevertheless, despite the inevitable challenges that will be encountered in the future, there are a great deal of exciting possibilities for the future of education that incorporates AI. A technology revolution is on the verge of occurring in the field of education. On the horizon are technologies like artificial intelligence-powered virtual reality experiences and predictive analytics that are designed to improve student accomplishment.

9.2 Understanding Artificial Intelligence

In order to fully appreciate the far-reaching implications that artificial intelligence (AI) has had in the area of education, it is essential to have a basic understanding of the innovative concepts that underpin this revolutionary technology. In its most basic form, artificial intelligence (AI) is the product of decades of research and development that aimed to imitate and ultimately surpass human intellect via the use of computer algorithms.

One of the most fundamental aspects of artificial intelligence is machine learning, which allows computers to autonomously gain information from data without the need for explicit programming commands. Through the process of sifting through vast amounts of data in search of patterns, machine learning algorithms are able to make predictions and decisions that are very accurately correct. This ability is essential for a number of applications of artificial intelligence, most notably those in the area of education.

The ability of computers to perceive, analyse, and synthesise language that is comparable to that of humans is made possible by natural language processing (NLP), which is another fundamental component of artificial intelligence. From language translation tools that allow international cooperation to virtual assistants who are able to participate in conversational discussion, the area of natural language processing (NLP) has a great deal of potential for the improvement of educational experiences.

It is useful to have a fundamental understanding of these concepts in order to have a better understanding of how artificial intelligence systems may adapt, learn, and replicate human cognitive processes in educational contexts (Blikstein, 2013). Two examples of how artificial intelligence (AI) is transforming the educational landscape are the introduction of personalised learning and the implementation of automated assessment. These technological advancements are more than simply resources; they are revolutionary forces that will change the manner in which we teach the generations who will come after us.

By embracing the complexity of artificial intelligence, educators and other stakeholders may be able to open up new vistas in education, better equipping children to thrive in a world that is more international and complex. As we embark on this journey of discovery and creation, it is essential to keep in mind that artificial intelligence (AI) is more than just a technological advancement; it is a paradigm shift that has the potential to totally transform the means by which we comprehend and carry out educational practices.

9.3 AI Applications in Education

9.3.1 Personalised Learning

One of the most fascinating breakthroughs in the world of education is the use of artificial intelligence (AI) in the process of developing personalised lesson plans that are tailored to the particular interests, abilities, and limitations of each individual student. Traditional, one-size-fits-all methods of education sometimes fail to meet the needs of students since each kid learns at their own pace and in their own individual style. However, there is a game-changing option available in the shape of tailored learning systems that are driven by artificial intelligence. These systems make use of data-driven insights to personalise the curriculum and instructional strategies for each individual learner.

An essential component of tailored education is the use of intricate algorithms for the purposes of machine learning and data analysis. These algorithms sift through mountains of data, which includes metrics for student performance, their learning preferences, and patterns of behaviour, in order to draw meaningful insights about each individual member of the student body. It is possible that artificial intelligence systems will make use of this information in order to dynamically adjust the learning environment in accordance with the student's current skill level and chosen style of learning.

There are several benefits associated with personalised learning. The first and most important benefit is that it enhances student engagement by adapting the presentation of material to the specific interests and preferences of each individual student. When students' educational experiences are personalised to meet their own requirements and interests, they are more likely to take an active part in their own education and strive for greatness. This is because students are more committed to their own education.

By adapting to the specific learning route of each individual student, personalised learning not only boosts the level of comprehension but also the amount of information that is retained. Students are not required to adhere to rigid timetables and traditional curricula; rather, they are given the opportunity to learn at their own pace with the assistance of technologies that are driven by artificial intelligence. Because of this, students are able to investigate concepts in more depth or return to regions in which they have difficulties as often as they want. This adaptive technique not only takes into consideration different learning rates but also promotes them. Mastery learning is a pedagogical strategy that places an emphasis on thorough knowledge and competency before moving on to new subjects. This method provides assistance for mastery learning.

In addition, students who participate in personalised learning have access to interventions and support that are timely and relevant, which helps to create an environment that is favourable to healthy learning. This is because algorithms that are driven by artificial intelligence have the ability to rapidly spot problem areas or misconceptions and give answers, corrections, or more assistance to address them. Through this proactive approach to providing assistance, students are provided with the resources they need to overcome obstacles and develop a sense of self-assurance in their educational journey.

9.3.2 Intelligent Tutoring Systems

Intelligent tutoring systems (ITS) are a cutting-edge use of artificial intelligence (AI) in the field of education. Their primary objective is to enhance the learning outcomes of students by providing assistance that is interactive, individualised, and adaptable. In contrast to traditional methods of instruction, which often rely on unidirectional communication and the transmission of material in a static format, ITS dynamically customises training to meet the specific needs and learning styles of each individual student. This is accomplished via the use of AI algorithms.

Students' performance data is processed by complex artificial intelligence algorithms, which are the backbone of ITS. These algorithms evaluate the data in order to discover knowledge gaps and generate tailored lesson plans. Through the use of techniques such as cognitive modelling, natural language processing, and machine learning, these algorithms take on the role of human tutors in order to provide assistance to students who are struggling with challenging academic subjects.

ITS offers a number of advantages, one of the most important of which is the ability to provide individualised instruction on a massive scale. Artificial intelligence (AI) may be used by ITS systems to provide students with customised feedback and coaching, regardless of the size of the class or the kind of classroom. This allows the systems to accommodate a broad range of pupils. Student engagement, understanding, and the ability to remember information are all significantly improved by using this one-on-one strategy.

On top of that, ITS systems enable active learning via the provision of adaptive and interactive learning environments. Through the use of interactive learning experiences such as simulations, virtual experiments, and problem-solving exercises, students are able to strengthen their talents in critical thinking, problem-solving, and conceptual understanding. In addition, instructional technology systems have the capability to modify the speed and level of difficulty of courses in real time based on the feedback provided by students. This ensures that students are both challenged and supported as they learn.

According to a number of studies, ITS has the potential to enhance learning outcomes across a wide range of subject areas and grade levels. Platforms for information and communication technology (ITS) have a significant influence on students' academic performance, retention, and capacity to self-regulate their learning because of the targeted assistance and scaffolding that they provide.

It is also beneficial for educators to use ITS platforms since these platforms provide valuable data on the growth and performance of their pupils while they are in the classroom. The capabilities of data analytics and reporting enable educators to get a deeper understanding of the learning pathways of their students, identify their students' strengths and areas in which they may develop, and tailor their classes to meet the specific needs of themselves and their students.

9.3.3 Automated Assessment and Grading

The traditional methods of evaluating student performance are being drastically transformed by automated assessment and grading systems that are powered by artificial intelligence. These systems provide a significant increase in efficiency, accuracy, and tailored feedback with regard to student performance. Manual grading methods are often used in traditional classrooms, which are labour-intensive, subjective, and prone to accuracy errors. As a result of the widespread availability of artificial intelligence (AI), automated assessment tools have made grading a breeze. Additionally, instructors are able to monitor the progress of their pupils in real time owing to the feedback they get.

The foundation of automated grading and assessment systems is comprised of algorithms that are powered by artificial intelligence and make use of machine learning to analyse the responses of students in an impartial manner. When it comes to assessing tasks such as essays, problem-solving exercises, multiple-choice questions, and other types of evaluations, these algorithms are very quick and accurate assessments.

One of the most significant advantages of automated assessment is that it provides students with feedback in a quick and simple manner. When students do not have to wait days or weeks for manual grading, they are able to immediately evaluate their learning strengths and areas in which they need development. Instead, they get instant feedback on their performance, which allows them to make more informed instructional decisions.

Furthermore, students have the opportunity to get feedback that is tailored to their individual learning style and progress via the use of automated assessment tools. There is a possibility that artificial intelligence systems may examine the responses of students in order to identify patterns. Based on the issues and misconceptions that are discovered, these systems would then provide recommendations on how to improve.

The use of automated grading systems is beneficial to educators since it allows them to increase the efficiency and effectiveness of assessment processes while simultaneously freeing up time and resources. If routine grading processes are automated, teachers will have more time to produce engaging courses, provide one-on-one assistance to students, and utilise the outcomes of assessments to influence the planning of classes.

Teachers are able to have a better understanding of how their pupils are progressing in terms of learning outcomes and development via the use of automated assessment technologies. Instructors are able to track the progress of their students over time, recognise patterns and trends in their learning, and pinpoint areas in which they may benefit from further assistance when they make use of data analytics and reporting tools.

However, it is essential that we should not overlook the fact that automated assessment approaches come with their own set of constraints and challenges. Ethical problems like algorithmic bias and assessment fairness need to be properly addressed in order to guarantee that automated grading systems do not aggravate existing inequities or cause harm to certain student groups.

9.4 Language Processing for Language Learning

The provision of learning environments that are engaging, individualised, and effective is made possible by artificial intelligence (AI)-driven language processing technologies. These tools provide language learners the opportunity to experience transformational potential. Getting proficient in a foreign language has always been a challenging endeavour that required a significant amount of time and effort from those who are learning the language. On the other hand, there are now language processing systems that are driven by artificial intelligence and employ cutting-edge techniques to assist individuals in learning languages in ways that no one could have envisioned.

Artificial intelligence systems that are up to date and capable of comprehending, interpreting, and producing human language are now at the forefront of language processing for the purpose of language learning. Because these algorithms employ natural language processing (NLP) to investigate linguistic patterns, semantics, and grammar, students are able to engage with language in meaningful ways. This is made possible by artificial intelligence.

As a technology that utilises language processing, speech recognition software is an essential resource for those who are learning other languages. Speech recognition systems that are powered by artificial intelligence are able to perform a number of functions, including accurate transcription of spoken language and rapid feedback on tone, fluency, and pronunciation. With the help of this tool, students are able to practice their communication skills in an environment that is both secure and interesting and that simulates genuine conversation.

Additionally, language processing technologies that are driven by artificial intelligence provide a wide range of customised and adaptive learning experiences that are customisable to meet the specific needs and preferences of each individual learner. One example that illustrates this notion is that language learning applications that are equipped with natural language processing capabilities have the ability to personalise their courses for each individual student by taking into consideration the student's existing level of knowledge and their goals. An example of this would be the development of interactive exercises, vocabulary drills, and explanations of grammatical concepts.

Both virtual language environments and conversational bots are examples of additional ways in which language processing technology may be used to create immersive educational experiences. Chatbots and virtual tutors that are powered by artificial intelligence communicate with students in real time for the purpose of answering their questions and helping them through their schoolwork. These interactions take place via conversations that are conducted in natural language. Using these immersive interactions, students have the opportunity to practice their language abilities in real-life scenarios, which helps them become more confident and fluent in their language skills.

Language processing systems aid language learning in a number of different ways, one of which is via the provision of translation and interpretation services, respectively. Learners now have the ability to readily access information and resources in different languages owing to translation systems powered by artificial intelligence (AI) that can translate text across languages in a quick and accurate manner. In a similar vein, students have the opportunity to engage in conversations and exchanges in more than one language by using interpretation tools that are powered by artificial intelligence (Singh et al., 2024a).

Despite the fact that language processing technologies hold a great deal of potential for language learning, it is necessary to give some consideration to the ethical concerns and challenges that are associated with these technologies. Finding solutions to problems relating to student privacy, data security, and algorithmic bias is necessary in order to ensure that language processing technology powered by artificial intelligence is used in classrooms in a fair and suitable manner.

9.5 Adaptive Learning Platforms

Using adaptive learning systems that are driven by artificial intelligence is a revolutionary new method to educate. Individualised instruction is provided by these systems, which take into account the specific circumstances, interests, and strengths and limitations of each individual student. Traditional, one-size-fits-all educational techniques sometimes fail to satisfy the requirements of pupils because of the variability in their learning styles and rates of comprehension. Adaptive learning systems, on the other hand, make use of artificial intelligence algorithms to personalise the educational experiences of each student based on their unique development (Singh et al., 2024b).

Intelligent artificial intelligence algorithms are the foundation of adaptive learning systems. These algorithms analyse student data, including performance indicators, learning preferences, and behavioural patterns, in order to develop individualised educational programs for each student. Using techniques from machine learning, these algorithms evaluate the strengths and weaknesses of the students, adjust the level of difficulty of the content, and intervene in a manner that is especially designed to assist the students in learning.

In particular, adaptive learning systems stand out due to their ability to adjust courses according to the strengths and limitations of each individual student, as well as their preferred learning method and current skill level. Instead of adhering to a predetermined curriculum or pacing guide, adaptive learning systems alter the level of difficulty and the order in which learning activities are performed based on the feedback and assessment data that are received in real time. By using this tailored approach, each and every student receives the appropriate amount of assistance and challenge to enable them to realise their greatest potential.

Additionally, learners have the flexibility and independence to go through the content at their own pace, go deeper into specific areas, or review previously covered information as required; all of these options are made possible by adaptive learning systems. Students are encouraged to take responsibility for their own education and to work towards achieving subject mastery via the use of adaptive learning systems, which are designed to accommodate a wide range of learning preferences and speeds.

A further benefit of adaptive learning systems is that they stimulate engagement and motivation by presenting content in a variety of ways that are tailored to the individual learner. Through the use of gamified exercises, real-world applications, and interactive multimedia resources, students are able to successfully learn and become more involved in the learning process.

Through the use of adaptive learning systems, educators have the potential to get a great deal of valuable information on the growth and performance of their students. Through the use of data analytics and reporting technologies, educators are able to monitor the development of their pupils, identify areas in which they are having difficulty or are being misinterpreted, and ensure that they get tailored assistance (Singh & Singh, 2023a). It is possible that teachers who adopt this method would be better able to respond to the particular requirements of their students and make decisions based on facts.

It is of the utmost importance to address the ethical implications and challenges associated with adaptive learning platforms, notwithstanding the fact that these platforms have substantial potential to improve the learning outcomes of students. It is imperative that adaptive learning platforms give priority to educational equality and provide support for the accomplishments of all learners. This may be accomplished by giving careful thought to issues like data privacy, algorithmic bias, and fairness in access to technology.

9.6 Emotion Recognition for Student Wellness

Educators now have the ability to identify and respond to the emotional states of their pupils in real time because of the development of emotion detection technology that is driven by artificial intelligence (AI). This ground-breaking method contributes to the overall well health of the pupils (Anderson & Whitelock, 2010). The identification and fulfilment of the emotional requirements of students in traditional classrooms may prove to be challenging due to the fact that signs are often disregarded or misinterpreted. On the other side, emotion detection systems that are powered by artificial intelligence provide instructors with vital information on the mental health of their pupils. This information may be used by teachers to intervene and assist their students when they need assistance.

For the purpose of ensuring the well-being of students, emotion detection is dependent on sophisticated artificial intelligence algorithms. These algorithms analyse students' facial expressions, voice intonations, and other physiological indications in order to comprehend their emotional states. By using machine learning techniques to discover patterns and relationships between facial expressions and emotions, these algorithms are able to accurately identify a wide range of emotional states, such as happiness, sadness, irritation, and worry.

One of the most significant benefits of emotion detection technology is that it enables educators to get instant feedback on the emotional states of their pupils while they are engaged in doing learning activities. The facial expressions of students may be monitored by artificial intelligence algorithms via the use of camera- or video-based systems. These algorithms can identify signs of discomfort or engagement in order to alert instructors when it is important to intervene. It is possible for teachers to assist students in achieving academic and personal success by taking the effort to deliver suitable words of encouragement and support at relevant times.

Additionally, the emotional requirements of students may be satisfied via the implementation of tailored therapies that are made feasible by technology that detects emotions. In the course of analysing the emotional responses of students over a period of time, artificial intelligence algorithms may disclose the students' coping methods and emotional triggers. The information provides educators with the ability to develop individualised interventions and support programs that are capable of effectively meeting the specific needs of each individual kid.

A further benefit of emotion recognition technology is that it enables students to become more self-aware and build skills in self-regulation (Singh & Singh, 2023b). This is accomplished by providing students with feedback on how they are feeling in connection to learning activities. Students may be able to better understand and control their emotions, as well as acquire techniques for dealing with stressful circumstances, via the use of interactive dashboards or feedback systems. Students benefit from this practice of self-monitoring and reflection because it helps them build emotional resilience and gives them more control over their own emotional health.

Additionally, the use of emotion detection technology facilitates open and honest conversation between educators, students, and parents about the best ways to provide emotional support to their children, therefore enhancing both communication and cooperation. When teachers communicate their results from emotion recognition data with parents and children in order to develop tailored assistance plans, they may be better able to satisfy the academic and emotional requirements of their pupils.

In order to fully achieve the potential of emotion detection technology in terms of improving the well-being of students, it is necessary to address issues pertaining to ethics and privacy difficulties. It is essential that data security, permission, and confidentiality be rigorously preserved in order to ensure that emotion recognition systems respect the private rights of students and provide a safe and encouraging learning environment.

9.7 Gamification and AI

A fresh and engaging approach to learning is produced when gamification ideas and artificial intelligence technologies are brought together. The dynamics of games are used in this approach to encourage students, maintain their attention, and enhance the outcomes of their learning opportunities (Singh & Singh, 2016). Gamification has the potential to transform educational activities into interactive and entertaining experiences by incorporating game mechanisms such as points, badges, leaderboards, and challenges into settings that are not traditionally associated with games.

One of the most important aspects of gamification and artificial intelligence is the use of AI algorithms to personalise and improve gaming experiences in accordance with learner preferences, performance data, and learning goals. The use of artificial intelligence (AI) gamification systems allows for the monitoring of student behaviour, the identification of trends, and the instantaneous modification of game rules in order to maintain player engagement and interest.

One of the most significant benefits of gamification and artificial intelligence is their ability to boost the intrinsic motivation and engagement of students. By incorporating elements of competition, success, and development into educational activities, gamified experiences are able to spark the attention of kids and stimulate their natural curiosity. Artificial intelligence algorithms are continually modifying the game mechanics to accommodate the preferences and skill levels of each individual learner. This is done to ensure that the learning experience is both challenging and practical.

Additionally, gamification and artificial intelligence foster active learning by providing students with opportunities to experiment with new things, find solutions to issues, and explore in a safe environment. Through the use of interactive challenges, quest-based activities, and immersive simulations, students engage in experiential learning that encourages creative thinking, collaboration, and critical thinking.

Additionally, the combination of gamification with artificial intelligence makes it possible to develop personalised learning regimens that are tailored to the particular needs and preferences of each individual learner. In order to assist students in accomplishing their educational objectives, artificial intelligence algorithms analyse the data pertaining to their performance, identify the areas in which they excel and the areas in which they need development, and then change the game's content and difficulty levels accordingly. This tailored approach makes it possible for students of all different kinds of backgrounds and skills to get the assistance they need in order to realise their greatest potential.

As an additional benefit, gamification and artificial intelligence inspire students to collaborate with one another and participate in social interactions by fostering a sense of belonging and a shared objective inside the classroom (Singh et al., 2013). Participating in team-based competitions, working together on group projects, and playing multiplayer games are all ways in which students may improve their social and emotional abilities as well as their understanding of the subject covered in the course.

Although gamification and artificial intelligence hold a great deal of potential for enhancing educational opportunities, we must not dismiss the ethical considerations and concerns that are associated with their use. The rigorous protection of data privacy, consent, and transparency is one of the most important factors in ensuring that gamified learning environments are conducive to the development of a safe and supportive learning environment for all students.

9.8 Smart Content Creation

The process of developing instructional materials is undergoing a revolution as a result of automation of procedures, enhancement of creativity, and optimisation of information dissemination for maximum impact. This is made possible by smart content creation that is driven by artificial intelligence. Traditional techniques for content generation often involve a lot of manual labour, long processes, and a limited amount of resources. The advancement of artificial intelligence, on the other hand, has made it possible for intelligent content creation platforms to automate operations that were previously arduous, to be open to fresh ideas, and to produce interactive instructional materials that are suited for a broad variety of student expectations.

Through the use of machine learning, deep learning, and natural language processing (NLP), intelligent content creation is able to assess, develop, and improve instructional materials. This is accomplished via the utilisation of artificial intelligence algorithms. With the assistance of these algorithms, it is possible to automate tasks such as the generation of text, the identification of pictures, the editing of videos, and the production of multimedia. This gives educators more time and energy to focus on instructional design and pedagogical innovation.

The ability to generate adaptable and tailored learning materials that are unique to the preferences, needs, and goals of each individual learner is a significant asset of smart content production. These tailored courses are designed by algorithms that are driven by artificial intelligence. These algorithms sift through student records of performance, patterns of learning, and comments in order to satisfy the demands of students who come from a variety of backgrounds, have diverse interests, and have various skill levels. Because of this individualised approach, each and every student is certain to get material that is not only fascinating but also helpful and supportive of their educational journey.

By providing educators with access to tools that enable them to experiment with various methods of presenting knowledge, intelligent content generation systems have the potential to foster originality and creativity in the classroom. Through the use of content production systems that are powered by artificial intelligence, it is possible to develop interactive simulations, virtual labs, gamified exercises, and multimedia presentations. Learners find these tools fascinating, and they stimulate active participation, which ultimately leads to improved learning results.

In addition, intelligent platforms for content generation not only improve the delivery of information but also quickly respond to the preferences, context, and performance of learners. This allows artificial intelligence systems to continuously monitor user behaviour, comments, and engagement metrics in order to fine-tune the speed, complexity, and presentation of material in order to achieve the highest possible level of challenge, motivation, and learning effectiveness. Through the whole of the course, this versatile approach maintains students' attention, keeps them challenged, and keeps them motivated.

In addition, smart content production platforms make it possible for educators to collaborate and share their expertise by providing them with access to a vast collection of curated, crowdsourced, and user-generated information. With the assistance of recommendation systems that are driven by artificial intelligence, educators are able to locate, remix, and customise information to meet their specific needs. These systems surface relevant, high-quality resources and learning materials based on the educators' interests, preferences, and instructional goals.

There are significant ethical problems that need to be solved before smart content production can be employed on a widespread scale. This is despite the fact that it has a great deal of potential for enhancing the materials that are used in educational settings. Preserving a culture of honesty, respect, and responsibility in the content creation processes while simultaneously following legal and ethical obligations requires that copyright, credit, and intellectual property rights be respected. This is vital for preserving a culture of respect.

9.9 Predictive Analytics for Student Success

We may be able to identify children who are at risk of experiencing academic issues and provide them with individualised treatments to assist them in achieving their goals by using data-driven, proactive techniques that are powered by artificial intelligence. In conventional methods of evaluating and intervening with students, it is usual practice to wait for students to demonstrate signs of difficulties before offering support to them. On the other hand, predictive analytics makes use of data from the past as well as data from the present in order to anticipate issues and take action at an earlier stage, so improving the possibility that students will be successful in school.

The purpose of advanced artificial intelligence algorithms is to draw predictions about the future performance of students by analysing a multitude of data points. These data points include students' academic accomplishments, attendance records, engagement levels, demographic information, and socio-economic variables. These algorithms make use of machine learning to identify patterns, correlations, and trends in the data in order to assist instructors in identifying students who may have difficulties in the classroom and in taking preventive actions.

The ability to identify students who are at risk of encountering academic challenges at an early stage is a key benefit that may be gained from the use of predictive analytics. The use of predictive analytics algorithms allows for the generation of risk scores or probability estimates for individual students. These scores are developed by analysing historical data and identifying patterns that are associated with academic success or failure. The identification of pupils who might benefit from further assistance or intervention is facilitated by this.

There is also the possibility that educators may make use of predictive analytics in order to better meet the specific requirements of children who are at risk by developing individualised plans of action and support services. It is possible for educators to develop comprehensive intervention programmes that address the academic, social, emotional, and behavioural challenges that children are experiencing by analysing the factors that contribute to the risk profiles of the children. Interventions such as academic support programmes, counselling, one-on-one mentoring, and tutoring are all examples of programmes that might potentially assist students in achieving academic success.

Data-informed decision-making is also made possible by predictive analytics, which provides educators with actionable information on the areas in which their pupils excel and those in which they might need development. Dashboards, reports, and other data visualisation tools provide instructors the ability to monitor the development of their students, evaluate the effectiveness of the interventions they are implementing, and adjust their support strategies as necessary. When educators use this data-driven approach, they have the ability to prioritise interventions, utilise data to decide which interventions are the most effective, and make the most of the student support resources available to them.

The use of predictive analytics helps educators see patterns and trends in student data across time, which in turn encourages them to adopt an attitude of continuous development. Teachers have the ability to systematically improve student outcomes by analysing longitudinal data in order to identify the factors that lead to success, the areas in which students are falling short, and the ways in which they might implement strategies that are backed by research.

Predictive analytics has a great deal of potential for assisting students in achieving their academic goals; nevertheless, prior to its broad use, there are significant ethical and privacy concerns that need to be addressed. It is essential that data security, consent, and confidentiality be carefully maintained in order to ensure that student data is used in a way that is both ethical and responsible and that contributes to the accomplishment of students.

9.10 AI-Based Virtual Labs

Through the provision of students with realistic laboratory experiences that are interactive, immersive, and hands-on, virtual labs that are driven by artificial intelligence (AI) are revolutionising the teaching of science and engineering. The potential for conventional laboratory-based learning experiences is limited by a number of hurdles, such as a scarcity of resources, concerns over the safety of students, and practical issues. Traditional laboratories have their limitations, but virtual labs driven by artificial intelligence employ cutting-edge technology to create realistic simulations that are so close to the real thing that they are nearly indistinguishable from what they are.

The sophisticated artificial intelligence algorithms that run AI-powered virtual labs are the brains of these laboratories. They enable real-time interaction and feedback in response to the actions and inputs of users. These algorithms make use of a variety of techniques, including machine learning, physics-based modelling, and data analytics, in order to model complex scientific processes, simulate experimental environments, and provide findings that are convincing.

Through the use of virtual labs that are driven by artificial intelligence, students are able to participate in learning that is both active and inquiry-based within an environment that is both secure and controlled. The students gain the ability to think critically, find solutions to issues, and participate in scientific inquiry via the use of virtual equipment, conducting experiments, and evaluating data through the use of physical practice.

In addition, students have the ability to access learning materials anytime they want, from any place, and at their own pace via the use of virtual labs that are developed using artificial intelligence. Students have the opportunity to engage in self-directed learning and autonomy when they participate in virtual laboratories since they are not constrained by the time or place constraints that are present in traditional laboratories.

Additionally, virtual labs that are driven by artificial intelligence make it possible to create personalised courses that are tailored to match the particular needs and objectives of each individual student. By assessing the students' performance data, learning preferences, and feedback, virtual lab activities may be adapted to meet the specific needs of each individual student in terms of their degree of expertise and preferred method of learning. Because of this, it is possible to adjust the level of difficulty, the tempo, and the material. Regardless of where a student begins in terms of their knowledge or abilities, this personalised technique ensures that they will have the required scaffolding to achieve success in their studies.

The use of artificial intelligence to power virtual labs makes laboratory-based instruction more cost effective and scalable. This is because virtual laboratories eliminate the need for actual equipment, supplies, and facilities. Virtual laboratories are able to accommodate a large number of students at the same time because of the utilisation of cloud computing and simulation technologies (Sangrá et al., 2012). This is accomplished without the hassle and expense that are associated with traditional laboratories.

It is imperative that serious consideration be given to the ethical problems and challenges that are associated with the broad use of AI-powered virtual labs, despite the fact that these laboratories have the potential to transform STEM education. It is of the utmost importance to take measures to protect the confidentiality, safety, and authenticity of student information in virtual laboratories in order to ensure that it may be exploited in a responsible manner.

9.11 Conversational Agents for Learning Assistance

By conducting one-on-one conversations with students and providing them with feedback, guidance, and assistance in a natural language environment, conversational agents, which are powered by artificial intelligence, provide a novel and interesting approach to assisting students in their educational pursuits. Due to the lack of participation and responsiveness, students often find it difficult to acquire timely and individualised assistance from conventional learning aid strategies such as textbooks, lectures, and tutorials. This is because these kinds of methods are not very responsive. Conversational bots, on the other hand, make use of artificial intelligence (AI) capabilities like machine learning and natural language processing (NLP) in order to engage in genuine discussions with students, clarify complicated concepts, and provide quick assistance.

The ability of conversational agents to comprehend natural language, manage discussions, and create replies is made possible by complex artificial intelligence algorithms. These capabilities are crucial for conversational agents to possess in order to provide assistance with learning. These algorithms simulate human interactions by assessing user inputs, understanding user intent, and creating appropriate responses (Means et al., 2010). This allows them to provide students with individualised assistance and guidance from a learning perspective.

There is a significant benefit associated with the utilisation of conversational agents, which is the capability to provide students prompt feedback and help whenever and wherever they want it. Through the use of conversational agents that interact with chat interfaces or voice assistants, students have the ability to circumvent the need to wait for assistance from their teachers or classmates by simultaneously asking questions, requesting clarification, and obtaining instant responses.

Conversational agents also provide individualised instruction that is tailored to each student's specific areas of interest, strengths, and areas in which they may develop. A student's skills, limitations, learning style, and previous interactions with the conversational agent are evaluated by algorithms powered by artificial intelligence. Based on this information, the conversational agent may tailor the level of difficulty, speed, and subject matter that is covered. Through the use of this approach, every single student, irrespective of where they begin in terms of their knowledge or abilities, will get the tailored attention that is necessary for them to flourish.

Through the facilitation of learning experiences that are conversational and interactive, conversational agents also promote active learning and involvement. Instead of just receiving information, students engage in conversation with the conversational agent (Sweller et al., 1998); they interact with the agent by asking questions, investigating ideas, and receiving instant responses. When students are given the opportunity to take an active part in their own education, this approach inspires them to engage in critical thinking and helps them build their ability to solve problems.

Accessibility and inclusivity are also promoted by conversational agents since they are able to accommodate a broad variety of learning styles and capabilities. Conversational agents make use of adaptive interfaces, multimodal interactions, and assistive technologies in order to accommodate students who have unique learning requirements, language difficulties, or disabilities. This is done in order to guarantee that all students have equal access to learning assistance and support (Paas et al., 2003).

There are significant ethical problems and challenges that need to be solved before conversational agents may be used, despite the fact that they hold a great deal of potential for enhancing learning aids. Within the context of conversational agent interactions, the appropriate utilisation and protection of student data is contingent upon the implementation of severe safeguards to ensure data privacy, security, and transparency.

9.12 Enhancing Teacher Efficiency

The purpose of boosting the productivity of instructors via the use of artificial intelligence technology is to maximise instructional practices, tailor education, and speed up administrative work by using innovative tools and solutions. It is now possible for educators to spend more time and effort on high-impact activities that improve student learning outcomes. This is made possible by the potential of artificial intelligence to automate tedious jobs, generate insights that can be put into action, and improve instructional decision-making.

Among the primary ways in which artificial intelligence improves the productivity of teachers is via the automation of administrative tasks such as grading, tracking attendance, and lesson preparation. When teachers employ grading systems that are powered by artificial intelligence to analyse student responses and deliver quick feedback, they may spend less time manually grading students and more time providing support to students and providing tailored instruction. Along the same lines, attendance monitoring systems that are powered by artificial intelligence could be able to automate the process of taking roll, which would free up more time for instructors to dedicate to actually teaching the curriculum (Yukselturk & Bulut, 2009).

Personalisation of instruction is also made possible by artificial intelligence (AI), which enables the analysis of student data and the giving of practical insights into individual learning preferences and needs (Lajoie, 2000). Through the use of machine learning algorithms, artificial intelligence systems have the capability to assess student performance data, identify patterns, and develop individualised learning pathways that are tailored to each individual student's strengths, weaknesses, and preferred method of learning. Teachers have the ability to maximise student engagement and achievement via the use of differentiated instruction, targeted interventions, and individualised instruction.

Moreover, artificial intelligence assists in the process of instructional decision-making by providing educators with access to data-driven insights and ideas (Plass et al., 2015). With the use of data analytics and reporting technology, teachers are able to analyse trends in student performance, track progress towards learning objectives, and identify areas in which there is room for growth. The utilisation of predictive data created by AI algorithms, which can also forecast future student outcomes, enables educators to participate in proactive interventions and provide assistance to students in terms of their performance.

By making curated resources, best practices, and opportunities for professional development more available, artificial intelligence also makes it simpler for educators to collaborate with one another and share what they know. A recommendation system that is driven by artificial intelligence may be able to assist educators in locating, modifying, and implementing research-based approaches in the classroom by surfacing appropriate instructional materials, research papers, and lesson plans (Vygotsky, 1978). This is accomplished by taking into consideration the interests, preferences, and learning goals of the educators.

There are significant ethical challenges and concerns that need to be addressed before artificial intelligence (AI) can be used in a broad manner, despite the fact that technology holds a great deal of potential for making schools more efficient. A stringent commitment to data privacy, security, and transparency requirements is required in order to protect student data and make appropriate use of it to help in teaching and learning.

9.13 Teacher-Student Collaboration

Through the collaborative efforts of both students and instructors, the traditional roles that have been established within the educational system are undergoing a fundamental recomposition. When going into this section of the text, the following are some important points to keep in mind:

9.13.1 Learning Experiences That Are Co-Created

In a setting that encourages collaborative learning, artificial intelligence may assist students and teachers in working collaboratively to construct courses. It is the responsibility of teachers to act as guides, ensuring that students have access to knowledge and tools that are powered by artificial intelligence and assisting them in exploring topics that are of interest to them.

9.13.2 *Tailored Paths to Education*

Teachers and students may collaborate to develop individualised classes that are tailored to the specific interests, preferences, and learning styles of each individual student (Squire & Jan, 2007). It is possible that artificial intelligence systems might be of significant aid in the areas of resource curation, the development of personalised suggestions, and the customisation of teaching methods to fit students with diverse needs.

9.13.3 *Feedback and Reflection*

Students and teachers are able to offer and receive continual feedback on how well they are learning when they collaborate. This allows for a more effective learning experience. Teachers who utilise data acquired by artificial intelligence on student performance to deliver constructive feedback are more likely to foster a growth attitude and a commitment to continuous progress in their students.

9.13.4 *Co-Created Assessments*

In order to more accurately represent learning outcomes, it is possible for instructors and students to collaborate on the creation of assessments. Assessment technologies that are driven by artificial intelligence and give insights on student knowledge and mastery may be used to assist the process of co-creating assessment activities that foster deeper learning and deeper critical thinking.

9.13.5 *Project-Based Learning*

It is encouraged via the use of collaborative projects, which provide students and teachers the chance to work together on challenges and undertakings that are relevant to real-world situations (O'Donnell & Dansereau, 1992). The use of artificial intelligence in project-based learning has the potential to improve the overall learning experience and foster the development of skills related to collaboration by simplifying the processes of research, data analysis, and the presentation of findings.

9.13.6 Students Should be Encouraged to Express Themselves Freely and Independently

Students have greater agency and decision-making power in their learning when they collaborate with their instructors on projects, topics, and learning routes. This is because students are given more opportunities to make decisions. AI systems have the potential to empower student agency in a number of ways, including by providing autonomy, flexibility, and opportunities for self-directed exploration.

9.13.7 Establishing Ties and Establishing Trust

Students and teachers who collaborate on projects together develop a sense of trust and enhance their relationships, which ultimately results in the formation of a community that provides support for learning. Artificial intelligence technologies augment human interactions in the classroom by enhancing communication, fostering empathy, and fostering a sense of community among students.

9.13.8 Collaborative Learning Communities

The work that students and instructors perform together extends beyond the confines of the classroom and into wider communities of practice outside school boundaries (Park & Choi, 2009). AI technology is being used by educators to facilitate ongoing professional development and the exchange of information. This is accomplished via the sharing of best practices, resources, and insights acquired from collaborative experiences.

9.13.9 Inspiring Individuals to Value Learning

Through the use of collaborative learning activities, both teachers and students have the opportunity to cultivate the behaviours, attitudes, and skills that are essential to continue learning throughout their whole lives (Sitzmann et al., 2006). When students learn in environments that are enabled by artificial intelligence and encourage curiosity, creativity, and critical inquiry, they are better equipped to tackle complex challenges and exploit opportunities in a world that is becoming more driven by artificial intelligence.

Partnerships between teachers and students raise significant ethical concerns around the protection of student data, equity, and the protection of student privacy (White & Walmsley, 2006). For the purpose of ensuring that the rights and well-being of all individuals are prioritised in collaborative interactions, it is of the utmost importance to establish clear rules and norms for the ethical use of artificial intelligence.

It is possible that this chapter will shed light on the revolutionary potential of collaborative learning experiences in encouraging engagement, empowerment, and meaningful learning outcomes for all parties involved. This will be accomplished by exploring the collaboration between teachers and students within the context of artificial intelligence in education.

9.14 Parental Engagement

The engagement of parents is critical to the academic success of their children, and artificial intelligence has the potential to significantly enhance and simplify this process (Riconscente, 2013). It is important to keep the following in mind while discussing the role of grandparents in this chapter.

When a parent uses communication systems that are powered by artificial intelligence, they have the ability to get up-to-the-minute information on their child's academic achievement, behaviour, and attendance details. These media make it simple for teachers and parents to maintain communication with one another, which in turn supports a collaborative effort to assist pupils in their educational pursuits (Karsenti & Bugmann, 2015).

9.14.1 Tailored Analysis

With the assistance of AI analytics technology, parents may be able to get tailor-made insights that demonstrate the areas in which their child excels, the areas in which they want improvement, and the learning styles that they like. Parents now have the ability to better steer their child's school selections and give personalised help at home due to the additional information they have acquired.

9.14.2 The Relationship between the Parents and the Children

Artificial intelligence-driven learning systems have the potential to facilitate the connection between parents and their children's schools by providing them with access to educational resources, activities, and curricular materials. This link may be used by parents in order to engage in meaningful conversations with their children about the concepts that they have learnt at school and to reinforce these concepts at home (Luckin et al., 2012).

9.14.3 Some Ways in Which Parents Can Assist Their Children in Learning

The use of artificial intelligence may increase parental involvement in their children's education by suggesting activities that are associated with school activities. These activities may include interactive games, educational videos, and hands-on projects. Engagement of this kind contributes to the establishment of a home environment that is favourable to learning and highlights the significance of education among members of the family.

Artificial intelligence systems have the potential to collect feedback from parents on their child's educational path, personal preferences, and any particular needs that may be necessary. As a result of this feedback loop, educators are able to swiftly react to concerns and queries raised by parents by modifying classes and developing personalised learning plans.

9.14.4 The Availability of Resources and Assistance Virtual Assistants

They are powered by artificial intelligence and have the potential to link parents to support groups, parenting services, and educational resources (Lester et al., 1997). The purpose of these virtual assistants is to aid parents in the social-emotional and intellectual development of their children by answering questions, providing advice, and connecting mothers and fathers to various resources.

9.14.5 Parental Empowerment

All of the resources that parents need to be active participants in their child's education are made available to them with the assistance of artificial intelligence technology. This, in turn, enables parents to provide more effective assistance to their kid throughout the learning journey. As a consequence of this empowerment, students develop a greater sense of self-assurance, and the partnership between the instructor, parents, and students becomes more robust.

Language and cultural considerations should be taken into account by designers in order to ensure that artificial intelligence (AI) systems are accessible to all families and can be understood by them. It is feasible to increase parental engagement across a wide range of cultures and backgrounds by using artificial intelligence interfaces that are bilingual, content that is culturally relevant, and communication strategies that are customised (Wood et al., 1976).

9.14.6 Ensuring the Privacy and Security of Information

When it comes to employing parental engagement tools that are powered by artificial intelligence, privacy and data security should be the top priority (Laffey et al., 2006). There should be well-defined norms, transparent and honest processes, and secure means for handling data in order to protect sensitive information and win the trust of parents. This will allow for the protection of sensitive information.

Through the use of artificial intelligence, educators are able to improve their ability to create a favourable learning environment for their students, strengthen the relationship between the home and the school, and encourage more engagement from parents.

9.15 Ethical Considerations

In the context of the educational application, artificial intelligence (AI) plays a very important role. In spite of the fact that artificial intelligence has a great deal of potential for enhancing learning experiences and boosting student performance, the deployment of AI in educational settings must be carried out in a responsible and equitable manner, which necessitates careful evaluation of the many ethical issues that it poses. The important ethical considerations are presented below.

9.15.1 Personal Information and Data Protection

Artificial intelligence systems are dependent on enormous amounts of data, including student performance metrics, personal details, and behavioural data. It is important to protect these types of data and information. Protecting sensitive information from being misused, accessed without authorisation, or exploited is of the highest significance in order to fulfil the goal of maintaining the privacy of students (Kirschner & van Merriënboer, 2013). The establishment of strong data privacy legislation, the encryption of all student data, and the implementation of additional cybersecurity protections are all necessary steps that must be taken by politicians and educators in order to preserve student information and guarantee that students have it.

9.15.2 The Importance of Fairness in Algorithms

Algorithms that are used in artificial intelligence have the potential to exhibit biases that are formed by the data that they are taught. This might potentially result in the persistence of educational gaps and prejudice (Conole, 2010). The training of artificial intelligence systems on datasets that are diverse and representative, the performance of periodic audits to ensure fairness, and the transparent disclosure of their decision-making processes are all essential elements in the process of minimising algorithmic bias. Instructors must be on the watch for biases in artificial intelligence systems and work towards eliminating them in order to ensure that all students have equal opportunities.

9.15.3 Maintain Transparency and Accountability

Artificial intelligence systems whose decision-making and recommendation-making processes are not always straightforward to understand. One of the most important things that can be done to promote accountability and transparency in AI-driven operations is to provide a clear and accessible explanation of the capabilities of the algorithms, the criteria for making decisions, and any biases that may exist (Wenger, 1998). It is imperative that stakeholders be honest about the dangers and constraints posed by artificial intelligence systems in order for them to be able to make well-informed decisions and hold developers and educators accountable.

9.15.4 Inclusion and Equity

Artificial intelligence has the potential to exacerbate existing educational inequalities and disparities if it is not used with caution. As a result of the fact that every student has a unique set of needs, experiences, and abilities, it is essential that interventions powered by AI take this into consideration (Russell & Plati, 2000). Legislators and educators have a responsibility to make it a top priority to guarantee that all students, particularly those who come from disadvantaged circumstances, have equitable access to the tools, resources, and opportunities that are associated with artificial intelligence (AI).

9.15.5 *Humans in Mind*

When designing artificial intelligence systems, it is essential to keep humans in mind throughout the design process. This entails prioritising everyone's requirements, including those of the kids and the instructors (Graesser & McNamara, 2010). It is of the utmost importance to design, develop, and evaluate artificial intelligence technologies in order to guarantee that they satisfy the needs, preferences, and values of stakeholders. Stakeholders consist of individuals such as students, teachers, parents, and members of the community. In order to enhance the outcomes of learning, educators should advocate for artificial intelligence solutions that prioritise the needs of users and offer students agency.

The incorporation of ideas of responsible AI governance into educational processes and the consideration of these ethical issues may lead to improvements in learning experiences, student performance, and the promotion of education that is ethical, egalitarian, and inclusive for all individuals. The transformational potential of artificial intelligence may then be harnessed by stakeholders in this manner. It is necessary to maintain regular communication, collaboration, and vigilance in order to overcome the ethical challenges and ensure that artificial intelligence (AI) is beneficial to children, educators, and society as a whole. This is because technology is always improving (Schnotz & Kürschner, 2007).

9.16 Overcoming Challenges

There is a need for a commitment to effectively addressing implementation hurdles, proactive strategies to integrate artificial intelligence into education, and collaboration among stakeholders. The following are examples of significant obstacles and possible solutions:

9.16.1 *Aversion to Change*

The presence of change aversion is a common barrier that prevents the incorporation of new technology, such as artificial intelligence, into educational settings. There is a possibility that educators are hesitant to make use of new resources because they are concerned about altering the status quo or potentially losing their jobs if they do so (Dede et al., 2013). Teachers need chances for professional development and training to gain competence and self-assurance in utilising AI technologies effectively if we are to overcome resistance to change in the classroom. Administrators and lawmakers should inform educators of the benefits of adopting artificial intelligence, address their concerns, and include them in decision-making processes in order to further encourage buy-in and ownership of research projects involving artificial intelligence.

9.16.2 Educators Must Have Access to Training and Professional Development Opportunities

For AI to be successfully used in the classroom, teachers must have the opportunity to learn how to use AI in their own teaching (Kennedy, 2014). To tackle this issue, schools and governments should fund extensive professional development programmes for teachers that help them integrate AI into their lessons through practical advice, real-world examples, and continuous support. Educators can be better prepared to use AI to its full potential in the classroom if professional development programmes are developed in conjunction with experts in the field, those working in educational technology, and researchers in artificial intelligence.

9.16.3 Educational Equity and Access

To avoid further widening of existing educational gaps, it is crucial to guarantee that all students have equal opportunity to utilise AI tools and resources. Educational and policy leaders should make it a top priority to ensure that underprivileged and disadvantaged communities have fair access to artificial intelligence (AI) resources, including tools, infrastructure, and support services, in order to eliminate access disparities (Reeves & Reeves, 1997). Schools and districts that are short on funds or infrastructure may be the recipients of financial aid, technical support, or even just outreach initiatives. Teachers should also be aware that AI systems could be biased and work to promote inclusive and culturally responsive practices that take into account the unique backgrounds and needs of their students.

9.16.4 Incorporation into Lessons and Coursework

Educators, curriculum designers, and AI specialists must work together to incorporate AI into lessons and coursework after extensive planning that is in line with learning goals. The best way for schools to help students integrate AI into their studies is to create interdisciplinary curricula that cover a wide range of topics and grade levels (Fisher et al., 2014). It is important to encourage educators to create real-world projects that students can work on as part of a class so that they can learn about AI in a practical setting and improve their analytical, problem-solving, and computational abilities. To further promote AI fluency and literacy, educational games, virtual labs, and adaptive learning platforms powered by AI can boost student engagement and learning outcomes.

9.16.5 Addressing Ethical and Legal Considerations

Managing the ethical and legal aspects of artificial intelligence in the classroom calls for careful discussion, well-defined regulations, and continuous supervision. Transparency, accountability, justice, and privacy protection should all be tenets of any ethical frameworks and standards that educational institutions develop for the use of AI. Responsible AI practices should be promoted in the classroom and in administrative decisions, and educators should be informed about the ethical concerns surrounding AI use. Data security, student privacy, and ethical AI governance are all areas that need to be addressed by lawmakers through new laws and regulations (Heidig et al., 2015).

Stakeholders can use AI's transformative potential to improve education, increase educational equity and inclusion, and equip students for digital success by working together to proactively address these challenges (Baker, 2016). To overcome obstacles and realise the potential of AI in education, lawmakers and educators must engage in strategic planning, launch capacity-building programmes, and commit to the responsible and ethical use of AI.

9.17 Future Trends and Possibilities

The incorporation of artificial intelligence (AI) into educational settings presents a significant opportunity to enhance the teaching and learning process, improve the results for students, and define the direction that education will take in the future. The following are some significant trends and potentials:

9.17.1 Personalised Learning Pathways

Artificial intelligence technology will continue to provide personalised learning experiences that are tailored to the exact needs, interests, and learning styles of each individual learner (Hone & El Said, 2016). The use of adaptive learning platforms, intelligent tutoring systems, and content recommendations powered by artificial intelligence will make it possible for teachers to design individualised learning pathways that improve student engagement, understanding, and the ability to remember material and skills.

9.17.2 The Integration of Augmented Reality (AR) and Virtual Reality (VR)

The technologies of AR and VR will play an increasingly major role in immersive learning experiences. These technologies will enable students to explore virtual environments, execute simulations, and engage in hands-on activities in realistic settings. Increasing student engagement, fostering collaboration, and fostering a deeper grasp of challenging concepts across a wide range of subject areas will be accomplished via the use of interactive learning environments and virtual teachers driven by artificial intelligence.

9.17.3 Natural Language Processing (NLP) and Conversational Agents

Students will be able to engage in natural language dialogue with AI-powered tutors, virtual assistants, and educational chatbots thanks to the implementation of natural language processing (NLP) technologies and conversational agents (Mason, 2011). These technologies will make it possible for students to have more interactive and conversational learning experiences. Students will get tailored assistance, guidance, and feedback from these conversational bots in real time, which will result in improved learning outcomes for the students and will encourage self-directed learning experiences.

9.17.4 Data Analytics and Predictive Modelling

The utilisation of data analytics and predictive modelling techniques will make it possible for educators to utilise large-scale data sets in order to recognise patterns, trends, and insights that can be used to guide instructional decision-making, predict student outcomes, and drive continuous improvement in teaching and learning practices (Mayer & Fiorella, 2014). The early diagnosis of children who are at risk, the implementation of tailored intervention strategies, and the provision of targeted support services will be made possible by predictive analytics algorithms, which will maximise student performance and retention.

9.17.5 Lifelong Learning and Professional Development

Learning platforms and adaptive training systems that are powered by artificial intelligence will support lifelong learning and professional development for educators. These platforms will enable personalised learning experiences, competency-based assessments, and just-in-time training opportunities that are tailored to the specific needs and career goals of educators. Mentoring programs, collaborative learning communities, and virtual coaching sessions that are powered by artificial intelligence will encourage educators to continually develop their skills, be creative, and encourage one another to work together.

9.17.6 Ethical Governance of Artificial Intelligence and Responsible Innovation

As the use of artificial intelligence technology becomes more widespread in the educational sector, there will be a growing emphasis on ethical governance of AI, responsible innovation, and transparency in the use of AI. The development of ethical norms, standards, and frameworks for the use of artificial intelligence in education will be a collaborative effort between educators, policymakers, and technology developers (Mok, 2014). This will ensure that AI technologies continue to uphold the ideals of justice, accountability, transparency, and privacy protection.

9.17.7 Global Cooperation and Information Sharing

Platforms that are enabled by artificial intelligence and online learning communities will improve global cooperation and information sharing among educators, researchers, and policymakers. This will be accomplished by transcending geographical boundaries and cultural barriers (Tversky et al., 2002). Through the use of artificial intelligence technology, educators will be able to collaborate on cross-cultural exchanges, collaborative research projects, and virtual conferences, which will enable them to share best practices, co-create innovative ideas, and find answers to common challenges in education.

Stakeholders have the opportunity to harness the transformative potential of artificial intelligence by embracing these future trends and possibilities in AI-driven education (Johnson et al., 2016). This will allow them to construct learning environments that are more inclusive, equitable, and student-centred, therefore preparing students for success in the twenty-first century. There is the opportunity for educators and policymakers to unleash the full potential of artificial intelligence (AI) to improve education and empower learners to succeed in a world that is rapidly changing. This can be accomplished via wise investments, innovative partnerships, and a commitment to ethical and responsible use of AI.

9.18 Conclusion

There is a tremendous amount of promise for artificial intelligence (AI) to revolutionise education since it has the ability to enhance student outcomes and affect educational policy. The employment of cutting-edge artificial intelligence technologies and approaches by educators in today's diverse classrooms may allow for the creation of more engaging classes that cater to the specific requirements of each individual student while also stimulating critical thinking, creativity, and coordination among students.

Learning may be done at your own pace with the assistance of technologies driven by artificial intelligence, such as conversational agents, intelligent teaching systems, and tailored learning platforms. Individualised support, guidance, and feedback are provided to pupils via the use of these technologies. Tools for data analytics, augmented reality experiences, and virtual labs powered by artificial intelligence are all examples of data-driven learning opportunities that provide students the opportunity to study in an immersive and engaging manner while also assisting them in better comprehending complex concepts.

In addition, technologies that are driven by artificial intelligence make it simpler for educators to continue their education throughout their careers. These tools include features such as competency-based assessments, online learning communities in which educators can collaborate with one another to improve their trade, and tailored training programs. By embracing ethical AI governance principles, being responsible innovators, and working together on a global scale, stakeholders have the ability to ensure that artificial intelligence technologies uphold educational values of justice, openness, and equality.

In the future, there is an almost infinite number of ways that artificial intelligence might be used in the subject of education. It is possible that educators and policymakers may harness the revolutionary potential of artificial intelligence to construct learning environments that are more inclusive, equitable, and student-centred in order to better prepare students for success in the context of the digital era. The full potential of artificial intelligence (AI) to revolutionise education and equip learners to thrive in a world that is changing at a rapid pace may be unlocked via strategic investments, collaborative efforts, and a commitment to the ethical and responsible use of AI for educational purposes.

References

Anderson, T., & Whitelock, D. (2010). Deep learning for social inclusion: A pedagogically driven approach. *Australasian Journal of Educational Technology, 26*(6), 729–740.

Baker, R. S. (2016). Stupid tutoring systems, intelligent humans. *International Journal of Artificial Intelligence in Education, 26*(2), 600–614.

Blikstein, P. (2013). *Digital fabrication and 'making' in education: The democratization of invention* (pp. 101–126). FabLabs: Of Machines, Makers, and Inventors.

Clark, R. E., & Mayer, R. E. (2016). *E-learning and the science of instruction: Proven guidelines for consumers and designers of multimedia learning*. John Wiley & Sons.

Conole, G. (2010). *New schemas for mapping pedagogies and technologies*. Ariadne. (62).

Dede, C., Mishra, P., & Voogt, J. (2013). Science, technology, engineering, and mathematics (STEM) education. *Handbook of Research on Educational Communications and Technology, 3*, 151–164.

Fisher, A. V., Godwin, K. E., & Seltman, H. (2014). Visual environment, attention allocation, and learning in young children: When too much of a good thing may be bad. *Psychological Science, 25*(7), 1362–1370.

Graesser, A. C., & McNamara, D. S. (2010). Self-regulated learning in learning environments with pedagogical agents that interact in natural language. *Educational Psychologist, 45*(4), 234–244.

Heidig, S., Müller, J., & Reichelt, M. (2015). Emotional design in multimedia learning: Differentiation on relevant design features and their effects on emotions and learning. *Computers in Human Behavior, 44*, 81–95.

Hone, K. S., & El Said, G. R. (2016). Exploring the factors affecting MOOC retention: A survey study. *Computers & Education, 98*, 157–168.

Johnson, L., Adams Becker, S., Cummins, M., Estrada, V., Freeman, A., & Hall, C. (2016). NMC horizon report: 2016 higher education edition. The New Media Consortium.

Karsenti, T., & Bugmann, J. (2015). Quels sont les facteurs d'efficacité des classes inversées en formation initiale des enseignants? *Formation et profession, 23*(3), 71–74.

Kennedy, G. E. (2014). Strategies for improving learning: Insights from cognitive science for computer tutors. *Educational Psychology Review, 26*(2), 245–269.

Kirschner, P. A., & van Merriënboer, J. J. (2013). Do learners really know best? Urban legends in education. *Educational Psychologist, 48*(3), 169–183.

Laffey, J., Lin, G. Y., Lin, Y. T., & Macchiarella, N. D. (2006). Analysis of online participation in a knowledge-building environment using electronic discourse analysis. *Journal of Educational Computing Research, 34*(2), 187–213.

Lajoie, S. P. (2000). Computers as cognitive tools: No more walls: Theory change, paradigm shift, and their influence on the use of computers for instructional purposes. *Journal of Educational Computing Research, 23*(2), 115–125.

Lester, J. C., Converse, S. A., Kahler, S. E., Barlow, S. T., Stone, B. A., & Bhogal, R. S. (1997). The persona effect: Affective impact of animated pedagogical agents. In Proceedings of the SIGCHI conference on human factors in computing systems (pp. 359–366).

Luckin, R., Bligh, B., Manches, A., Ainsworth, S., Crook, C., & Noss, R. (2012). *Decoding learning: The proof, promise and potential of digital education*. Nesta.

Mason, R. (2011). Learning technologies for adult continuing education. *Studies in Continuing Education, 33*(1), 11–24.

Mayer, R. E., & Fiorella, L. (2014). Principles for reducing extraneous processing in multimedia learning: Coherence, signaling, redundancy, spatial contiguity, and temporal contiguity principles. The Cambridge handbook of multimedia learning, 279–315.

Means, B., Toyama, Y., Murphy, R., Bakia, M., & Jones, K. (2010). *Evaluation of evidence-based practices in online learning: A meta-analysis and review of online learning studies*. US Department of Education.

Mok, H. N. (2014). Teaching tip: The flipped classroom. *Journal of Information Systems Education, 25*(1), 7–11.
O'Donnell, A. M., & Dansereau, D. F. (1992). *Scripted cooperation in student dyads: A method for analyzing and enhancing academic learning and performance*. Springer.
Paas, F., Renkl, A., & Sweller, J. (2003). Cognitive load theory and instructional design: Recent developments. *Educational Psychologist, 38*(1), 1–4.
Park, S., & Choi, H. J. (2009). Factors influencing adult learners' decision to drop out or persist in online learning. *Educational Technology & Society, 12*(4), 207–217.
Plass, J. L., Homer, B. D., & Kinzer, C. K. (2015). Foundations of game-based learning. *Educational Psychologist, 50*(4), 258–283.
Reeves, T. C., & Reeves, P. M. (1997). Effective dimensions of interactive learning on the World Wide Web. In *Web-based instruction* (pp. 59–66). Educational Technology Publications.
Riconscente, M. M. (2013). Effects of game-based learning on students' mathematics achievement: A meta-analysis. *International Journal of Educational Research, 58*, 79–93.
Russell, J. E., & Plati, T. (2000). The influence of web-site usability factors on students' perception of educational web sites. *Journal of Educational Computing Research, 23*(2), 139–162.
Sangrá, A., Vlachopoulos, D., & Cabrera, N. (2012). Building an inclusive definition of e-learning: An approach to the conceptual framework. *International Review of Research in Open and Distance Learning, 13*(2), 145–159.
Schnotz, W., & Kürschner, C. (2007). A reconsideration of cognitive load theory. *Educational Psychology Review, 19*(4), 469–508.
Singh, K. K., & Singh, A. (2016). Detection of 2011 Sikkim earthquake-induced landslides using neuro-fuzzy classifier and digital elevation model. *Natural Hazards, 83*, 1027–1044.
Singh, A., & Singh, K. K. (2023a). YORES: An Ensemble YOLO and Resnet Network for vehicle detection and classification.
Singh, A., & Singh, K. K. (2023b). FedDDR: A federated improved DenseNet for classification of diabetic retinopathy. Proceedings http://ceur-ws.org ISSN, 1613, 0073.
Singh, K. K., Mehrotra, A., Nigam, M. J., & Pal, K. (2013, April). Unsupervised change detection from remote sensing images using hybrid genetic FCM. In 2013 Students Conference on Engineering and Systems (SCES) (pp. 1–5). IEEE.
Singh, K. K., Rho, S., Singh, A., & Sergei, C. (2024a). Big data analytics and knowledge discovery for urban computing and intelligence. *Complex & Intelligent Systems, 10*(1), 1–2.
Singh, A., Dhull, A., & Singh, K. K. (Eds.). (2024b). *Blockchain and deep learning for smart healthcare*. John Wiley & Sons.
Sitzmann, T., Kraiger, K., Stewart, D., & Wisher, R. (2006). The comparative effectiveness of web-based and classroom instruction: A meta-analysis. *Personnel Psychology, 59*(3), 623–664.
Squire, K., & Jan, M. (2007). Mad city mystery: Developing scientific argumentation skills with a place-based augmented reality game on handheld computers. *Journal of Science Education and Technology, 16*(1), 5–29.
Sweller, J., Van Merrienboer, J. J., & Paas, F. G. (1998). Cognitive architecture and instructional design. *Educational Psychology Review, 10*(3), 251–296.
Tversky, B., Morrison, J. B., & Betrancourt, M. (2002). Animation: Can it facilitate? *International Journal of Human-Computer Studies, 57*(4), 247–262.
Vygotsky, L. S. (1978). *Mind in society: The development of higher psychological processes*. Harvard University Press.
Wenger, E. (1998). *Communities of practice: Learning, meaning, and identity*. Cambridge University Press.
White, J., & Walmsley, S. (2006). *Inclusion: Changing practice*. Routledge.
Wood, D., Bruner, J. S., & Ross, G. (1976). The role of tutoring in problem solving. *Journal of Child Psychology and Psychiatry, 17*(2), 89–100.
Yukselturk, E., & Bulut, S. (2009). Gender differences in self-regulated online learning environment. *Educational Technology & Society, 12*(3), 12–22.

Chapter 10
ChatGPT in Academia and Research: A Comprehensive Review of Integrating AI in Higher Education

Aashka Thakkar, Andinet Asmelash Fentaw, and Habtamu Ditta Hirpo

Abstract The ChatGPT, which is also known as generative pre-trained transformer, is a technique that was developed specifically to generate information in a manner similar to conversation based on data inputs. This is designed to resemble natural language and can be used for a variety of purposes, including chatbots to communicate with virtual assistants and language translator systems. Thanks to its intensive training on large-scale text datasets and intelligent computational intelligence techniques, ChatGPT is able to generate remarkably sophisticated responses based on user-provided data. The study examines the potential benefits, challenges, and ethical dilemmas associated with integrating ChatGPT into teaching methodologies. It investigates the ways in which ChatGPT impacts levels of learners of engagement, creative thinking, and problem-solving abilities. Additionally, it investigates the ways it could foster imaginative thinking and provide chances for tailored instruction. The aim of this article is to take an extensive examination of these challenges as well as examine ChatGPT's potential uses in educational settings. Ambiguities about security and the copyright predispositions in content created by artificial intelligence are additionally addressed in the paper, as well as the importance of finding an appropriate equilibrium amongst the use of artificial intelligence (AI) and traditional teaching methods.

Keywords GPT · ChatGPT · Chatbots · AI

A. Thakkar (✉)
Faculty of Management Studies, Parul University, Waghodia, Vadodara, India
e-mail: aashka.thakkar@paruluniversity.ac.in

A. A. Fentaw
Accounting and Finance Department, New Global Vision College, Addis Ababa, Ethiopia

H. D. Hirpo
Advisory Department, Berhan Bank SC, North Addis District, Ethiopia

A. Singh, K. K. Singh (eds.), *Multimodal Generative AI*,
https://doi.org/10.1007/978-981-96-2355-6_10

10.1 Introduction

The rapid technological developments in the area of artificial neural networks have led to the establishment of progressively more advanced AI-driven educational resources. The aforementioned tools may provide individualised education, adapt to the particular needs of every learner, and can also grade exercises. Artificial intelligence (AI) has become prevalent in educational environments nowadays in a broader spectrum, which covers focused individuals training, automated methods for handling admin duties, teaching, learning, tutoring, and mentoring, and beyond. A growing array of colleges and universities have begun to implement hybrid learning strategies—which integrate both conventional and technological instruction—into practice in an effort to increase students' academic attainment and performance. The rising popularity of distant and online education is driving this upward trajectory. In this case, helping learners finish their homework through the use of AI tools like ChatGPT ends up being a practical and beneficial choice. However, improper use of such instruments could lead to moral dilemmas and intellectual deception. Higher education institutions, as well as other educational bodies, must establish rules that regulate the way students behave when using those instruments to acquire knowledge in order to uphold academic standards of integrity and morality. As the technology evolves, we should anticipate seeing more innovative uses of AI for educational purposes in the course of time thereafter.

OpenAI has also introduced the GPT-4, which is the latest and most recent large-scale multidimensional model language programmes, in tune with the ongoing advancement and replication of artificial technological devices. In addition to accepting language and visual input, the GPT-4 version offers significant improvements and advances in the chatbots over the GPT-3.5 in terms of thinking and response, comprehending complicated problems, critical analysis, and generating ideas and solutions to problems. Simultaneously, GPT-4 has made advances in the recognition of images, putting in text constraints, response precision, and numerous other areas. GPT-4 can process instructions with greater detail, produce a wider variety of imaginative writings, and operate with greater dependability and originality. ChatGPT is regarded as an insightful instrument in higher education, reinforcing the significance of retaining interpersonal interaction and logical thinking abilities in the course of education. ChatGPT extends beyond the constraints of current indexation and extraction by precisely comprehending the context and motive of enquiries, offering organised and cohesive human-like observations, and adapting solutions to reflect feedback from users. This sets it apart from traditional Google and other search engines and smart chatbots that only provide automatic answers built on keywords searched. ChatGPT fulfilled 92.5% of each assignment of the mind exam. Instructors will need to follow along, picking up knowledge while they go and reiterating previously acquired skills like ethical information usage, critical thinking, and the verification and examination of references. ChatGPT can be used to create anything, from covering letters to brief overviews of widely recognised works of fiction. There could be significant disruptions to the educational system, and educational institutions might have to pay more. The speed at which AI has been used recently has shocked multiple sectors, including education. The popularity of ChatGPT has been unbelievable, which can be analysed from the following data.

Duration of online platforms to reach one million users

Online platform	Year of launch	No. of years/months/days
Netflix	1999	3.5 years
Kickstarter	2009	2.5 years
Airbnb	2008	2.5 years
Twitter	2006	2 years
Foursquare	2009	13 months
Facebook	2004	10 months
Dropbox	2008	7 months
Spotify	2008	5 months
Instagram	2010	2.5 months
ChatGPT	**2022**	**5 days**

Source: Statista

OpenAI's hit chat programme with computational intelligence, ChatGPT, has attracted one million users in just a period of 5 days. In merely 5 days, the OpenAI-developed chatbots platform garnered nearly a million subscribers, sparking intense worldwide fascination and conversation. A global innovation fever was triggered by the publication of ChatGPT, leading organisations in the technology and online sectors, together with conventional and real enterprises, to be drawn into the ranks of varied software products being created, driven by ChatGPT. ChatGPT, which was introduced in November 2022, accomplished 100 million users in just 2 months, which shows the popularity and usage of the platform among the users. Reaching a million customers took a span of 4 years for Facebook, 2 years for Instagram, and nearly 1 year for Google. The growing popularity of ChatGPT has led to a divisive discussion on the pros and cons of the programme. Some businesses actively welcomed it, but several others—especially educational institutions—have persistently avoided it out of fear that the students might copy content. Despite the fact that it is necessary to have an understanding of the various issues at hand for the purpose of ChatGPT to function to the fullest extent possible. It's critical to acknowledge ChatGPT's issues, especially for educators.

ChatGPT is poised to become an excellent instructor and learner together. Additionally, just as scientists did using math calculators in times gone by, educators as well as learners may be able to expand their capabilities as well as possibilities with the use of technology like artificial intelligence (AI). Like a logical educational staff member, chatbots powered by artificial intelligence can be employed to give students swift responses to their most frequently asked questions. Individuals who continued their studies outside of and after the lecture may find this support useful. Artificial intelligence-powered instructional assistants may improve search and provide personalised recommendations on content and other instructional materials.

This article aims to tackle the biggest problem facing the world of education, which is "Can be ChatGPT a useful tool to feed education?" But these days, we have to know the manner in which to make the most of technology judiciously. The potential benefits of ChatGPT and GPT-4 over other linguistic systems such as BERT, RoBERTa, and XLNet tend to be significant (Liu et al., 2022). GPT-4's scalability serves as one of its main advantages; with trillions of characteristics and quick access to enormous volumes of data, it's capable of accurately accomplishing a broad spectrum of language-related tasks. Additionally, the ChatGPT applications can be configured for a range of activities, including questions/answers, resolving problems, communicating with and synthesising them, as well as translators. This makes them highly versatile. It also uses data efficiently and generates high-quality results with relatively little preparation abilities. Additionally, ChatGPT might generate language that is close to natural language, making it challenging to discern between proofs the evidence has been manufactured and its original source. ChatGPT is a powerful AI tool that has the potential to completely transform the educational landscape. Its ability to perform knowledge-driven and cognitively engaging tasks, such as grading assignments and student counselling, has the potential to drastically alter how education is provided. Considering today's environment, integrated instruction is required to create smaller groups that provide proper interpersonal distance in addition to lowering physiological attendance, However, it is also to go back to frontal instruction as a primary instructional method, with technological advances now further enhancing the educational process (Thakkar 2023).

When given suggestions, LLMs have shown an astonishing ability—and occasionally an inability—to generate language. Certain LLMs can leverage current insight by poring over the published material and recommending certain enquiries or perspectives on a particular subject or inquiry. The potential applications of LLM will increase with the new GPT-4's ability to extract information from images, particularly in the fields of higher education, the medical field, and scientific investigation, wherein visual aids are essential for generating or improving knowledge. For ChatGPT operational management and digital empowerment in educational organisations, the present investigation performed a SWOT evaluation. The outcomes of the SWOT analysis are used to identify a number of concerns that will have an impact on ChatGPT's numerous stakeholders in education. Furthermore, a series of suggestions are provided for education sector practitioners who plan to utilise this technology. The use of AI for educational purposes must strike a balance with maintaining the human element and interpersonal interactions, which are crucial to the transfer of information. Just within a couple of years, ChatGPT, a freely accessible platform constructed around Horizon 2020 gadget's innovations, is quickly becoming well-known. The emergence of ChatGPT, as well as additional AI-powered technologies, carries tremendous responsibility along with its immense potential. Given this, moral concerns need to be taken into account. Algorithmic linguistics: Emily Bender has pointed out that while ChatGPT's extensive internet dataset provides a multitude of viewpoints and views, diversification is not always ensured by scale (Bender et al., 2021). The open-source programme software produces a somewhat accurate imitation of human-written content and is becoming ubiquitous in various fields, including optimising websites for search engines and binding contracts in written form. It is crucial to thoroughly investigate the technology in order to ascertain the expected and unforeseen rewards and implications when incorporating emerging AI tools into teaching methods or altering pedagogical practices on the basis of AI tools (Heath et al., 2022). In the portion of the article that follows, we go over possible applications, dangers, abuses, and prospects for ChatGPT as a means of assisting administrators in thinking intelligently about what kind of effects this device may have on the classroom. The paper examines the published literature with the intent of covering ChatGPT studies along with examining their methodologies. The study to critically explore how artificial intelligence (AI) systems influence academic outcomes in diversified learning environments. Based on an extensive evaluation of a multitude of study findings, the present research investigates whether or not fueled by artificial intelligence initiatives could strengthen academic performance, instructional adaptability, and involvement among students (Thakkar, 2024)

The research has investigated some of the following enquiries into research:

- In what manner are ChatGPT studies conducted in educational writing?
- How might ChatGPT be used in educational writing?
- What restrictions and difficulties does ChatGPT face according to educational writings?

10.2 Literature Review

ChatGPT is a sophisticated chatbot that is capable of processing text much like an individual's brain. While individuals are hopeful regarding its potential application, others are somewhat fearful. Kasneci et al. (2023) outlined the dangers and difficulties of utilising LLMs in teaching, emphasising both operational and ethical issues. While Qadir (2022) pointed out potential biases and ethical considerations, Tlili et al. (2023) expressed concerns regarding ChatGPT's accuracy, equitable treatment, and confidentiality of user data. Plagiarism and scamming are further issues. Computing programmes known as chatbots converse along with users in real time (Clarizia et al., 2018). They engage in a variety of aspects of higher education, including motivating, useful for learning, peer, and instructional drivers (Kuhail et al., 2022). Chatbots have the ability to combine data obtained from several sources, respond to enquiries from students right away, as well as inspire those (Okonkwo & Ade-Ibijola, 2021). Nonetheless, critiques include the absence of user-centred architecture, fictitious dialogues, and moral dilemmas (Kuhail et al., 2022; Murtarelli et al., 2023). Due to the significant shifts that the use of artificial intelligence (AI) is bringing about in the workforce, which is one of the educational primary purposes, questions concerning what and how best to educate future generations have been brought up. These concerns highlight the need for future citizens to have an education that is going to equip their children with the knowledge and abilities needed to survive in an ecosystem that is evolving rapidly (Zhai, 2021). Approximately 89% of the United States students in college use ChatGPT to finish their assignments, with 53% of them employing the application to write essays, according to a latest poll. Furthermore, according to McGee (2023), 22% of students utilise ChatGPT to create assignment summaries, and 48% of participants apply it while assessments. Not only does ChatGPT provide an additional comfort, but it can also produce data. AI has the potential to be a useful tool for both teachers and students when applied properly. The GPT can assist educators in developing individualised lesson plans that consider each student's distinct SWOC preferences in order to give participants a better-focused and successful educational journey (Ausat et al., 2023). The customized Region-based neural network with convolution was created as a machine learning method for analyzing footage for vehicle recognition. Video surveillance cameras placed on the roadways record traffic footage, which is then examined to identify the vehicle in a certain shot (Singh et al. 2021; Singh et al. 2024a, Singh et al. 2024b)

Even though some research revealed that ChatGPT could potentially be inspiring and beneficial for learning, additional investigations raised issues about how it might affect equal access to education and academic credibility (Yan, 2023; Shoufan, 2023). The large number of publications about ChatGPT that have been published often examine its advantages in-depth and caution against any possible drawbacks. However, we discover that in order to fully comprehend ChatGPT's actual capabilities and execution, its study circumstances must be revisited. Current reviews examine ChatGPT algorithms' attributes (Roumeliotis & Tselikas, 2023). Halaweh (2023) argues in favour of the employing of ChatGPT in the classroom and predicts that soon it will be necessary to employ techniques that facilitate the use of ChatGPT in the classroom in an enlightened and moral way. According to Karaali (2023), analytical skills and reasoning are not as significant as the fundamental ability to read and write. Thus, students' usage of AI technologies like ChatGPT shouldn't discourage them from developing these kinds of literacy skills. As per Ali et al. (2023), certain students stated that ChatGPT enhanced their abilities in reading and writing, whereas it did not improve in speaking or paying attention. Thurzo et al. (2023) investigate how AI is used in the field of dentistry education, but they conclude that AI cannot fully comprehend the human body's structure. AI seems to be able to learn everything by heart; nevertheless, it doesn't seem to have the identical sense of purpose. In a study intended to investigate if software that detects plagiarism was capable of recognising essays created with ChatGPT, Khalil and Er (2023) discovered that 40 of the 50 articles examined had a score for similarity of 20% or lower, indicating a high level of uniqueness. As a result, ChatGPT can design lessons with their analytical thinking at the forefront by offering questions or suggestions that will help them develop their higher-level analytical skills.

A topic of discussion might revolve around whether teachers ought to persist in taking on some traditional responsibilities like grading assignments on coherence and formatting given the significant shift that AI programmes like ChatGPT have brought to the educational landscape (Emenike & Emenike, 2023). Trying to figure out where to draw the divide regarding using these programmes properly and misusing them appears to be the main source of conflict here. Some people completely reject the idea of using these programmes as a remedy, while other people are more upbeat and advocate for their careful application.

10.3 ChatGPT in Education

The opinions of educators concerning ChatGPT have been divided. Some teachers are concerned as students have been employing the latest technologies to take shortcuts on reports, tests, and projects. Some who take a more liberal perspective on it believe it has the power to completely transform education by facilitating kids in developing as intellectuals, writers, and educators. Prior to the invention of calculators, the solutions were usually all that remained. Nevertheless, as computations were available, it evolved into crucial to show how the issue had been solved. To guarantee that their educational experience is successful and valuable, it is imperative that educators educate students on ways to use these devices and applications appropriately. It is preferable to incorporate these artificial intelligence (AI) devices into the educational system so that students can learn and utilise them in a proper and accountable manner rather than forbidding them from utilising them in order to safeguard their time and energy. To ensure that learners use these resources appropriately and responsibly, it is crucial to strike a balance between their autonomy as well as the demands of integrity in education. (Villaseñor, 2023). Researchers have suggested that something similar might occur with educational papers, whereby students are currently assessed on how efficiently they amend as well as enhance a text produced by artificial intelligence (AI) in addition to everything they express. These tasks encompass text creation, reiterating paraphrasing, computerised translation, as well as resolving questions. In the appropriate setting, interactive artificial intelligence (AI) tools like chatbots as well as AI assistants may additionally be effectively employed.

- Intelligent computing (AI) has the potential to revolutionise academia by providing students with personalised educational opportunities. The process of customising educational materials and knowledge via technological devices to each student's unique needs, interests, and talents is referred to as 'personalised education'. Programmes for education using computational intelligence (AI) capabilities can evaluate a student's progress and adjust the informational level in an instant to make sure the learner is receiving enough knowledge. Personalised education is the technique of using automated technology to change the content and degree of complexity of education based on a learner's performance. The goal of this approach is to provide students with customised training so they can understand material more quickly and effectively.

- Specifically tailored educational tool suggestions guided by artificial intelligence could be generated by looking at student's preferences and desired study pattern. Personalised solutions can be used in educational settings to help students track down new materials or assignments that are tailored to their specific interests and needs. These suggestions could be based on the student's prior learning outcomes, chosen teaching style, or extra factors like their interests or ambitions. ChatGPT can be integrated into platforms to provide interactive learning environments where participants can engage in discussions to strengthen their understanding of a variety of disciplines.
- Additionally, AI can pinpoint academic areas in which learners may be falling short and offer specialised assistance in order to assist them in catching up. It can finish every one of the data entry tasks in a matter of seconds, when professors would often need many days to accomplish them. As a result, educators have less homework to complete and more time to work directly with their pupils.
- ChatGPT has the potential to be incorporated into shared-learning scenarios to encourage peer communications and conversations.
- The GPT can create business simulations built around real-world occurrences that are needed for my beginning microeconomics course. Offer remedies for the problem of the mismatch between supply and want. It can include scenarios that are applicable to the participants' day-to-day experiences. Write down the educational goals for the course of study and make sure they are the first thing you look at in each instance.
- ChatGPT can be used to enhance language proficiency by engaging in dialogues with learners. It is capable of simulating real-life conversations, aiding learners in building fluency and boosting confidence in a language.
- Serving as a virtual teaching assistant ChatGPT can manage tasks, address common queries, and provide details on schedules or course materials.
- ChatGPT offers feedback on assignments, essays, or coding tasks, aiding students in recognising errors and gaining insights from them. It can also streamline the grading process for educators, saving them time.
- ChatGPT might prove particularly advantageous in an area such as economic analysis, wherein students must work using datasets throughout the course of study. It makes it possible to spend less time on time-consuming chores like transferring data and debugging, which may frustrate students. It also helps the students gain a greater comprehension of the subject matter while permitting a professor to concentrate even more on the in-depth examination of scientific models.
- ML (machine learning) and automation software have made various administrative tasks in the field of education less complicated and more automated as well. Alongside storing and processing the information, creating schedules, taking attendance, grading assignments, and taking care of scholarships, these duties may encompass administrative responsibilities for teachers very easily. The goal of automating the administrative procedures in universities and schools aims to boost accuracy and effectiveness while saving time and resources for other vital tasks like teaching, learning, and mentoring. By simplifying and reducing the

burden of administrative duties, teachers and administrators may focus on the primary objective of teaching, which is to provide students with exceptional learning experiences.

- ChatGPT's accessibility at all points in time ensures the learners worldwide can consistently receive support and assistance for their educational endeavours. Students have the ability to use ChatGPT whenever it seems most beneficial to them, regardless of constraints on time. Time and again the system's adaptability to a variety of learning styles will make it a perfect educational buddy for pupils (Rahman and Watanobe 2023 and Rai et al. 2020). Additionally, its permanent accessibility guarantees students can get assistance when they stumble across difficulties. This promotes education that takes place outside of the educational setting. When deploying these kinds of tools, it's important to pay close attention to issues like any possible prejudices, confidentiality of data, and the ethical application of AI in education. Teachers can use these tools to build competitive games, evaluations, and a multitude of other imaginative endeavours that will help them to further enhance their conventional methods of instruction.
- The research project approach may have been conceptualised with the use of ChatGPT. Scholars are able to chat about their goals, the methodology of the study, gathering information strategies, along with information processing methodologies by interacting using ChatGPT. Because ChatGPT can interpret natural speech, it can offer recommendations and analysis depending on its extensive expertise as well as research methodology expertise. While ChatGPT can help spark recommendations and offer direction, it is crucial for investigators to assess the recommendations seriously and modify them to suit their unique requirements and desired outcomes.
- Teachers will also gain a deeper understanding of the technology and its possible applications in the classroom. Administrators wishing to incorporate technological advances into their educational programmes so they can make them more fun and engaging for pupils, regardless of their level of education or subject area, may find ChatGPT to be helpful. Prior to using ChatGPT in the educational setting, educators ought to ensure they completely comprehend the principles of the software ChatGPT as a platform for creating and revising texts. Incorporating ChatGPT into the curriculum fosters the ability to think critically about the material covered during instruction as well as technology literacy, which is important in today's constantly tech-dependent society. Despite the fact that successful writing for academic purposes is crucial to the integrity and performance of academic publications, it still remains one of the biggest impediments that worldwide postdocs and students have to conquer.
- Investigators of all flecks, but especially non-native English speakers, can benefit from using LMMs as a publishing and editing tool to improve the calibre of scholarly work. With a period of time, software-based editing instruments have seen tremendous developments. Modern software-based editing instruments are full of capabilities to fix grammatical errors while improving textual lucidity, ranging from simple spellings checks carried out by text editors to paid internet

services like Grammarly, Scribbr, and QuillBot. Nevertheless, these instruments generally possess a predetermined collection of assessments of the written sample and generate an analysis in accordance with these assessments. Teachers can utilise ChatGPT to help them study for examinations. It can offer worksheets, quizzes, and practice enquiries so that participants can assess their understanding and pinpoint regions on which they must prioritise their study efforts. Better grades and improved performance in exams are possible outcomes of this customised methodology. It may assist students in writing better and conveying their thoughts more clearly, which can facilitate their success with assignments like writing.

10.4 Challenges of Using ChatGPT in Education and Research

Considering the fact that ChatGPT can significantly impact the education of learners, using it in educational settings needs to be done in a responsible and ethical manner. In order to do this, ongoing research and observation are required. However, there are a number of ethical and sociological concerns associated with using AI in education. For example, it might perpetuate ingrained bias and intolerance, violate individuals' private rights by shrivelling them more frequently, erode autonomy for students, and discriminate against learners that are consistently treated unfairly in the classroom. Although ChatGPT appears to be able to write almost anything, it doesn't come without stumbling blocks. Since ChatGPT was developed using data collected from the Internet before 2021 and fails to have any access to the Internet, its algorithms are unable to generate trustworthy responses on any information or events created after 2021. There are a lot of constraints using ChatGPT for education and research. The mechanism can produce information that can be technically incorrect but conceptually precise, which is actually the biggest challenge of using AI in education. Additionally, ChatGPT has a propensity to provide prejudiced findings, much like every AI programme (Fuchs, 2022). The complicated structure and intricacies of natural speech may be difficult for the technology to comprehend, and this could result in errors of judgement and inaccurate responses. Additionally, the calibre and variety of data for training, which were put into creating the templates, might serve as a cause of errors. It is also unable to access the Internet or forecast upcoming occurrences (yet). It also only has the ability to respond via text. For instance, it can create a radio show script but not an audio recording for the programme. Moreover, users are unable to upload video with ChatGPT to evaluate. The primary threats associated with ChatGPT for educational purposes are as follows: an over-reliance on technology; biased or untrustworthy data sources; ethical concerns about copied content; privacy issues; the improper use of information; lack of face-to-face contact; fading optimism for the acquisition of new knowledge; and lack of ability to complete certain assignments due to unresolved technical issues, for instance, server crashes and malfunctions; and safety concerns.

Assignments requiring logical thinking, specific expertise, or up-to-date data might not be appropriate using ChatGPT. It appears from the past research that ChatGPT has difficulty with complicated mathematical operations or algebraic operations, suggesting that these models do not fully understand logical reasoning. Furthermore, it is not possible to update such massive linguistic models in the identical manner as we upgrade knowledge bases—that is, by just adding or changing entities. Since the majority of the instructional text in the present model comes from data that was made freely accessible prior to 2021, these models are unable to produce correct results quickly. Compared to the search giant's Google, ChatGPT creates a more extended description featuring visual illustrations and prevalent dangers. Yet again, ChatGPT's invasion of educational facilities across the entire globe astonished them. The context rather than the content is extremely important in research investigation. In the absence of it, it is impossible to envision artificial intelligence (AI), or any other technology, offering an insightful response to the question. Providing the background information is the first step when applying a chatbot to assist you, editing an aspect of your work for lucidity. What exactly is the subject of your research project, and precisely what is the major point you make?

For instance, the following threats are mentioned in relation to implementations for higher education and general investigations as well as instructional assistance:

- Due to too much use of ChatGPT, scholars who are investigators, teachers, and students might lose their creative powers along with the absence of diversification in the studies they undertake. If learners see that ChatGPT is doing their work for them, they may lose interest in studying, and there will be a loss of creativity and critical thinking due to overreliance on chatbots.
- The remarks that ChatGPT makes can be a problem, and its training data could accidentally encourage biases or preconceptions. This can lead participants astray and produce misleading recommendations or one-sided information, which also reinforces existing prejudices in research.
- However, while ChatGPT can assist and provide feedback, technology cannot replace individualised instruction from a classroom teacher or coach, who is required for children's social-emotional development.
- The fact that it might be challenging to comprehend exactly they ways ChatGPT comes along with the replies and generates remedies for the issues, it can be challenging to look at the findings drawn by the system or determine whether they are incorrect or defective. Despite ChatGPT's wide range of capabilities, its relevancy in specific research scenarios remains limited due to the fact that it cannot always be perceived as adequately equipped enough to navigate more complex commitments or understand how particular tasks are completed.
- One of the most common problems with automated technologies is the technical glitch or issue. The equivalent technological developments, like ChatGPT, may experience troubles that might involve glitches, network outages, or incompati-

bilities with specific software or the datasets. All of these issues could have a significant impact on research, its outcomes, and the teaching-learning process.
- The boundaries and constraints within which ChatGPT operates that may give rise to moral and legal issues when employed in scholarly research, encompassing matters like data ownership, privacy, and security consequences. Moreover, researchers possess the risk of taking advantage of ChatGPT by putting literature into it without first properly citing it, which could result in the act of plagiarism.
- There is a greater chance of hacking attacks and an alarming threat to customer safety when sensitive data, such as test scores, grades, and other protected details, are stored on ChatGPT.
- Minimal disclosure of data and difficulty managing intricate research work can be the biggest challenges for the researchers. While figuring out ChatGPT's methodology could turn out complicated, confirming its conclusions or detecting inaccuracies could present a challenge. While ChatGPT is capable of handling a wide range of duties, its applicability in some inquiry circumstances is limited because it cannot be considered to be able to cope with more sophisticated activities or comprehend methods to carry out particular duties.
- For artificial intelligence to be effective in learning environments, educators must be ready to adopt and apply technology. However, a few educators may be sceptical about adopting AI, either because they are unfamiliar with the technology or because they are worried concerning how it can impact their responsibilities and also dilute their role as teacher. There are several challenges to overcome when implementing artificial intelligence (AI) in the classroom to gain an edge. It becomes crucial for investigators to remember that ChatGPT is only a tool and that, therefore, other recommendations and guidelines need to be examined, combined with their own interpretations and suggestions for the problem, rather than using it exclusively. It's also important to keep in mind that ChatGPT conclusions should always be carefully evaluated prior to being used in any kind of academic research investigation.

10.5 Artificial Intelligence: The Double-Edged Sword

Despite the fact that artificial intelligence (AI) innovations are here to stay, these are still being developed to benefit people. Consequently, it is preferable to utilise computational intelligence instead of letting it take command in order to combat the worry that it would eventually replace people. With technologies developing at a rapid pace, worries about artificial intelligence's potential implications on education and the long-term prospects of academia have grown. Experts argue that relying excessively on artificial intelligence (AI), for example in ChatGPT, could result in a future generation of children whose minds are reluctant to think critically or independently. ChatGPT's Effect on Student: An in-depth investigation was carried out by the British Institute in Egypt to investigate the contributions made by its undergraduates. Turnitin was implemented to find that articles included a stupendous quantity of artificial intelligence (AI)-created information. Investigations were conducted in nine departments, where a minimum of a single course was checked for plagiarism as well as usage of artificial intelligence (AI). Ayman et al. (2023) considering a certain group of worries that AI programmes like ChatGPT might result in students who depend heavily on technology, empirical evidence suggests that AI could serve as a helpful tool in the educational environment. Given that educators have to continue to stress on critical thinking and human interaction. AI has the potential to enrich and enhance the whole student's educational endeavours. Keeping in mind copyright issues becomes crucial while using ChatGPT or any other translation framework. Original pieces of writing, including plays, songs, novels, and artwork, are protected against unauthorised use sans the owner's permission by copyright law. As such, using any material that is copyrighted in ChatGPT exchanges might infringe on the intellectual property holder's rights of infringement. It is imperative to acknowledge that the results generated by ChatGPT do not inherently confer freedom of usage. These published findings could be protected by copyright issues, just like various other kinds of information. Because of this, getting prior consent from the owner of the intellectual property rights may be required before using the information obtained in any way (Kocoń et al. 2023; Hill-Yardin et al., 2023). This is crucial to remember that even with such an effective AI device, there are drawbacks. Among them is the possibility of compromised academic standing, skewed assessments of acquiring knowledge, factual errors, and an excessive dependence on AI that may impede the acquisition of life skills that are essential. Therefore, in order to enable the successful application of an artificial intelligence (AI) device for learning and research, these challenges need to be rectified. Given the dangers associated with using ChatGPT, a few suggestions have been placed in order. In the beginning, despite the fact that ChatGPT has been restricted by certain educational organisations, studies and instruction should nevertheless take into account this cutting-edge AI technology. To make the most of ChatGPT, researchers, educators, and students deserve to be urged to experiment with its application. Nonetheless, great care must be taken to ensure that ChatGPT is used in an open, fair, visible, and moral manner. Rahman et al. (2023) Al Ahmed and Sharo (2023) examine ChatGPT's prospective benefits and risks regarding education as a whole from the viewpoints of teachers and students. Additionally, researchers examine the way ChatGPT assists learners in honing their technical abilities. They used ChatGPT to perform several coding-related tests, such as generating codes from problem narratives, code rectification, and code snippets extraction of programmes from writings.

Zhai's (2022) research proposes the suggestion that educational objectives should be modified. As opposed to emphasising generic skills, higher education should concentrate on developing students' ability to think critically and imaginatively. Participants ought to become able to employ artificial intelligence (AI) tools to complete activities related to their subject matter. Investigators should create AI-involved educational endeavours that engage participants in solving real-world issues in order to meet the learning objectives. Concerns about individuals outsourcing assessment work are also raised by ChatGPT (Rahman et al., 2023). The main objective of this investigation is to illustrate how ChatGPT can be used in educational settings by providing a useful example and some suggestions. Publicly available papers, blogs, social media profiles, and graphical and numerical representations were used to collect data for this study. At one moment, ChatGPT is regarded as an effective instrument that raises the motivation of learners to acquire knowledge and a sense of self-confidence. Learners with specific requirements and/or other students who might find it difficult to learn using typical classroom techniques will especially benefit from this. ChatGPT's interactive approach enhances the educational process. The imaginations of learners can be unleashed by ChatGPT, which can also offer individualised tutoring and assist with preparing them for potential employment with artificial intelligence (AI) (Memarian & Doleck 2023). Students' productivity and performance can be enhanced by using this technology to better match their learning demands. Scholars at the school of Hong Kong, for example, are strictly forbidden from using ChatGPT or any other AI programmes without prior authorisation from their professors. Failure to do so can be considered plagiarism. The aforementioned measures are being taken to make sure college students are using ChatGPT technologies appropriately, to prevent overuse and excessive dependence, as well as to protect the standard of their higher education and their academic credibility (Chan & Hu, 2023). Creating ethical standards for education and training has recently become more essential than it was in the past (Farrokhnia et al., 2023). Taking on several portrayals and evaluating the morality of its created notions versus the standards of such diverse identities is an unexplored possibility for ChatGPT. Plagiarism in various forms may be becoming more common because of technologies like ChatGPT. Because of this, increasingly intricate equipment is required to identify and address unanticipated forms of higher education malpractice (Geerling et al., 2023). In order to take into consideration the variety of copyright infringement that may occur in both research and teaching methods, this could lead to a vicious cycle wherein hardware and processing capacity become increasingly scarce. It is challenging to unravel the steps involved in arriving at answers and solutions as well as the underlying assumptions of derived answers owing to the highly complex nature of artificial intelligence systems. This can make it very challenging for learners to figure out why they receive certain answers and what they can do to correct the errors they made. For instance, when a student uses wrong syntax, the programme might propose fallacious modifications, but that student may struggle to figure out why and how the recommendation makes sense or what they can do to refrain from making the same errors again (Alabool 2023; Biswas 2023; Laupichler et al., 2022).

Due to ChatGPT's extensive usage, some educational institutions across the globe have put restrictions on or outright banned it. For instance, ChatGPT use was prohibited in Seattle public institutions starting in January 2023, while the College of Sciences in Paris stressed that researchers were not allowed to utilise ChatGPT or any other AI technology for disguise reasons (Zhou et al., 2023). Students are not permitted to use ChatGPT in order to finish assignments, take examinations, or perform experiments, according to a stringent prohibition enacted by Bangalore's RV University (Yadava, 2023). Teachers can enhance their ability to modify courses and deliver instruction by using ChatGPT's support in monitoring and assessing their students. To guarantee that educators and students use ChatGPT appropriately and freely, as well as to help them develop the self-sufficient thinking and innovative skills needed to meet new challenges and seize opportunities, it is crucial to acknowledge the potential hazards and constraints connected with this type of technology as well as put in place regulations that are appropriate (Ienca, 2023). Participants are encouraged to interact more thoroughly with the content because of its capacity to provide individualised educational opportunities and quick suggestions, which enhances retention of the knowledge. Furthermore, ChatGPT's accessibility at all hours ensures the learners with hectic routines are not deprived of any opportunities for learning, and this is a monumental advantage. This is especially important for students from families with modest incomes because their commitments may conflict with regular classes. Studies have demonstrated that ChatGPT might be a helpful tool for academic and scientific studies during their idea generation phase. However, the researchers can encounter certain challenges when assessing data, describing problems, including enough study citations, and integrating the existing literature. Previous to contending whether or not artificial intelligence (AI) should continue to be allowed to operate in educational settings, it is imperative that all of us carefully analyse the challenges confronting higher education presently and the significant influence that AI has on higher education. From a worldwide standpoint, there are currently a number of issues with higher education, such as inequitable allocation of resources, unsteady quality of instruction, inadequate material, antiquated methodologies for instruction, a great deal of work for students, and a flawed system of education assessments. These issues seriously limit the quality and equity of education, impeding its advancement.

10.6 Conclusion

Teachers and students can benefit from tailored training, advanced text help, and guidance in thinking critically thanks to ChatGPT. Nevertheless, the use of new technologies raises many questions about their abuse. Teachers have expressed concern about bias and misinformation in ChatGPT's responses as well as cases of plagiarising and duplicity. It is important that teachers, professionals, and policymakers take preventive measures towards educating themselves together with the students they serve on the right and ethical uses of these tools. It is also very necessary that all teachers be aware that AI tools do not come without restrictions, as every kind of technology has its own set of pros and cons, but inborn dangers are associated with each one. Rathore (2023) reveals that Educational organisations have shown great enthusiasm for ChatGPT because it can provide personalised educational experiences across disciplines. However, there is a lack of real-life studies looking into how ChatGPT was used within formal learning settings including integrative ones, despite what researchers think about this occurrence. Scholars who are searching for directions are thereby recommended to visit ChatGPT first because it is a great place to start.

It can also be used to gather basic information, schedule the response, and accumulate resources from the internet related to a particular subject. It's just the start of it, though. In an effort for the field of education to remain significant, humans have to keep striving to do better than AI. As a result, academicians must accept AI as an instrument for managing time that opens up new opportunities. Instructors are in an exceptional position to explicate to learners how to effectively use artificial intelligence (AI) for both their educational assignments and future careers. Teachers may use a variety of strategies to counteract any potential negative effects of ChatGPT. Choosing to ensure that students are conscious of the possibility of inadequate or biased information regarding promoting engagement and instruction among peers to make up for a lack of interpersonal cooperation, establishing specific requirements and standards regarding the utilisation of artificial intelligence (AI) to address moral as well as security threats, and having backup plans prepared to cope with any technological issues that might pop up are some of these situations. Emphasising the importance of analysis and analytical skills will assist in preventing any decrease in creative thinking and determination. In conclusion, the extensive adoption of artificial intelligence (AI) technology into higher education will significantly alter the nature of the educational system in the years to come. This will enhance its efficiency and efficacy, give learners enhanced educational resources, and enable them to more effectively adapt to the demands of a world that is constantly evolving (Ellis and Slade 2023; Gozalo-Brizuela & Garrido-Merchan, 2023). Human beings are depending progressively more on AI (artificial intelligence) for performing a variety of jobs as a result of its rapid growth. Instead of squandering time and effort on monotonous tasks that automated systems can perform with ease, humans should embrace newer technologies in order to foster future developments in technology. Instead of trying to outpace continuously evolving instruments and technology, people ought to leverage them intelligently to create new discoveries (Yu 2023).

References

Al Ahmed, Y., & Sharo, A. (2023, June). On the education effect of CHATGPT: Is AI CHATGPT to dominate education career profession?. In *2023 International Conference on Intelligent Computing, Communication, Networking and Services (ICCNS)* (pp. 79–84). IEEE.

Alabool, H. M. (2023, August). ChatGPT in education: SWOT analysis approach. In *2023 International Conference on Information Technology (ICIT)* (pp. 184–189). IEEE.

Ali, J. K. M., Shamsan, M. A. A., Hezam, T. A., & Mohammed, A. A. (2023). Impact of ChatGPT on learning motivation: Teachers and Students' voices. *Journal of English Studies in Arabia Felix, 2*(1), 41–49.

Ausat, A. M. A., Massang, B., Efendi, M., Nofirman, N., & Riady, Y. (2023). Can chat GPT replace the role of the teacher in the classroom: A fundamental analysis. *Journal on Education, 5*(4), 4. https://doi.org/10.31004/joe.v5i4.2745. *(1) (PDF) The Impact of ChatGPT on*

Student Learning/performing. Retrieved April 6, 2024, from https://www.researchgate.net/publication/372481501_The_Impact_of_ChatGPT_on_Student_Learningperforming

Ayman, S. E., El-Seoud, S., Nagaty, K., & Karam, O. (2023). The impact of ChatGPT on student learning/performing. https://doi.org/10.13140/RG.2.2.28890.11205.

Bender, E. M., Gebru, T., McMillan-Major, A., & Shmitchell, S. (2021). On the dangers of stochastic parrots: Can language models be too big? In *Proceedings of the 2021 ACM Conference on fairness, accountability, and transparency* (p. 610–23). Virtual Event Canada: ACM. https://doi.org/10.1145/3442188.3445922

Biswas, S. Role of Chat GPT in education (February 25, 2023). Available at SSRN: https://ssrn.com/abstract=4369981

Chan, C. K., & Hu, W. (2023). Students' voices on generative ai: Perceptions, benefits, and challenges in higher education. *arXiv*. https://doi.org/10.48550/arXiv.2305.00290

Clarizia, F., Colace, F., Lombardi, M., Pascale, F., & Santaniello, D. (2018). Chatbot: An education support system for student. *Cyberspace Safety and Security*, 291–302. https://doi.org/10.1007/978-3-030-01689-0_23

Ellis, A. R., & Slade, E. (2023). A new era of learning: Considerations for ChatGPT as a tool to enhance statistics and data science education. *Journal of Statistics and Data Science Education, 31*(2), 128–133.

Emenike, M. E., & Emenike, B. U. (2023). Was this title generated by ChatGPT? Considerations for artificial intelligence text-generation software programs for chemists and chemistry educators. *Journal of Chemical Education, 100*(4), 1413–1418.

Farrokhnia, M., Banihashem, S. K., Noroozi, O., & Wals, A. (2023). A SWOT analysis of ChatGPT: Implications for educational practice and research. *Innovations in Education and Teaching International*, 1–15.

Fuchs, K. (2022). The importance of competency development in higher education: Letting go of rote learning. *Frontiers in Education, 7*, 1004876. Frontiers.

Geerling, W., Mateer, G. D., Wooten, J., & Damodaran, N. (2023). ChatGPT has aced the test of understanding in college economics: Now what? *The American Economist, 68*(2), 233–245.

Gozalo-Brizuela, R., & Garrido-Merchan, E. C. (2023). ChatGPT is not all you need. A State of the Art Review of large Generative AI models. *arXiv preprint arXiv:2301.04655.*

Halaweh (2023). ChatGPT in education: Strategies for responsible implementation.

Heath, G. A., Ravikumar, D., Hansen, B., & Kupets, E. (2022). A critical review of the circular economy for lithium-ion batteries and photovoltaic modules–status, challenges, and opportunities. *Journal of the Air & Waste Management Association, 72*(6), 478–539.

Hill-Yardin, E. L., Hutchinson, M. R., Laycock, R., & Spencer, S. J. (2023). A Chat(GPT) about the future ofscientific publishing. *Brain, Behavior, and Immunity, 110*, 152–154. https://doi.org/10.1016/j.bbi.2023.02.022. *(1) (PDF) The Impact of ChatGPT on Student Learning/performing*.

Ienca, M. (2023). Don't pause giant AI for the wrong reasons. *Nature Machine Intelligence, 5*(5), 470–471.

Karaali, G. (2023). Artificial intelligence, basic skills, and quantitative literacy. *Numeracy, 16*(1), 9.

Kasneci, E., Seßler, K., Küchemann, S., Bannert, M., Dementieva, D., Fischer, F., Gasser, U., Groh, G., Günnemann, S., Hüllermeier, E., Krusche, S., Kutyniok, G., Michaeli, T., Nerdel, C., Pfeffer, J., Poquet, O., Sailer, M., Schmidt, A., Seidel, T., Kasneci, G. (2023). Chatgpt for good? On opportunities and challenges of large language models for education. https://doi.org/10.35542/osf.io/5er8f.

Khalil, M., & Er, E. (2023). Will ChatGPT get you caught? Rethinking of plagiarism detection. *arXiv*. https://doi.org/10.35542/osf.io/fnh48.

Kocoń, J., Cichecki, I., Kaszyca, O., Kochanek, M., Szydło, D., Baran, J., … Kazienko, P. (2023). ChatGPT: Jack of all trades, master of none. Information Fusion, 101861.

Kuhail, M. A., Alturki, N., Alramlawi, S., & Alhejori, K. (2022). Interacting with educational Chatbots: A systematic review. *Education and Information Technologies, 28*(1), 973–1018. https://doi.org/10.1007/s10639-022-11177-3

Laupichler, M. C., Aster, A., Schirch, J., & Raupach, T. (2022). Artificial intelligence literacy in higher and adult education: A scoping literature review. *Computers and Education: Artificial Intelligence, 3*, 100101.

Liu, J., Shen, D., Zhang, Y., Dolan, W. B., Carin, L., & Chen, W. (2022). What makes good in-context examples for GPT-3?. Proceedings of deep learning inside out (DeeLIO2022): The 3rd Workshop on Knowledge Extraction and Integration for Deep Learning Architectures, 3, 100–114. *(1) (PDF) ChatGPT and a New Academic Reality: Artificial Intelligence-Written Research Papers and the Ethics of the Large Language Models in Scholarly Publishing*.

McGee, R. W. (2023). Is Chat Gpt biased against conservatives? an empirical study. *An Empirical Study (February 15, 2023)*.

Memarian, B., & Doleck, T. (2023). ChatGPT in education: Methods, potentials, and limitations, computers in human behavior. *Artificial Humans, 1*(2), 100022., ISSN 2949-8821,. https://doi.org/10.1016/j.chbah.2023.100022

Murtarelli, G., Collina, C., & Romenti, S. (2023). Hi! How can I help you today? Investigating the quality of chatbots–millennials relationship within the fashion industry. *The TQM Journal, 35*(3), 719–733.

Okonkwo, C. W., & Ade-Ibijola, A. (2021). Evaluating the ethical implications of using chatbot systems in higher education. *digiTAL, 2021*, 68.

Qadir, J. (2022). Engineering education in the era of chatgpt: Promise and pitfalls of Generative AI for education. https://doi.org/10.36227/techrxiv.21789434.v1.

Rahman, M., & Watanobe, Y. (2023). ChatGPT for education and research: Opportunities, threats, and strategies. https://doi.org/10.20944/preprints202303.0473.v1.

Rahman, M. & Terano, H. J., Rahman, M., Salamzadeh, A., Rahaman, M., & Saidur (2023). ChatGPT and academic research: A review and recommendations based on practical examples. 3. 1–12. https://doi.org/10.52631/jemds.v3i1.175.

Rai, A. K., et al. (2020). Landsat 8 OLI satellite image classification using convolutional neural network. *Procedia Computer Science, 167*, 987–993.

Rathore, B. (2023). Future of AI & generation alpha: ChatGPT beyond boundaries. *Eduzone: International Peer Reviewed/Refereed Multidisciplinary Journal, 12*(1), 63–68.

Roumeliotis, K. I., & Tselikas, N. D. (2023). Chatgpt and open-ai models: A preliminary review. *Future Internet, 15*(6), 192.

Shoufan, A. (2023). *Exploring students' perceptions of ChatGPT: Thematic analysis and follow-up survey*. IEEE Access.

Singh, K. K., Yadav, P., Singh, A., Dhiman, G., & Cengiz, K. (2021). Cooperative spectrum sensing optimization for cognitive radio in 6 G networks. *Computers and Electrical Engineering, 95*, 107378.

Singh, A., Dhull, A., & Singh, K. K. (Eds.). (2024a). *Blockchain and deep learning for smart healthcare*. John Wiley & Sons.

Singh, K. K., Rho, S., Singh, A., & Sergei, C. (2024b). Big data analytics and knowledge discovery for urban computing and intelligence. *Complex & Intelligent Systems, 10*(1), 1–2.

Thakkar, A. (2023). Blended learning: The new normal. *The Blended Teaching and Learning, 22*.

Thakkar, A. (2024). An empirical study on the impact of use of ict in higher education-with special reference to MBA students. *Board of Editors*, 52.

Thurzo, A., Strunga, M., Urban, R., Surovková, J., & Afrashtehfar, K. I. (2023). Impact of artificial intelligence on dental education: A review and guide for curriculum update. *Education Sciences, 13*(2), 150.

Tlili, A., Shehata, B., Adarkwah, M. A., Bozkurt, A., Hickey, D. T., Huang, R., & Agyemang, B. (2023). What if the devil is my guardian angel: Chatgpt as a case study of using Chatbots in education. *Smart Learning Environments, 10*, 1. https://doi.org/10.1186/s40561-023-00237-x

Villaseñor García, Elio. (2023) (2023). Applying neural networks analysis to assess digital government evolution. *Government Information Quarterly., 40*, 101811. https://doi.org/10.1016/j.giq.2023.101811

Yadava, O. P. (2023). ChatGPT—A foe or an ally? *Indian The Journal of Thoracic and Cardiovascular Surgery, 39*, 217–221. https://doi.org/10.1007/s12055-023-01507-6

Yan, D. (2023). Impact of ChatGPT on learners in a L2 writing practicum: An exploratory investigation. *Education and Information Technologies*, 1–25.

Yu, H. (2023). Reflection on whether chat GPT should be banned by academia from the perspective of education and teaching. *Frontiers in Psychology, 14*, 1181712. https://doi.org/10.3389/fpsyg.2023.1181712

Zhai, X. (2021). Practices and theories: How can machine learning assist in innovative assessment practices in science education. *Journal of Science Education and Technology, 30*(2), 139–149. https://doi.org/10.1007/s10956-021-09901-8

Zhai, X., ChatGPT user experience: Implications for education (December 27, 2022).

Zhou, C., Li, Q., Li, C., Yu, J., Liu, Y., Wang, G., et al. (2023). A comprehensive survey on pretrained foundation models: a history from BERT to ChatGPT. *arXiv*. https://doi.org/10.48550/arXiv.2302.09419.

Further Reading

Sharma, A., Bahl, S., Bagha, A. K. et al. (2022). Blockchain technology and its applications to combat COVID-19 pandemic. *Res. Biomed. Eng. 38*, 173–180. https://doi.org/10.1007/s42600-020-00106-3.

Chapter 11
Exploring Multi-modal Hate Speech Detection Using Machine Learning and Deep Learning Models

Shefali Khera, Anuradha, Akansha Singh, and Krishna Kant Singh

Abstract The growing social media use has caused a troubling issue of hate speech occurrence. As more individuals engage on these platforms, the harmful and offensive expressions towards certain groups have also escalated. Due to this, it is imperative to analyse this multi-modal data available online and eliminate toxic data to curb hate crimes on a global level. This book chapter presents an overview of multi-modal hate speech detection and publicly available datasets, followed by a discussion about the effectiveness of various machine and deep learning techniques used for multi-modal hate speech detection. Social media platforms now leverage AI technologies to filter out toxic content and combat hate crimes. Machine and deep learning techniques are gaining traction for analysing such data. The data collected on such social media sites include text, visual, and audio, leading to multi-modal data collection. Multi-modal data is used for improved accuracy and adaptability to get better results. Thus, the survey in this research delves into definitions of discriminatory speech, the rationale for discovery, and standard written content analysis strategies. It explores cutting-edge hate speech identification methods, including multi-modal ones, discussing their advantages and drawbacks. The paper also highlights datasets, their challenges, and their performance metrics and categorisation ratings of popular hate speech detection approaches. In conclusion, the paper provides insights into the current landscape, offering comparisons, addressing challenges, and proposing future research directions in multi-modal and multi-lingual hate speech detection, contributing to AI-driven social media analysis advancements.

S. Khera · Anuradha
Department of CSE, The NorthCap University, Gurugram, India

A. Singh (✉)
School of Computer Science Engineering and Technology, Bennett University, Greater Noida, Uttar Pradesh, India
e-mail: akansha1.singh@bennett.edu.in

K. K. Singh
Delhi Technical Campus, Greater Noida, Uttar Pradesh, India

A. Singh, K. K. Singh (eds.), *Multimodal Generative AI*,
https://doi.org/10.1007/978-981-96-2355-6_11

Keywords Hate speech detection · Machine learning · Deep learning · Multi-modal Data

11.1 Introduction

The increased utilisation of social networking sites has made them the primary source of communication globally. A substantial amount of textual, audio, and visual data in diverse formats like images, videos, and written content is present online. This multi-modal data contributes to misleading information and offensive speech, creating an urgent demand for more effective control measures.

11.2 Overview of Hate Speech

Hate speech has been articulated by many researchers as a language that targets societies and individual members based on their culture, colour, sex, and religion (Schmidt & Wiegand, 2017). Certain definitions as listed in Table 11.1 refer to the platforms of hate speech and discuss community perspectives. Following a comprehensive analysis, hate speech can be defined as a harmful and offensive attack on an individual's identity on different grounds that can incite violence.

11.3 Multi-modal Data

Multi-modal information, such as text, images, and sound, offers a comprehensive perspective for understanding online interactions as depicted in Fig. 11.1. The escalation in social networking interaction has significantly expanded the range of data available for analysis, highlighting multi-modal data in the last 4 years. Data processing and model architecture present challenges in integrating visual elements. Overcoming these obstacles requires the collaboration of experts in computer

Table 11.1 Different forms of hate speech

Forms	Definitions
Cyberbullying	Characterised as a deliberate demonstration completed by a social occasion or individual using electronic stages (Chen et al., 2012)
Discrimination	Interaction via a distinction and afterward utilised as the premise of unreasonable treatment (Thompson, 2016)
Toxic comments	Conveying disrespectful content, abusive, unpleasant, and harmful (Risch & Krestel, 2020)
Abusive language	The term abusive language seeks to diminish or humiliate some person or group (Nobata et al., 2016)

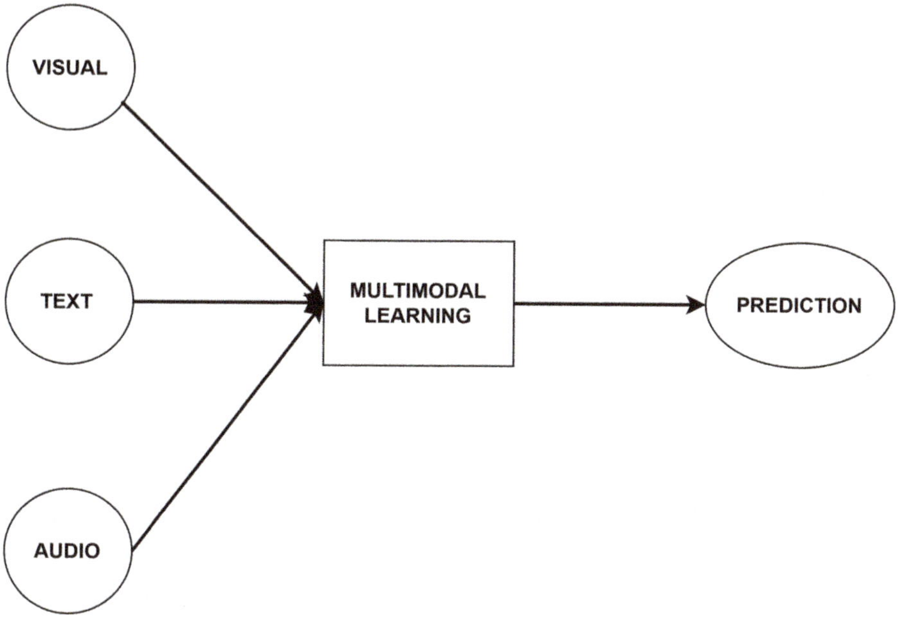

Fig. 11.1 Forms of multi-modal data

vision, natural language processing, and audio processing. Moreover, ethical concerns surrounding multi-modal hate speech detection, especially concerning privacy and potential biases in visual data interpretation, require thorough examination.

This evolving situation calls for employing multi-modal strategies for combating hate speech. The significance of employing multi-modal data to improve accuracy and adaptability has been highlighted by recent studies. Research articles have grown, as shown in Fig. 11.2. It emphasises the necessity of using visuals in hate speech identification since they frequently offer important context needed for precise analysis. Given the speed at which the digital landscape is changing, it is important that models for identifying hate speech online must be able to process a variety of material types. This requires an understanding and application of multi-modal data.

11.4 Multi-modal Hate Speech Detection

Both humans and automated systems find it challenging to recognise and comprehend hate speech; examples of several types of hate speech include those shown in Fig. 11.3. International minority associations define hate crimes as criminal acts targeting specific groups, while the European Union Commission defines hate speech as publicly inciting aggression based on particular traits. These and other organisations' definitions add diversity to the conversation. Academics such as

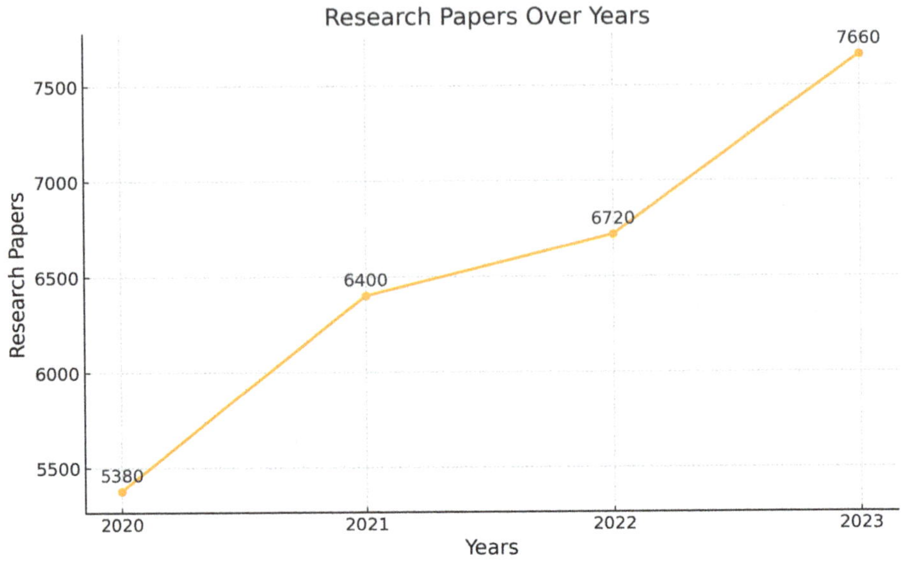

Fig. 11.2 Research papers on multi-modal hate speech detection in the preceding 4 years

Fig. 11.3 Forms of hate speech on social media

Nobata et al. (Fortuna & Nunes, 2018) concentrate on language that criticises an individual or groups people together based on contradictory.

Prominent social media sites like Facebook and Twitter have their definitions emphasising protected characteristics and prohibiting violence or threats based on specific attributes. Exploring these definitions reveals common themes like religion, gender, race, and violence, along with complexities like cursing, disability, property damage, age, and serious diseases. Figure 11.4 shows how the internet talks about the urgent need to curb hate speech and how it has become the topic of discussion today. This research aims to synthesise these diverse perspectives to enhance our understanding and improve hate speech detection.

Fig. 11.4 Examples of the urgent need to curb hate speech on social media

11.5 Highlights of Work Done in Multi-modal Hate Speech Study

The comprehensive exploration of hate speech detection in online textual content is recently surveyed by Schmidt and Wiegand (Schmidt & Wiegand, 2017), covering evolving terminology, features, datasets, and approaches. However, the field lacks a consistent dataset and standardised evaluation, hindering effective method comparisons. Saleem et al. (Saleem, 2021) compare classification methods for Reddit hate speech, while Wassem and Hovy (Waseem & Hovy, 2016) focus on Twitter, providing a manually annotated dataset. Davidson et al. (Davidson et al., 2017a) evaluate hate speech detection on Twitter, and Malmasi and Zampieri (Malmasi & Zampieri, 2017) enhance this dataset with more features. ElSherief et al. (ElSherief et al., 2018) propose a substantial hate dataset on Twitter. Zhang et al. (Zhang et al., 2018) advance hate speech detection using a CNN and a GRU over Word2Vec embeddings.

Examining existing datasets, examples include RM (Davidson et al., 2017b) (2435 tweets on refugees and Muslims), DT (Park & Fung, 2017) (24,783 tweets annotated for hate/offensive language), WZ-LS (Davidson et al., 2017a) (18,624 tweets on racism/sexism), and semi-supervised (Schmidt & Wiegand, 2017) (27,330 general hate speech Twitter tweets). Despite the prevalence of images in modern social media, limited contributions leverage visual information. Zhong et al. (Zhou et al., 2015) classify Instagram images for cyberbullying, with marginal improvement from pre-trained CNN features. Hosseinmardi et al. (Hosseinmardi et al., 2015) address cyberbullying on Instagram using pre-trained CNN features, while Yang et al. (Yang et al., 2019) at Facebook explore multi-modal feature fusion strategies for user-reported data, and Sabat et al. (Sabat et al., 2019) detect hate speech memes using both images and text.

The survey has been performed as follows: The introduction has been discussed in Sect. 11.1, followed by data acquisition, pre-processing, and formation of structured data in Sect. 11.2. Section 11.3 covers different AI/ML models, like deep learning and hybrid techniques, that can be used for hate speech detection. Section 11.4 covers the dataset training and performance. Finally, Sections 11.5 and 11.6 talk about the challenges, conclusion, and future advances, respectively.

11.6 General Framework for Hate Speech Detection

Figure 11.5 shows the broad structure outlining hate speech detection. Determining the main source—such as social media channels like Twitter and Facebook, where active attempts are made to suppress hate speech—is one of the first stages in combating hate speech. The subsequent phases include using AI/ML approaches to address the issue of hate speech shortly after pre-processing the gathered data to present it in an organised manner. The detection of hate speech in multi-modal information is presented in this research along with a discussion of several deep learning and machine learning algorithms taking into account various kinds of data.

11.6.1 Corpus Formation?

Corpora are representative of systematically compiled datasets containing diverse textual, visual, or auditory content that is extracted from numerous online sources for hate speech identification.

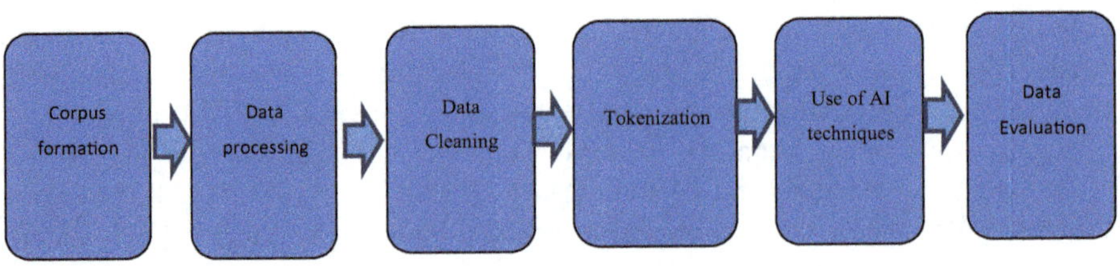

Fig. 11.5 Stages of hate speech identification

To adequately teach and analyse a hate speech identification model, an accurately curated corpus must be developed. The shifting linguistic patterns and contextual complexities included in hate speech, which is a snapshot of the complexity of real-world online interactions, can be better understood by analysing several datasets. The relevance of a thoughtfully constructed corpus lies in its capacity to accurately represent the diverse and dynamic nature of online content, thereby making model training and assessment easier to implement. Considering insights from contemporary research, this study highlights how it is imperative to design and leverage corpora that accurately reflect the intricacies of online communication so that hate speech detection approaches are relevant and impactful.

11.6.2 Types of Corpus

Within the sphere of hate speech detection, employing diverse types of corpora is paramount to comprehensively addressing the intricacies of online interactions. Recent studies highlight the significance of tailoring corpus types to capture varied dimensions of hate speech dynamics. Textual corpora, drawn from social media posts, comments, and forum discussions, facilitate a thorough examination of linguistic patterns and textual markers associated with hate speech. Visual corpora, inclusive of images and memes disseminated online, offer valuable insights into the visual elements intertwined with hate speech expressions. Auditory corpora, comprising hate speech embedded in audio content such as podcasts or voice messages, extend the analysis to the spoken domain. Multi-modal corpora, integrating text, images, and audio, present a holistic approach, recognising the multi-faceted nature of hate speech across different modalities. By diversifying corpus types, researchers can gain a more nuanced interpretation of hate speech manifestations in the ever-evolving landscape of web-based interaction, paving the way for the establishment of stable and adaptable hate speech detection techniques.

11.6.3 Data Pre-Processing

The initial and fundamental step in data preparation is data pre-processing, a comprehensive assignment such as data cleaning, tokenisation, data integration, conversion, and stop word removal, with the normalisation as shown in Fig. 11.6. Researchers are often faced with complex and heterogeneous datasets in the modern big data landscape, which necessitates careful treatment to extract valuable insights. The intrinsic ability of data preparation to successfully handle issues like missing data, outliers, noise, and discrepancies within the set of data highlights the critical relevance of this process. This meticulous curation contributes to the validity and robustness of the research findings by ensuring that based on clean and trustworthy data, the following analyses may be carried out.

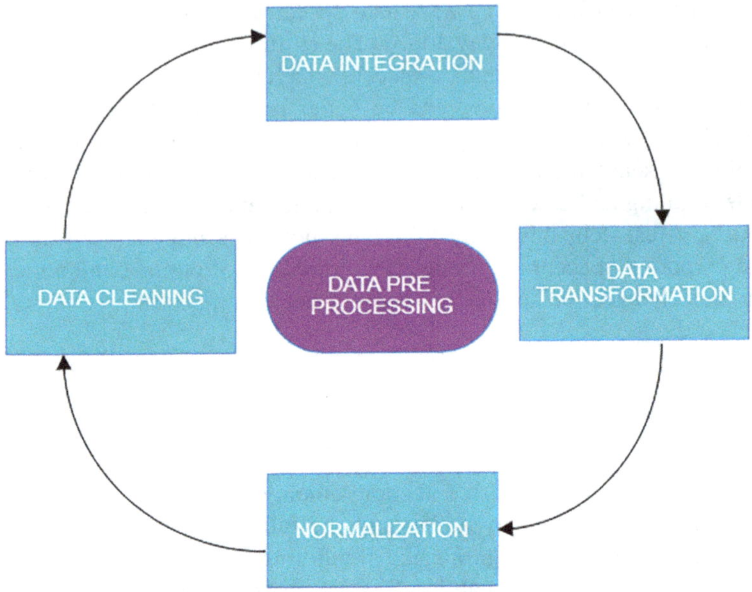

Fig. 11.6 Steps for data preprocessing

11.6.4 Data Cleaning

Inside the sphere of natural language processing (NLP), data cleaning addresses non-essential factors like hyperlinks, punctuation symbols, hashtags, and numerical characters. Punctuation and hashtags are often excluded from conventional practices, but it is not the most effective method. For example, punctuation marks can be used for emojis, conveying the emotions of customers, and hashtags can be used to identify abusive comments.

In addition to data cleaning, integrating data is also crucial to creating a more comprehensive dataset by combining numerous sources of information. This integration allows the model to be advanced in a variety of verbal frameworks and settings so that it can identify hate speech with greater accuracy in a variety of contexts. In addition, data transformation strategies are used to translate uncooked text into a format suitable for evaluation, and normalisation procedures ensure the consistency of records distributions. By combining these strategies with powerful data cleansing techniques, we can establish the foundation for advanced AI tactics that can identify hate speech on a mass scale and make certain record distributions consistent. As soon as paired with powerful data cleansing, those strategies provide a stable groundwork for the subsequent use of advanced AI tactics in hate speech recognition.

11.6.5 Tokenisation

An NLP version development process is facilitated by segmenting text information into phrases or sentences, or tokens, in order to comprehend the context and simplify the process. By assessing the linguistic order, tokenisation contributes to the understanding of the general content of a text. An exceptional criteria, like punctuation marks or whitespace, may be used to segment the remarks. The records are further polished by removing noise phrases, such as formatting tags, numerals, pronouns, prepositions, conjunctions, and auxiliary verbs. By normalising the textual content, variability can be reduced and the text can be aligned closer to a standard. In this way, data variation is minimised, which may enhance the performance of subsequent methods by reducing data variation.

11.6.6 Use of AI Techniques

After segmentation, more than one AI technique is used to analyse the facts to detect hate speech. The detection of hate speech has been significantly enhanced using modern artificial intelligence (AI) techniques such as system mastering, deep learning, hybrid models, and big language models. With machine learning algorithms, especially, it is possible to identify patterns and distinguish complicated factors in textual records, making it possible to develop effective hate speech detection models. Using deep learning, complex links within languages can be captured, enabling automatic acquisition of hierarchical representations and preventing hate speech from being misinterpreted as non-offensive. The extensive effectiveness of hate speech detection structures has been enhanced through the integration of a variety of AI techniques into hybrid models. As a result of the advent of huge language models, such as GPT-3, researchers have gained a tremendous amount of contextual expertise, which allows them to better identify hate speech in its context and specificity. With this aggregate of AI methods, hate speech identification has reached a crucial milestone in terms of accuracy, efficiency, and flexibility to enhance the converting landscape of online content.

11.6.7 Data Evaluation

To determine the accuracy and effectiveness of hate speech detection algorithms that are based on advanced artificial intelligence techniques, it is imperative that they are subjected to a useful and reliable analysis. Several conventional criteria are frequently used to determine whether a model is effective at detecting hate speech and controlling false positives, including accuracy, recall, and F1. Furthermore, receiver operating characteristic (ROC) curves and area under curve (AUC) measurements assess the version's ability to distinguish hate speech from non-offensive content material at different decision thresholds. The evaluation should also consider the system's real-global utility, including its processing speed, scalability, and generalisability to different types of datasets in addition to the above parameters.

11.7 Stages of Multi-modal Hate Speech Identification

In recent years, there has been immense focus on detecting hate speech content online, and machine learning algorithms play a vital role in this process. This review examines the use of different advanced computational methodologies like hybrid techniques, deep learning, and traditional learning to address the challenge of hate speech detection. Figure 11.7 shows diverse models that have been recently developed using various machine learning techniques for hate speech detection.

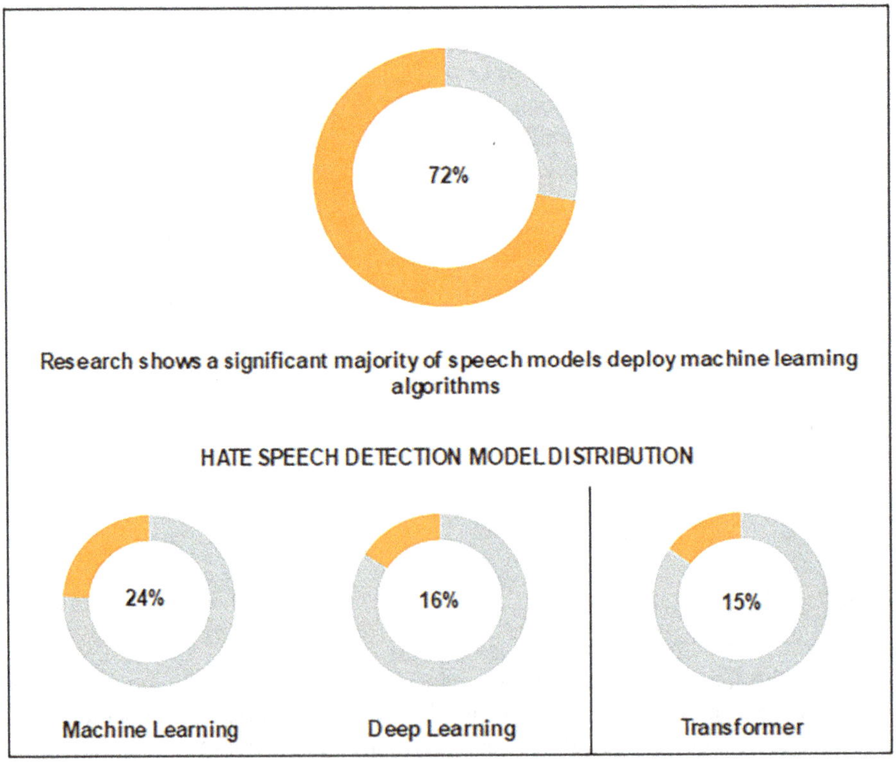

Fig. 11.7 Different hate speech detection models

11.8 Machine Learning Models

In the realm of hate speech detection, a traditional machine learning model plays an immensely fundamental role by taking into account tasks like categorising, predicting, and clustering to classify and counteract hate speech. Reliance on labelled data during training is crucial for ensuring the accuracy of classification tasks. There have been numerous studies that have stated that machine learning algorithms are effective. In particular, these studies highlight the advantage of training models on a diverse number of datasets rather than those that focus on a single dataset (Chiril et al., 2022). Supervised machine learning methods, including commonly employed methods like Support Vector Machines (SVM) and Random Forest, have proven successful in classifying social media data into hate or non-hate categories, demonstrating their efficacy across various languages. Moreover, researchers have also explored ways to improve the quality of text through the use of preprocessing techniques that retain essential text features without compromising the quality of the model itself to enhance its performance over time (Husain & Uzuner, 2022). In the pursuit of a more comprehensive understanding, efforts have been made toward multiclass classification, categorising online user comments into distinct classes (Nobata et al., 2016). Despite the efficiency of supervised learning in domain-dependent tasks, manual labelling challenges and limited execution times persist (Burnap & Williams, 2014).

11.9 Deep Learning

Traditional image encryption techniques employ rounds of diffusion and confusion to obtain an explicit trade-off between security and efficiency (Panwar et al., 2023). Deep learning strategies, which have become far more efficient at discovering hate speech than any conventional machine learning strategy, were used to accomplish a considerable change in a few years. Deep learning has revolutionised the field of machine learning by demonstrating its capacity to discern complex patterns from extensive datasets (Singh et al., 2024). In trendy language, artificial neural networks perform better thanks to their complex topologies, which provide a more in-depth analysis of textual patterns. Recurrent neural networks and convolutional neural networks are two models widely used in natural language programs to detect hate speech, despite the fact that their topologies are unable to execute straightforward operations naturally. These facts suggest that the use of deep learning methodologies is successful because of their excellent performance, which is frequently attributed to the availability of large datasets. However, the results obtained using this approach are sufficient for making any claim, which depends on the crucial choice of the algorithm, the configuration of hidden layers, and the techniques used for feature representation. Regarding the identification of hate speech, researchers have made significant progress by employing RNN models, exceeding the performance of other deep learning techniques. Deep learning techniques, such as Visio-Linguistic (VILIO) models, are highly successful in recognising hostile memes in the context of analysing hate pictures (Muennighoff, 2020). It is important to mention that, although deep learning approaches excel in some situations, they may not always surpass traditional techniques in terms of performance.

11.10 Hybrid Model

Hybrid learning models, seamlessly combining traditional and deep learning elements, have gained prominence in hate speech identification. These models competently leverage the strengths of both paradigms to reduce misclassification rates and enhance accuracy, finding applications in sentiment analysis and message characterisation (Yuan et al., 2016). Recent research, exemplified by Distil-BERT, XML-RoBERTa, and MuRIL, focuses on improving prediction rates and minimising misclassification instances, particularly in code-mixed Dravidian languages (Bölücü & Canbay, 2024).

Examples of up-to-date results that are based on deep and hybrid teaching paradigms, specifically in the hate speech subdomain, express the transformation process of the development and classification tasks in all languages. Long short-term memory is the only model that repeatedly outperforms baseline and up-to-date approaches for hate speech detection (Erico et al., 2020). The rules clustering method in Ayo et al. (2021) has better performance metrics. The semi-supervised multitask learning, which introduces the fuzzy ensemble process and Latent Dirichlet Allocation, has the difference of improving the field with subtle topic extraction and revealing hate speech in a form (Liu et al., 2019).

All of the mentioned three models have categorisations based on different techniques as mentioned in Fig. 11.8.

11.11 Testing and Performance of Dataset

Social media is the talk of the town nowadays and many keep uploading posts. All this makes a lot of videos like hate speech, etc., which are floating around. It is challenging to draw the correct amount of data across these vast online resources for researchers.

Social media sites make it easy for researchers to get data through their APIs. But it's not just about using APIs—there are different ways to collect data from social media, as shown in Fig. 11.9. Identifying hate speech has become a big deal all over the world. Videos are a big part of spreading content, reaching a lot of people, including young ones. For example, about 1 billion lengthy content videos are watched on YouTube each day.

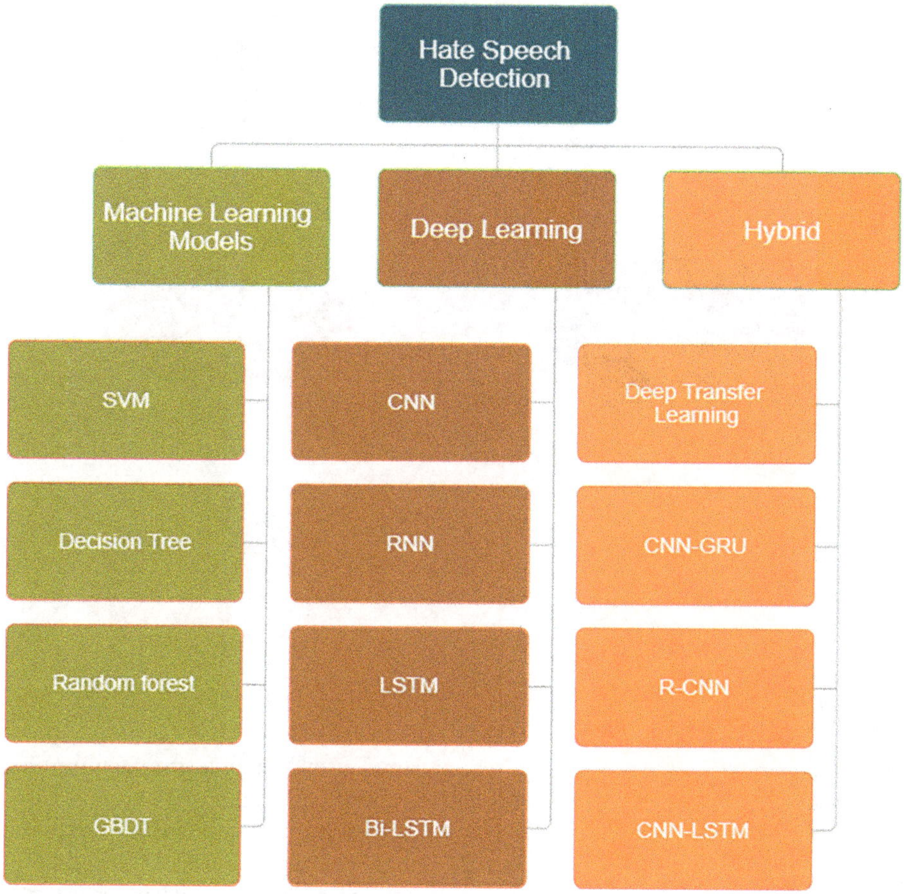

Fig. 11.8 Categorisation of hate speech detection across different models

Spotting hate speech is important to make sure young people are safe online and everyone has a good experience. Until now, most researchers focused on analysing text—things like social media posts, news comments, and tweets. Even though most methods for detecting hate speech mainly look at text, there's not much research on finding it in different kinds of content.

11.11.1 Dataset Evaluation

Measuring the impact of hate speech identification heavily depends on datasets. A properly normalised dataset is crucial for achieving the optimal performance of an algorithm. In this section, varied measurements, such as the F1-score, recall, and

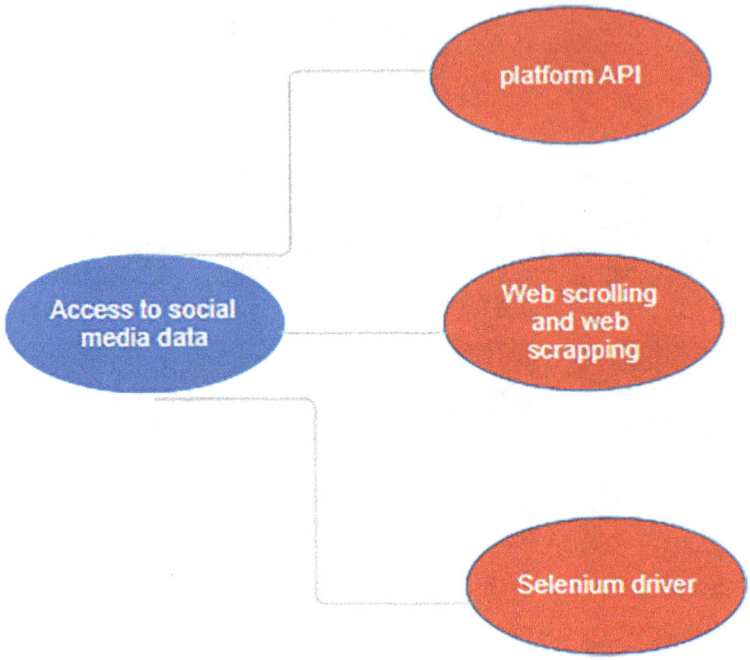

Fig. 11.9 Different ways to access data

precision, are utilised to evaluate the effectiveness of the machine and deep learning techniques. Accuracy is utilised as a key measure of performance.

The majority of the cutting-edge hate speech identification employed accuracy, recall, and F1-score as performance metrics; some, however, used AUC and accuracy because of dataset imbalances. Numerous research studies have consistently demonstrated that no single machine learning (ML) algorithm consistently outperforms others across diverse datasets. Consequently, it is imperative to conduct a comprehensive comparison of various ML algorithms to determine the most effective one for a given dataset. The experimental results, like those of Abro et al. (Abro et al., 2020), reveal that the SVM and AdaBoost classifiers exhibited superior performance, potentially attributed to SVM's utilisation of threshold functions for data separation, prioritising margin over the number of features. The independence of SVM from the feature count in the data was underscored (Cavnar & Trenkle, 1994; Zhang & Zhou, 2005). Additionally, SVM showcased its proficiency in handling non-linear data through the incorporation of kernel functions. AdaBoost's efficacy lies in its application of adaptive algorithms for iterative rule learning (Schapire, 2003) and its focus on minimising training errors.

Results from RF and LR discriminators, although marginally less than SVM and AdaBoost, surpassed those of NB, DT, KNN, and MLP. RF's diminished performance could be ascribed to the absence of relevant characteristics, leading to inaccurate predictions (Xu et al., 2012). The lower execution of LR might be attributed to its linear classification boundary, proving inadequate for effectively handling non-linear data (Eftekhar et al., 2005).

The least satisfactory performance was observed with NB, DT, MLP, and KNN classifiers. NB's dependence on dependent autonomy among characters negatively impacted its effectiveness as the complexity of conditional dependence increased with a higher number of features (Lewis, 1998). DT struggled in predicting hate speech due to the continuous data representation within the master features vector, posing challenges in determining the ideal threshold values for constructing a decision tree (Dreiseitl et al., 2001). MLP's suboptimal performance was linked to insufficient training data, rendering it a complex "black box" (Singh & Husain, 2014). KNN's poor performance was attributed to the laziness of its learning algorithm, proving inadequate for handling noisy data (Bhatia, 2010), underscoring its unsuitability for identifying hate speech in tweets.

11.12 Challenges Faced

Spotting instances of hate speech and opinion mining deals with several difficulties and complexities, which affect the effectiveness of models. One notable hurdle from the informal and inadequately written nature of social media messages is departing from formal structures and complicating the identification of text patterns. Another significant challenge lies in the laborious and costly task of constructing high-quality labelled datasets, especially for languages beyond English. This not only requires significant time and resources but also leads to a lack of diverse and accurately annotated data, especially for languages with limited resources, hindering the effectiveness of the models.

The incorporation of hate speech recognition and sentiment analysis in varied literary settings presents unique challenges related to language, such as varying linguistic structures, emotional dictionaries, and cultural expressions. The ongoing issue of distribution of data and imbalance presents difficulties in identifying significant trends, which in turn affects the accuracy of detecting hate speech and doing sentiment analysis. Moreover, the intrinsically subjective and environmentally sensitive nature of hate speech and expression of feeling poses a significant obstacle. Precise identification necessitates a subtle understanding of context and cultural subtleties, since comments considered hostile in one setting may not be interpreted similarly in another. The analysis of statements such as 'I'm dying to see you' in different cultural and linguistic settings highlights the complex difficulties in understanding intended meanings using algorithms. Indirect hate speeches increase the intricacy, underlining the need for complete contextual comprehension to differentiate between harmful and harmless words.

11.13 Conclusion and Discussion

Taking into account the dynamic nature of social media, the study sheds light on the pressing issue of hate speech requirements within this dynamic space. Increasingly, toxic conversations have emerged as a global issue as online communication grows, prompting social media platforms to utilise artificial intelligence tools to filter and prevent hate crimes on social media platforms. A fundamental shift is underway in the approach to addressing this challenge as a result of a comprehensive study and the integration of machine learning techniques into data analytics. As part of the study, we examined definitions of hate speech, motivations for conducting the research, and established methods for interpreting texts based on prior research. Notably, it contains an in-depth analysis of cutting-edge methods for identifying hate speech, emphasising the importance of multi-modal approaches and multilingualism, and describing their strengths and limitations.

In summary, this chapter provides valuable insights into the study of hate speech detection and its analysis in the current scenario. Through detailed analogy, addressing problems faced by hate speech detection and proposing future research directions, it significantly contributes in the realm of multi-modal and multi-lingual detections. The findings not only deepen our understanding of the intricacies involved but also establish the foundation for progress in AI-driven social media analysis. As the virtual realm continues evolving, this chapter acts as a cornerstone for ongoing efforts to refine and optimise hate speech detection mechanisms, fostering a secure and diverse online community.

References

Abro, S., Shaikh, S., Khand, Z. H., Zafar, A., Khan, S., & Mujtaba, G. (2020). Automatic hate speech detection using machine learning: A comparative study. *International Journal of Advanced Computer Science and Applications, 11*, 8.

Ayo, F. E., Folorunso, O., Ibharalu, F. T., Osinuga, I. A., & Abayomi-Alli, A. (2021). A probabilistic clustering model for hate speech classification in twitter. *Expert Systems with Applications, 173*, 114762.

Bhatia, N. (2010). Survey of nearest neighbor techniques. *arXiv preprint arXiv:1007.0085.*

Bölücü, N., & Canbay, P. (2024). Syntax-aware offensive content detection in low-resourced code-mixed languages with continual pre-training. *ACM Transactions on Asian and Low-Resource Language Information Processing.*

Burnap, P., & Williams, M. L. (2014). Hate speech, machine classification and statistical modelling of information flows on Twitter: Interpretation and communication for policy decision making.

Cavnar, W. B., & Trenkle, J. M. (1994). N-gram-based text categorization. In *Proceedings of SDAIR-94, 3rd annual symposium on document analysis and information retrieval.* Citeseer.

Chen, Y., Zhou, Y., Zhu, S., & Xu, H. (2012, September). Detecting offensive language in social media to protect adolescent online safety. In *2012 international conference on privacy, security, risk and trust and 2012 international confernece on social computing* (pp. 71–80). IEEE.

Chiril, P., Pamungkas, E. W., Benamara, F., Moriceau, V., & Patti, V. (2022). Emotionally informed hate speech detection: A multi-target perspective. *Cognitive Computation*, 1–31.

Davidson, T., Warmsley, D., Macy, M., & Weber, I. (2017a, May). Automated hate speech detection and the problem of offensive language. In *Proceedings of the international AAAI conference on web and social media* (Vol. 11, No. 1, pp. 512–515).

Davidson, T., Warmsley, D., Macy, M., & Weber, I. (2017b). *Automated hate speech detection and the problem of offensive language*. ICWSM.

Dreiseitl, S., et al. (2001). A comparison of machine learning methods for the diagnosis of pigmented skin lesions. *Journal of Biomedical Informatics, 34*(1), 28–36.

Eftekhar, B., et al. (2005). Comparison of artificial neural network and logistic regression models for prediction of mortality in head trauma based on initial clinical data. BMC medical informatics and decision making. 5(1).

ElSherief, M., Nilizadeh, S., Nguyen, D., Vigna, G., & Belding, E. (2018). *Peer-to-peer hate: Hate speech instigators and their targets*. ICWSM.

Erico, C., Salim, R., & Suhartono, D. (2020). A systematic literature review of different machine learning methods on hate speech detection. *International Journal of Informatics Visualization, 4*, 213–218.

Fortuna, P., & Nunes, S. (2018). A survey on automatic detection of hate speech in text. *ACM Computing Surveys (CSUR), 51*(4), 1–30.

Hosseinmardi, H., Mattson, S. A., Rafiq, R. I., Han, R., Lv, Q., & Mishra, S. (2015). Analyzing labelled cyberbullying incidents on the Instagram social network. *Lecture Notes in Computer Science, 9471*, 49–66.

Husain, F., & Uzuner, O. (2022). Investigating the effect of preprocessing arabic text on offensive language and hate speech detection. *Transactions on Asian and Low-Resource Language Information Processing, 21*(4), 1–20.

Lewis, D. D. (1998). Naive (Bayes) at forty: The independence assumption in information retrieval. In *European conference on machine learning*. Springer.

Liu, H., Burnap, P., Alorainy, W., & Williams, M. L. (2019, May). Fuzzy multi-task learning for hate speech type identification. In *The world wide web conference* (pp. 3006–3012).

Malmasi, S., & Zampieri, M. (2017). Detecting hate speech in social media. *arXiv Preprint, arXiv*, 1712.06427.

Muennighoff, N. (2020). Vilio: State-of-the-art visio-linguistic models applied to hateful memes. *arXiv preprint arXiv:2012.07788.*

Nobata, C., Tetreault, J., Thomas, A., Mehdad, Y., & Chang, Y. (2016, April). Abusive language detection in online user content. In *Proceedings of the 25th international conference on world wide web* (pp. 145–153).

Panwar, K., Kukreja, S., Singh, A., & Singh, K. K. (2023). Towards deep learning for efficient image encryption. *Procedia Computer Science, 218*, 644–650.

Park, J. H., & Fung, P. (2017). One-step and two-step classification for abusive language detection on twitter. *arXiv Preprint, arXiv*, 1706.01206.

Risch, J., & Krestel, R. (2020). Toxic comment detection in online discussions. *Deep learning-based approaches for sentiment analysis*, 85–109.

Sabat, B. O., Ferrer, C. C., & Giro-i-Nieto, X. (2019). Hate speech in pixels: Detection of offensive memes towards automatic moderation. *arXiv Preprint, arXiv*, 1910.02334.

Saleem, H. M. (2021). *Abusive language through the lens of online communities*. McGill University (Canada).

Schapire, R. E. (2003). The boosting approach to machine learning: An overview. In *Nonlinear estimation and classification* (pp. 149–171). Springer.

Schmidt, A., & Wiegand, M. (2017, April). A survey on hate speech detection using natural language processing. In *Proceedings of the fifth international workshop on natural language processing for social media* (pp. 1–10).

Singh, P. K., & Husain, M. S. (2014). Methodological study of opinion mining and sentiment analysis techniques. *International Journal on Soft Computing, 5*(1), 11.

Singh, A., Dhull, A., & Singh, K. K. (Eds.). (2024). *Blockchain and deep learning for smart healthcare*. John Wiley & Sons.

Thompson, N. (2016). *Equality, diversity and social justice* (6th ed.). Palgrave Macmillan.

Waseem, Z., & Hovy, D. (2016, June). *Hateful symbols or hateful people? Predictive features for hate speech detection on twitter* (pp. 88–93). In Proceedings of the NAACL student research workshop.

Xu, B., Ye, Y., & Nie, L. (2012). An improved random forest classifier for image classification. In 2012 IEEE International Conference on Information and Automation. IEEE.

Yang, F., Peng, X., Ghosh, G., Shilon, R., Ma, H., Moore, E., & Predovic, G. (2019). *Exploring deep multimodal fusion of text and photo for hate speech classification* (pp. 11–18). In Proceedings of the third workshop on abusive language online.

Yuan, S., Wu, X., & Xiang, Y. (2016, March). A two phase deep learning model for identifying discrimination from Tweets. In *EDBT* (pp. 696–697).

Zhang, M.-L., & Zhou, Z.-H. (2005). A k-nearest neighbor based algorithm for multi-label classification. *GrC, 5*, 718–721.

Zhang, Z., Robinson, D., & Tepper, J. (2018). Detecting hate speech on twitter using a convolution-GRU based deep neural network. *Lecture Notes in Computer Science*.

Zhou, B., Tian, Y., Sukhbaatar, S., Szlam, A., & Fergus, R. (2015). Simple baseline for visual question answering. *arXiv*.

Chapter 12
Multi-modal Generative AI for People with Disabilities

N. R. Raji, C. L. Biji, and V. Vineetha

Abstract With the advancement of technology, especially in the applications based on artificial intelligence, many new avenues opened to the persons with disabilities (PwD) in improving their quality of life. This population faces different challenges across various domains. This chapter tries to explore the possibilities of multi-modal generative artificial intelligence (GenAI) on refining the lives of PwD. It also discusses various challenges of these persons in the areas of communication, mobility, education, and emotional well-being. The potential of multi-modal GenAI can address these major issues through augmentative and alternative communication (AAC), visual assistance, mobility assistance, personalised accessibility solutions, communication, intervention, education and employment, independent living, mental well-being, and emotional support. This chapter also discusses the recent trends and challenges in the implementation of multi-modal GenAI applications. Data privacy, bias, and technological barriers are indeed major concerns to be addressed in this regard. Finally, the paper highlights the future scope of multi-modal GenAI in progressing accessibility, inclusion, and empowerment for PwD, underscoring the potential for further innovation and collaboration in this crucial field of research and development.

Keywords Multi-modal GenAI · Disability · Assistive technology · Artificial intelligence

N. R. Raji (✉)
Department of Computer Science, Degree (HI), National Institute of Speech and Hearing, Thiruvananthapuram, India
e-mail: rajinr@nish.ac.in

C. L. Biji
Department of Analytics, School of Computer Science and Technology, Vellore Institute of Technology, Vellore, Tamil Nadu, India

V. Vineetha
Verizon Wireless, Southlake, TX, USA

A. Singh, K. K. Singh (eds.), *Multimodal Generative AI*,
https://doi.org/10.1007/978-981-96-2355-6_12

12.1 Introduction

As per World Health Organisation (WHO) reports, 1.3 billion people, that is, 16% of the world population, has considerable disability. This population is a diverse population, and the different aspects, like gender, age, race, sexual orientation, financial condition, and their ethnicity, affect their life experiences and health needs. According to the United Nations (UN) fact sheet, 80% of PwD live in developing countries. Among that 80%, only 11% has higher education and 19% has lower education. Adverse socioeconomic outcomes, such as poor education, worse health outcomes, lower employment rates, and higher rates of poverty, are more common among people with disabilities. The Rights of Persons with Disabilities (RPWD) Act (2016), India has identified 21 conditions of disability. The identified disabilities are blindness, low vision, leprosy cured persons, deaf and hard of hearing, locomotor disability, dwarfism, intellectual disability, mental illness, autism spectrum disorder, cerebral palsy, muscular dystrophy, chronic neurological conditions, specific learning disabilities, multiple sclerosis, speech and language disability, thalassemia, haemophilia, sickle cell disease, multiple disabilities including deaf blindness, acid attack victim, and Parkinson's disease. The accommodations needed for each disability are well documented in this act. The challenges faced by a person with disability (PwD) vary depending mainly on the disability and the accommodations needed to overcome these challenges also vary.

To provide equal opportunities to all, the accommodations that can be given in each area of life need to be identified. Assistive technologies are gaining popularity as they help this population to lead an easy life. Artificial intelligence is the new popular word of the era, and it finds its application in almost all domains, like art, education, medicine, music, business, marketing, coding, designing, etc., making the works easier, more creative, and more accessible. By the end of the 2022, generative artificial intelligence (GenAI) captured popularity with its ability to create human-like content, which includes text, video, audio, and tabular data. With the introduction of ChatGPT and Dall-E, a milestone was set in the field of artificial intelligence, especially in content generation (García-Peñalvo & Vázquez-Ingelmo, 2023). The recent development in this area is multi-modal generative AI, which can accept input of various modalities and output content on various modalities. The power of GenAI is vast and continuously expanding. The basic multi-modal GenAI is represented in Fig. 12.1. It can be considered a fusion model that can accept different data type inputs, which include tabular data, images, text data, audio, and

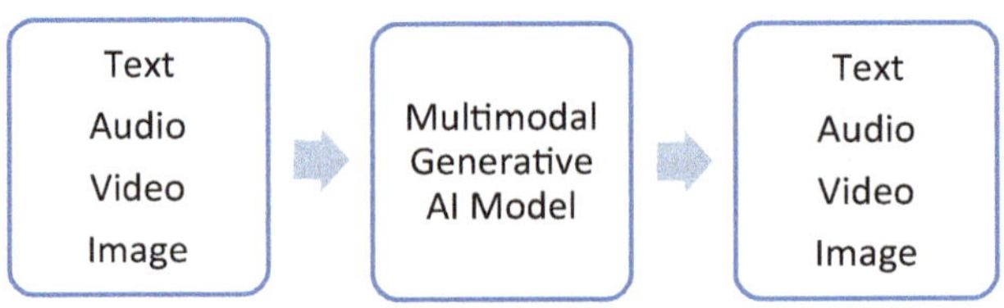

Fig. 12.1 Basic multi-modal generative AI model

videos. Though this technology is used in various domains, one of the areas, which it holds particular promise, is in applications, which can provide accessibility to PwD. This technology can be used to provide equal opportunities for all in all domains, including education, transportation, social interaction, communication, and employment irrespective of their physical challenges.

This chapter is an attempt to explore the possibilities with multi-modal GenAI to improve the lives of PwD with a focus on accessibility, inclusion, and empowerment. The next section outlines the disability aspects and challenges faced by PwD. The accessibility issues are also discussed in this section. Section 12.3 gives a general overview about multi-modal GenAI. Section 12.4 discusses the possible and current applications that can enhance the life quality of PwD. Section 12.5 discusses the challenges, and the last section (Sect. 12.6) concludes the chapter with throwing light on the future scope of multi-modal GenAI in the area.

12.2 Understanding Disability and Accessibility

According to the WHO, disability is any impairment, restriction, or limitation in one's ability to do activities that is primarily brought on by environmental and health-related factors. The report of disability mentioned earlier states that the chances of developing conditions like anxiety, depression, obesity, and asthma are twice as high for persons with disabilities when compared to others. It further states that the persons with disabilities suffer many inequities like stigma, discrimination, poverty, and barriers to education and employment. In India, according to the census of 2011, the distribution of disability across various disabilities is shown in Fig. 12.2. Considering the fact that 80% of PwD in the world live in developing countries, this can be considered a representation of statistics across various disabilities in developing countries. Here, the 20% of the total disabled population is people with mobility issues. The people with challenges in hearing and vision come next.

According to the United Nations International Children's Emergency Fund (UNICEF), 240 million children in this world are with disabilities, and the majority of these children are less likely to attend school. Most of the PwD have fragile health, and the majority are not able to afford treatment or buy assistive devices. The WHO report says that the PwD experience dependency, and their participation in society is much restricted. This statistic reveals the situation of PwD, which needs urgent attention. Recently, there has been an advancement globally in creating disability inclusion awareness. Many international organisations have shown keen interest in this area and taken initiatives to promote this agenda.

The challenges faced by PwD vary individually depending on the disability. The visually challenged persons face difficulties participating with respect to mobility, domestic life, interpersonal interaction and relationships, major life areas, and leisure activities (Salminen & Karhula, 2014). The communication and language difficulties are the challenges faced by the deaf and hard of hearing population

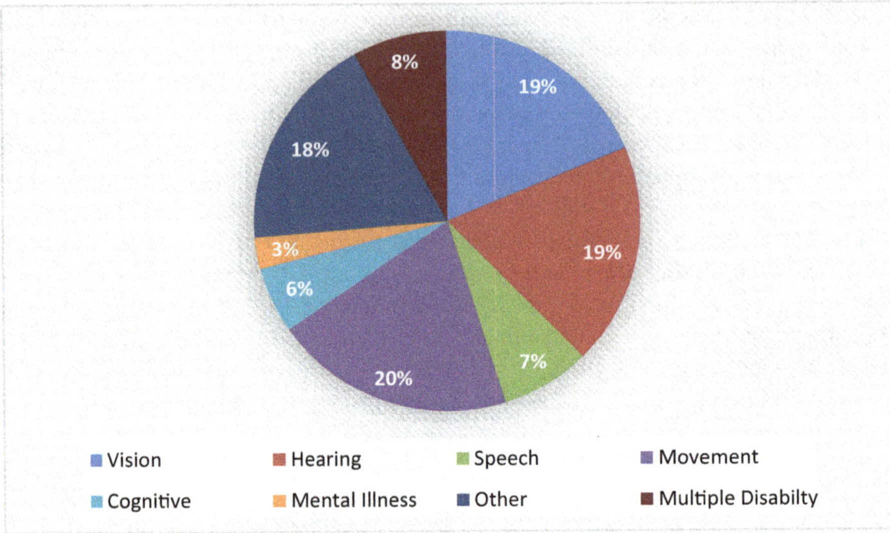

Fig. 12.2 Disability Census, India 2011

(Lederberg et al., 2013). Due to these challenges, they find difficulties in education, access to public services, and social interaction. The people with mobility challenges majorly find difficulty in transportation and navigation. The infrastructure accessibility is very crucial for them to access the facilities. Many of these buildings are not accessible to this population (Oladipo et al., 2008). The challenges faced by the PwD during the climate and environmental changes are discussed by the researchers, and it is observed that the challenges faced by the PwD populations in shelters, extreme conditions like heat waves and floods, are severe compared to the other people. The challenges identified include the lack of assistive devices, inaccessible facilities such as beds and bathrooms in shelters, and incidents of sexual violence (Kosanic et al., 2022).

The challenges faced in education by PwD also vary. The verbal instructions given by the teacher, language inadequacy, and proper classroom settings are some of the challenges faced by deaf students. The students with low vision also suffer with classroom settings with poor lighting and non-availability of accessible libraries and bathrooms (Lintangsari & Emaliana, 2020). The infrastructure accommodations are needed for persons with mobility challenges. During the pandemic, the students with disabilities faced difficulties in accessing online platforms (Clinton, 2020). In general, for providing better educational opportunities for PwD, the infrastructure, classroom setting, sensitisation of disability among teachers, and accessibility in all activities need to be ensured for full inclusion of this population (Moon et al., 2012). The mental health challenges experienced by autistic students at university need to be identified and highlighted, which may help faculty and support services to provide required accommodations (Scott & Sedgewick, 2021). The social, communicational, and behavioural characteristics of individuals with autism spectrum disorder are unique to each individual. Due to this, they face challenges in employment. By considering these characteristics and providing proper accommodations, they can be made employable (Hendricks, 2010).

The barriers that have been highlighted in the workplace include an inaccessible physical environment, the absence of appropriate assistive technology, and negative attitudes of people towards people with disabilities (Narayanan, 2018). According to research, women with disabilities frequently encounter difficulties when travelling to obtain specialised medical care and when attempting to access unfriendly physical health facilities. Other related barriers to access include the insensitivity and ignorance of healthcare professionals regarding the unique maternity care requirements of women with disabilities; unfavourable attitudes of service providers; the belief held by people in good health that women with disabilities ought to be asexual; and health information that is not specific enough in addressing the unique maternity care requirements of women with disabilities (Ganle et al., 2016). Some of the accessibility challenges faced by different PwD are represented in Table 12.1.

In all areas of life, the PwD faces challenges, and this exclusion can be eliminated in one way or another by providing more accessibility to this population. The advancement of technology has helped to overcome the challenges faced by the PwD population to a great extent. The wearable devices that help the blind to navigate freely and independently and text-to-speech conversion software help them to access the libraries. Similarly, the advanced digital hearing aids and speech-to-text conversion software help the deaf to overcome their barriers (Singh et al., 2021). The assistive devices that are commonly used by the deaf and hard of hearing are hearing aids, cochlear implants, assistive listening devices, real-time captioning systems, captioned telephones, and video relay systems. The popular assistive devices used by blind or low-vision people are screen readers, text-to-sound converters, magnifying tools, Braille converters and identifiers, and tactile graphics. AACs are the popular device among the individuals with autism. They also use visual schedulers, sensory integration apps, social skill training apps, etc. The people with mobility challenges use automatic wheelchairs, prosthetic limbs, and voice command-controlled devices. The studies in all areas of disability are going on with an aim of developing new technologies that make the world accessible to all.

12.3 Overview of Multi-modal GenAI

The newest development in AI technology is called generative AI (GenAI) (Goodfellow et al., 2014). Deep neural networks are utilised in the design of generative AI models to learn the structure and pattern of large training datasets in order to produce new data that is similar. GenAI can generate a variety of content types, including text, video, audio, images, and other types of data (Mariani & Dwivedi, 2024). Multi-modal GenAI can be considered as an extension of GenAI that can produce more exact predictions or more accurate decisions concerning "real-world"

Table 12.1 Accessibility challenges faced by PwD

Individuals with disabilities	Accessibility challenges faced
Deaf and Hard of Hearing	• Non-availability of sign language interpretation during public events • Non-accessibility to social services due to communication difficulties • Closed captioning not available for most of the videos • Non-accessible audio alerts and alarms • Inadequate sound amplification systems • Non-accessible massive open online courses (MOOCs) • Limited higher education opportunities due to inaccessible teaching resources • Lack of classroom accommodations like note takers or sign language interpreters or real-time captioning services
Visually Challenged	• Websites and online applications that are not compatible with screen readers • Inaccessible online visual content due to the absence of alternative texts • Difficult navigation • Unawareness of visual cues during social interactions • Unawareness of surroundings as a whole • Identification of poorly placed objects and obstacles • Colour contrast problem for people with poor vision • Non-availability of braille identification and translation to text to braille • Inaccessible learning resources in text format
People with Autism	• Bright environments or noisy environments that are overstimulating • Difficulty in understanding visual cues and non-verbal communication • Appropriate AAC devices for communication based on individual needs • Complex environments that may cause anxiety
Mobility Challenges	• Navigation difficulty due to inaccessible infrastructures • Wheelchair non-friendly environments • Inaccessible public places and transportation facilities • Dependence on others for operating computers and other electronic devices for bedridden patients

settings, scenarios, or challenges. The multi-modal GenAI models are trained on datasets having different data types that need to be interpreted, whereas the traditional GenAI focuses only on one type of data.

A paradigm shift in creativity, learning, personalisation, efficiency, and problem-solving is represented by multi-modal generative AI. Creating original content in various genres like poetry, music, coding, and realistic images is one of the popular and useful abilities of multi-modal GenAI. This ability changes industries like education, entertainment, and design by opening more creative avenues. The technological advances in the area of machine translation and sentiment analysis are improved by combining data of different modalities, as it improves the learning, training, and comprehension of models. The experiences can be customised and personalised depending on the preferences of each individual, which typically improvise many applications in the areas of marketing, healthcare, and education (Singh et al., 2022). The intervention of multi-modal GenAI in data analysis and content generation by automating the processes encourages faster innovation and efficiency. It also simplifies human tasks and saves time, which can be effectively utilised in more important activities. As the technology deals with different modalities, it helps in identifying relationships between various modalities and understanding hidden patterns and solutions.

The unique feature of the multi-modal GenAI is its ability to accept inputs of different modalities and deliver the output in various modalities after interpreting the hidden relations between the different modality input data. Each of these inputs is processed through a separate subnetwork and also allows the interaction between them. A simple block diagram on the same is represented in Fig. 12.3. The various inputs undergo different pre-processing and encoding processes. The methods for pre-processing and encoding vary depending on the input modality. The text input after encoding undergoes an embedding process that helps to represent the encoded text in fixed-size vector form, which captures the semantic meaning of the input text. The pre-processing of images includes resizing, normalisation, colour enhancement, or converting into greyscale and other similar process that enhances the digital quality of images. The audio is pre-processed with resampling, extracting features, normalising, etc., making the input compatible to the model. Extracting frames, resising, and normalising are the popular video pre-processing steps. After the encoding, all the inputs will be in numerical format. This is fed to a multi-modal fusion network. The numerical representations obtained from various modalities are combined here. For combining the inputs, various methods are used. Concatenation of inputs from various modalities is one common method, while another method is to find the element-wise sum or average of embeddings. Attention mechanisms are used to determine which modality or part of the modality should be given more importance. Along with the integration of these inputs, the model identifies the complex relations, interactions, and patterns that exist between these inputs. This makes the output contextually rich and diverse. The output generator generates the output based on the finding obtained from the previous block. This can be in the form of text, audio, video, image, or a combination of any of these.

Though the data needed for the training is large, its key features of contextual awareness and cross-modal understanding outrun the limitations. As the model learns across various existing relations among various inputs of different modalities, the processing capacity needed is also large. When the model is trained using inputs of one modality, extra measures need to be taken to reduce the bias. As the multi-modal generative AI considers various inputs, it has the potential to improve fairness and reduce bias by evaluating a variety of datasets and producing outputs that are less prone to bias from humans. The diverse needs of PwD can be addressed by creating specific tools and materials that suit their individual needs. Apart from these, the challenges faced by PwD need to be addressed in a creative way that may be beyond the human intelligence alone. By fostering creativity and the problem-solving abilities with modern technologies, novel viewpoints can be provided, which suggest new adaptive solutions to the challenges faced by PwD.

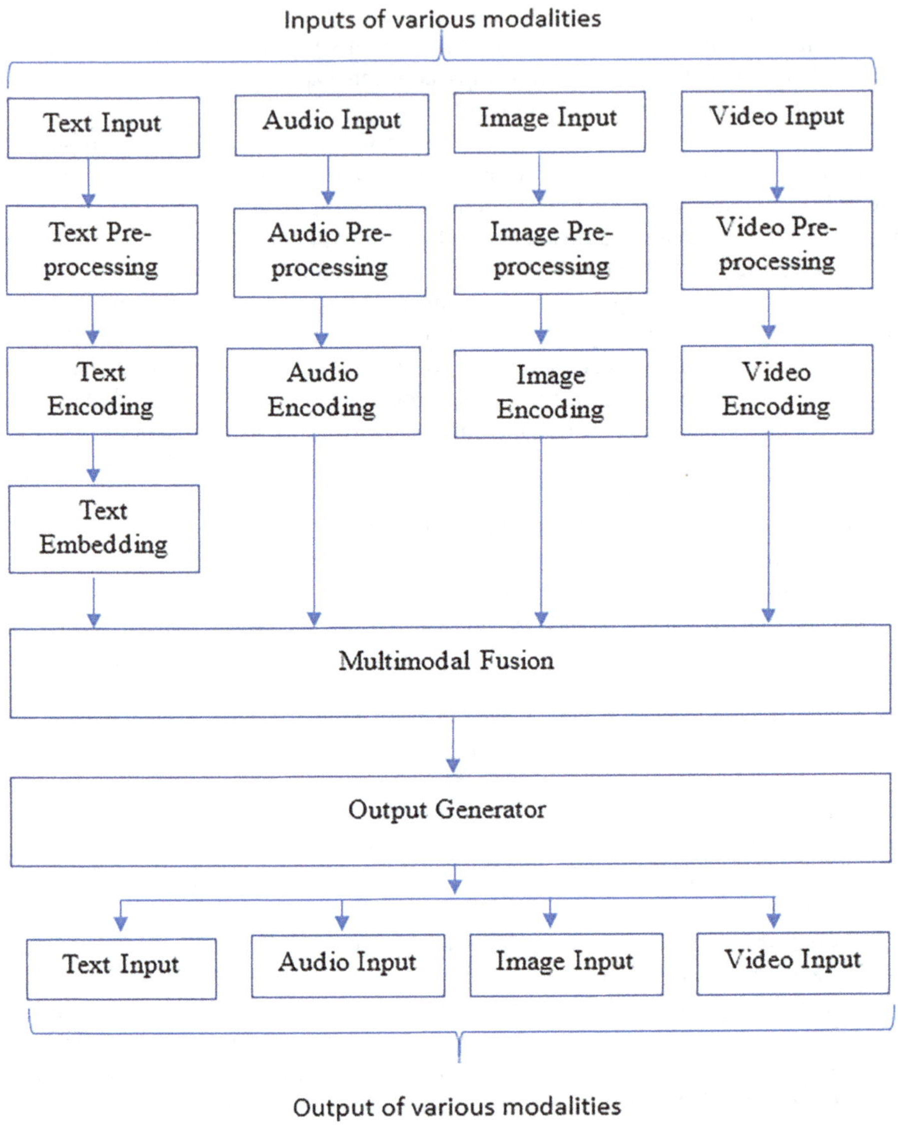

Fig. 12.3 Block diagram of multi-modal GenAI application with various input–output modalities

12.4 Applications of Multi-modal GenAI for the PWD

The recent developments in GenAI made it popular worldwide, and more studies are going on developing new applications in various domains. This book chapter aims to identify the possibilities and challenges of using multi-modal GenIA in the

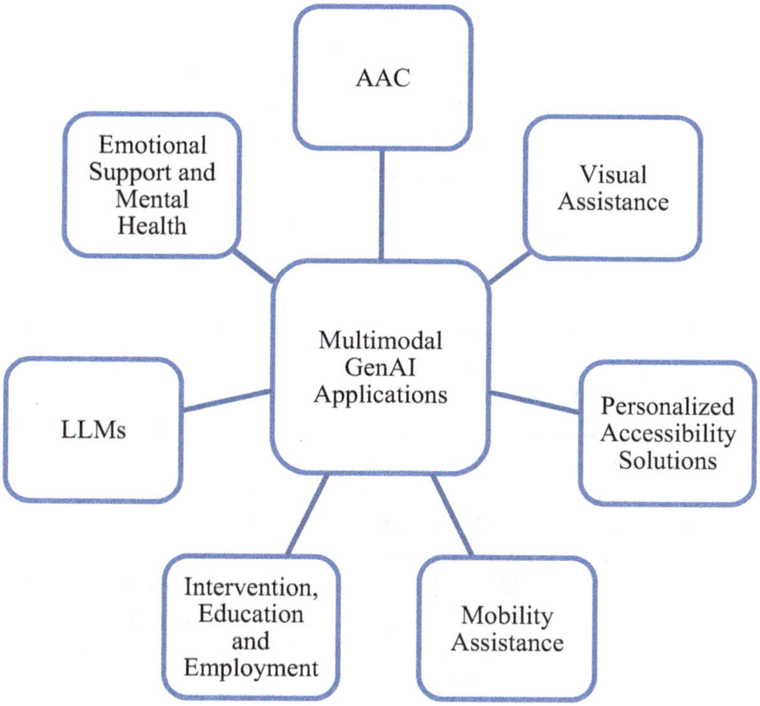

Fig. 12.4 Multi-modal GenAI application domains for PwD

domain of assistive technology, which in turn can improve the life of PwD. The various application domains are visually represented in Fig. 12.4.

12.4.1 Augmentative and Alternative Communication (AAC)

American Speech and Hearing Association (ASHA) describes AAC as "multiple ways of communication that can supplement or compensate (either temporarily or permanently) for the impairment and disability patterns of individuals with severe expressive communication disorders." AAC usually provides all ways of communication to people except talking. In many situations, the people who have communication difficulties are unable to express their competence, potential, and ability to think and learn. AAC aids such people to confidently communicate their needs. The assistive devices that are used for AAC include electronic boards, keyboards with text and images, tablets, and laptops. The text-based keyboards are generally used by the people who can type. For some people who cannot spell or read, symbol- or picture-based AAC devices are used. Clicking on the pictures in electronic boards generates sounds and also constructs sentences, which helps the people with speaking difficulty (Pereira et al., 2021).

Though speech is one of the most effortless modes of communication, the preference of communication mode varies according to many factors. The people with autism spectrum disorders vary widely, and they largely depend on these assistive technologies (Aydin & Diken, 2020). Apple revolutionised the communication boards by introducing easy-to-use apps that helps the kids with autism to put words and symbols together and eventually create sentences (Allen et al., 2016). Text input can be educated into natural-sounding voice using GenAI models. This allows people who have trouble speaking to interact by selecting or entering text on a device, which the AI model subsequently turns into voice. Image captioning-trained GenAI models are able to provide natural language descriptions of images. This can help by giving written explanations of their surroundings or objects of interest to people who struggle to perceive or describe visual information. The multi-modal GenAI can improve the current AAC devices by combining the text-to-speech feature and image captioning feature (Singh & Saravanan, 2024). The user can choose his preference of input and generate an image with sound and caption. Predictive text input can also be included in AI-powered AAC systems depending on the user's past and the context of the discussion. This minimises the effort needed to type or choose words, and it also speeds up conversation.

The currently available AACs with audio output can be enhanced by multi-modal GenAI techniques by incorporating sentiment analysis features. The system can add the user feelings to the generated audio output if the device is able to identify the emotion of the user. The ability of the GenAI model to learn and adapt can be useful in AACs to understand the unique communication pattern of the user. This can be useful in improving prediction, which may give clearer and more useful outputs.

Natural language understanding tasks, including sentiment analysis, intent detection, and question answering, can be trained with multi-modal GenAI models. This can facilitate more effective communication using AAC devices when people express their wants, feelings, or concerns. AAC systems are able to recognise and interpret user gestures by combining multi-modal techniques with vision-based artificial intelligence algorithms. This enables people to converse through gestures and sign language, which are subsequently converted into text or voice output. As sign language is considered as mother tongue of most of the deaf people, the conversion of text or audio into sign language will also be helpful for them for effective communication and also in academics. The various AAC tools and their benefits are shown in Table 12.2.

12.4.2 Visual Assistance

The popular assistance for the smooth navigation of the visually challenged are smart canes, mobile applications, e-canes, laser-based walker, and infrared cane (Khan et al., 2021). The short-range capability of ultrasonic and infrared rays is one of the limitations of these devices. The possibilities of multi-modal GenAI to improve these devices are yet to be explored. Deep learning techniques along with

Table 12.2 AAC tools and their advantages

Applications	Advantages
Converts text, images, and symbols to meaningful audio sentences	• User can effectively communicate his needs
Image description in natural language	• People with visual difficulty get descriptions of images
Sentiment analysis	• The behaviour analysis of the user can be done • Includes user's feelings in the audio output • Prediction based on individual communication patterns
Sign language translator that converts sign language to text and sound	• Enables the deaf to communicate easily

IoT are widely used in many advanced navigation applications. The multi-modal feedbacks from these applications are widely appreciated by the visually challenged population. Safety, ease–of-use, and overall experience are the factors expected by the users from the applications that assist in both outdoor and indoor navigation (Nair et al., 2022). Based on this perspective, prototypes for real-time location sharing, indoor and outdoor navigation, obstacle avoidance, and obstacle recognition are created (Khairnar et al., 2020). With the advancement of artificial intelligence, a variety of devices are designed with computer vision algorithms and deep learning technologies to assist the mobility of the visually challenged (Tapu et al., 2020; Manjari et al., 2020).

Images may be analysed using multi-modal GenAI models, which can then describe the scene, objects, and the people in them. This makes it possible for visually challenged people to comprehend their environment better. Using cameras or wearable technology, these models can identify and categorise items in real-time, whether they be images, videos, or any other sort of data. With the use of this knowledge, people will be able to explore both interior and outdoor surroundings with greater independence. The modern AR applications with multi-modal GenAI made real-time overlays of needed data into the direct field of view of the user possible. This helps the user to get a better understanding of their surroundings with explanations, cues for navigation, signage with descriptions, and identification of objects.

The people with visual impairment usually depend on text-to-sound conversion applications. These applications helped them to lead an independent life and improve the quality of life. This can be further improved by using multi-modal GenAI applications that can recognise non-verbal cues during social interactions by recognising gestures and facial expressions. The persons with low vision or colour blindness can benefit from this technology as it can recognise colours in the surroundings and describe the colours or modify the contrasts. These applications can be integrated on wearables, mobiles, or any other assistive gadgets that can provide seamless visual support to the users wherever they go. These visual aids with multi-modal GenAI capabilities can be designed or customised based on the preferences

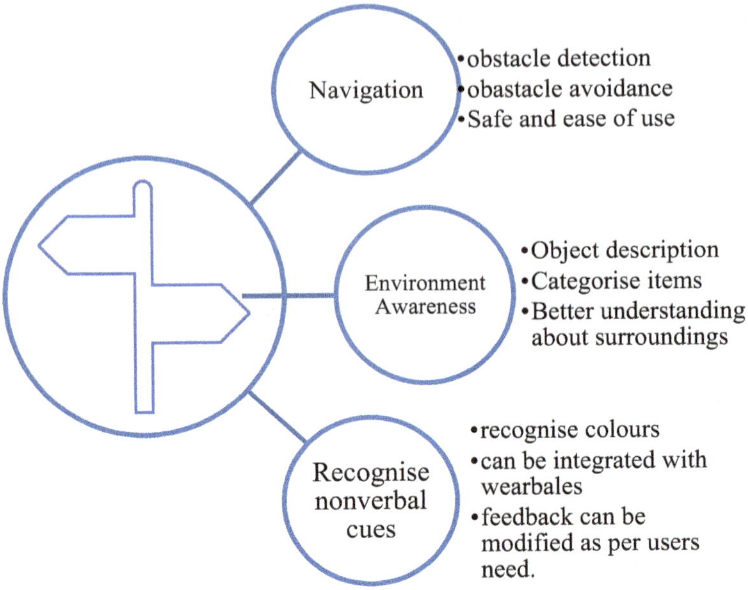

Fig. 12.5 Visual assistance applications and their advantages

or requirements of the users. The output modality of these applications can be modified according to the user's feedback, like auditory or tactile feedback. The visual assistance devices are summarised in Fig. 12.5.

12.4.3 Personalised Accessibility Solutions

Multi-modal GenAI's capabilities can be effectively utilised for developing customised accessibility solutions that meet the various needs and preferences of people with disabilities. The usage of input and output of different modalities, the multi-modal GenAI systems can offer more inclusive, flexible, and effective support in various scenarios and situations. Using this significant feature of multi-modal GenAI, the user needs and preferences can be comprehended in a more detailed way. This further helps to provide personalised assistance to the needy PwD. The outputs can be generated according to the personalised preference of the user. This way the accessibility to the output and input can be ensured. For instance, the deaf population is a very diverse population. The hard of hearing usually prefer subtitles, while the persons with profound deafness prefer output in sign language. That is, the program can modify the input and output according to the user's preference. These diverse data streams can be processed and integrated by multi-modal GenAI to give a comprehensive picture of the context and requirements of the user.

As communication is the biggest challenge for persons with deafness, navigation is the major difficulty for the visually challenged. By combining the user's data regarding his location, schedule, and communication history, timely reminders and navigational assistance can be provided. Multi-modal GenAI systems can be modified continuously by learning from user interactions and feedback. Through an iterative learning process, the system's comprehensive abilities about user preferences and requirements get improved. This can help the system to provide more effective navigation assistance to the user. To suit various skills and preferences of users and provide customised accessibility solutions, feedback needs to be provided through various sensory channels. To achieve this, the capability of multi-modal GenAI to produce various modality outputs can be used. According to user preference, the output can be text or speech or haptic feedback or visual cues or a combination of any of these. To achieve the best personalised accessibility solutions, collaboration between user and AI system is needed, which can be effectively achieved using multi-modal GenAI that can understand and react to human input in a variety of modalities. Table 12.3 represents the examples of personalised accessibility solutions.

12.4.4 *Mobility Assistance*

Multi-modal GenAI models can improve accessibility, independence, and quality of life by providing solutions for mobility challenges of persons with disabilities. These models can offer individualised route planning and real-time directing based

Table 12.3 Personalised accessibility solutions

Beneficiaries	Advantages
Deaf and Hard of Hearing	• Subtitles/transcripts for the hard of hearing • Sign language interpretation for deaf
Visually Challenged	• Navigation solutions based on user travel history, communication history, reminders, etc. • Feedbacks based on user preference
People with Autism	• Sensory inputs tailored to individual needs • Preference to individual communication mode and capabilities • Identification of one's emotions and response according to affective states like stress or anxiety • Visual schedules • Feedback according to user need and reinforcement • Customised social skill training
Challenges in Mobility	• Integration with assistive devices like wheelchairs, prosthetic limbs, etc. • Communication interfaces as per user needs • Voice command control • Improved navigation • Identification of accessible services and places

on mobility limitations and preferences of users. Based on accessibility options like wheelchair ramps or elevators, routes can be chosen. By combining data from GPS, indoor positioning systems, and environmental sensors, alerts can be given about risks and obstacles in the path. By integrating multi-modal GenAI-based apps into the mobility devices, like wheelchairs or prosthetic devices, their functionalities can be enhanced. Through continuous learning from the user movements, the devices can adjust according to the user movements, the terrain, and physiological markers. This can maximise the flexibility and safety. Also, these devices can then provide more effective stability aids, terrain detection, and predictive control.

When an AI system can read the user gestures or spoken cues and execute commands to change the mobility device's speed or direction or any other characteristics, it enhances the autonomy and convenience of the user. Multi-modal GenAI, with its gesture detection and voice instruction detection capabilities, helps the user in hands-free operations of mobility devices. This increases the independence of those with restricted movement navigating in their environment. Studies on multi-modal GenAI-driven assistive robots are progressing, which can help people with mobility problems by physical support and help with everyday life activities. These robots can respond to verbal instructions, gestures, facial expressions, etc., which in turn give greater independence and self-reliance to the people with mobility challenges.

The accessible transportation services are another challenge faced by the people with mobility issues. By combining the data from various sources in various modes, the multi-modal GenAI model can provide information on accessible routes, available vehicles, and transportation options. Through integration with ride-sharing services and transportation networks, it helps users find wheelchair-accessible cars and plan accessible routes, increasing mobility and fostering community involvement. The model can support VR-based therapies to improve mobility and coordination for those with physical limitations. Also, the social isolation decreases when the accessibility options increase. Using this advanced model-based application, the accessibility can be increased, which will motivate the public services also to adopt measures to provide accessibility in their services. The advantages of mobility assistance powered by multi-modal GenAI are shown in Fig. 12.6.

12.4.5 Intervention, Education, and Employment

The higher education opportunities for PwD are very limited. Most of the time, an inclusive environment does not cater to their special individualised needs. Multi-modal GenAI can play a major role in paradigm changes in education by including inclusivity, empowerment, and tailored learning experiences that value each learner's unique capabilities and potential. Providing education to all is one of the primary responsibilities of all countries, which in turn decides the development and growth of the country. Using multi-modal GenAI models, a more equal and accessible environment can be provided by removing all obstacles to growth and

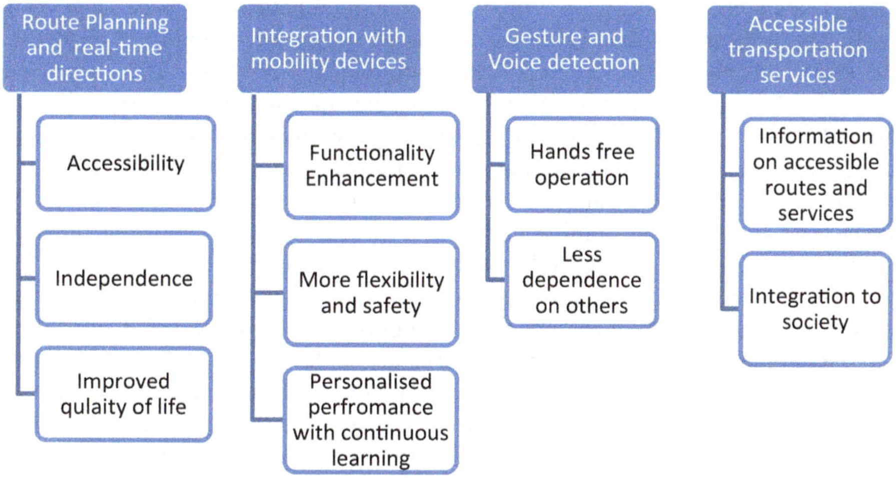

Fig. 12.6 Advantages of multi-modal GenAI powered mobility assistance applications

achievement. The capability of multi-modal GenAI in processing inputs in different modalities and providing outputs in different modalities ensures people with different kinds of disabilities—like visual, auditory, or cognitive challenges—interact with instructional materials in ways that best fit their own learning preferences and styles.

With the technological advancement and usage of online resources, many doors are open for all students to access numerous academic resources available across the world. The accessibility of these materials is still inadequate for PwD. The higher education opportunities of deaf and hard-of-hearing students are very limited. Though the massive open courses provide subtitles, for a deaf student the content is still not accessible because of his language difficulties. Many of the web images are without annotations, which makes inaccessible to a visually challenged student. Multi-modal GenAI-based applications can provide a universal solution for this by providing sign language interpretation along with audio and image descriptions. The visual representation of audio clues and the audio representation of visual clues can be very helpful for all the students. The teachers can use the customised and individual instructions and plans for students with disabilities according to their needs with the help of multi-modal GenAI-based applications.

Education and training are two important aspects for an individual to get employed. The employment improves the living conditions of an individual, which in turn improves the quality of his life. For a PwD, accessibility to the education and training is very important. Language learning is one of the major hurdles faced by the deaf due to the lack of auditory input. The GenAI models like ChatGPT play an important role in language learning. For the text input, it is able to generate images and simple explanations, which is useful for all who need to learn a language. In employment, the deaf employers always find it difficult to write proper official emails due to their language challenges. This hinders their career growth. The

GenAI aids in this writing proper emails without grammatical mistakes. The GenAI tools that offer more accessibility need to be encouraged by employers to include the PwD and encourage them to perform with their full potential. These tools can empower the students by motivating them. For example, a text or a story written by a blind student in Braille can be seen as an image or video that describes his world in a 3D environment that will be enjoyable for his classmates (Yang & Zhang, 2024).

During the era of the pandemic, most of the interventions that are needed for PwD were given through online methods. This led to the development of many applications that can be used by individuals for identification and therapies. Communication disorder identification is one of such applications. For stammering or stuttering identification and for its intervention therapies, the audio signals are not enough. During the therapy, the therapist also notes the expressions along with the audio signals. Most of the persons with stuttering or stammering need psychological intervention too. This is considered as part of the therapy. The multi-modal GenAI systems can automate such sessions by capturing the facial expressions, audio signals, and emotions of the user. This will be helpful in designing individualised therapy sessions for each person considering their individual needs. The advantages of using AI tools in intervention, education, and employment are tabulated in Table 12.4.

12.4.6 Empowerment with Large Language Models for Hearing Impaired

The advancements in large language models (LLMs) have been revolutionising various industries, including healthcare, finance, credit scoring, ecommerce, and cost-effective tutoring. The ability to model long-term dependencies has significantly improved, allowing for more coherent and contextually relevant text outputs. This is particularly beneficial in dialogue modelling, where maintaining context and

Table 12.4 Advantages of using GenAI-based tools in intervention, education, and employment

Domain	Advantages
Intervention	• Individualised and customised therapies • Automatic identification of individual needs from the behaviour • Feedback modality can be decided based on user need
Education	• Sign language interpretation • Audio to text and images • Simplified text for people with language difficulties • Accessibility to resources like online courses • Language learning support • Audio-to-Braille conversion and to digital format
Employment	• Improved accessibility • More participation in group discussions and meetings • Accessible tools for training • Improved communication

coherence over a series of exchanges is crucial. The role of powerful LLMs like ChatGPT in enhancing accessibility, communication, and inclusivity is important in empowering people with disabilities (Wei et al., 2023). LLMs are deep learning models trained on a large corpus of text data. Its success in processing natural languages is indeed inspiring to explore its use case in sign language recognition. Generally, communicating with hearing-impaired people seems difficult as the sign language itself is uniform throughout. Establishment to communicate between both parties is possible through a human translator, but this is challenging too. The lack of experts as translators adds to the barriers for the hearing impaired. Sign language recognition (SLR) systems are used to establish this communication link without a human mediator. Established state-of-the-art literature is mainly focused on alphabets/digits/static signs/few words and sentences. Therefore, an urgent need to focus on dynamic signs and non-verbal kinds of communication is required. Though many advanced solutions exist under the sign language recognition task, still there exists a gap in communication between the deaf and hearing. LLM-based applications for sign language learning processes become rewarding between the deaf and hearing individuals. As a learning tool, the LLM-based application has the potential to redefine our understanding of how we most effectively teach and engage with sign language, thereby providing wider opportunities to the 70 million deaf community across the globe.

Recently, sign language recognition (SLR) systems are used to establish this communication link. Since many people do not know the sign language, communication becomes a major obstacle and affects communication. Automated two-way communication between sign language and text or speech is an unsolved problem globally. Sign language recognition can be done in two ways: glove-based or vision-based recognition. Although some traditional desktop computer-assisted tools were created for encouraging active listening, the usage rate is not high. As mobile smart devices become widely popular among the deaf community, LLM-based translators will have wider acceptance.

The sign language recognition system using LLM is shown in Fig. 12.7, which includes four key modules as reported by Telmo et al.: (a) video dataset

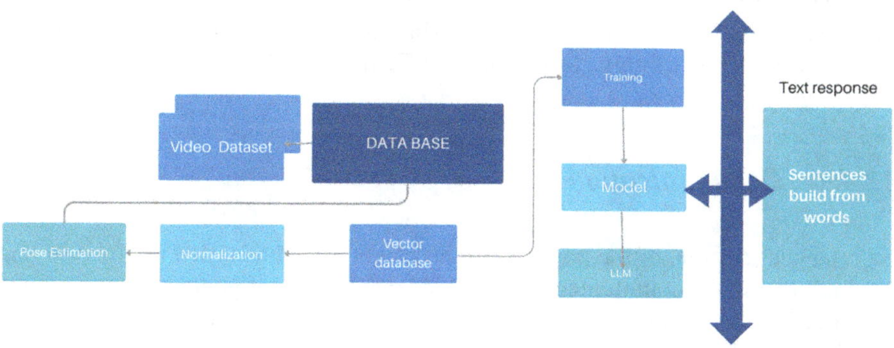

Fig. 12.7 Sign language recognition system using LLM

compilation, (b) video frame construction on database, (c) feature extraction and normalisation followed by storage in the vector database, and (d) user console for displaying the text (Adão et al., 2023). Jia Gong et al. proposed a novel SignLLM capable of generating sign token representation following the language-like characteristics. A vector quantised visual sign module is optimised through a context prediction to produce sign sentences. The Codebook Reconstruction and Alignment module empowers the conversion of character-level sign tokens into word-level sign tokens (Gong et al., 2024).

LLMS can be configured in three ways: (a) provide a prompt and get the response, (b) apply retrieval-augmented generation, and (c) additional training. Retrieved augmented generation of prompt is more reasonable in terms of resources. Initially, datasets need to be collected from authenticated sources. The collected documents may have images, textual data, or even scanned documents. Later, text splitters break documents into splits of specified size and will be stored in a vector store. Based on the prompt added, the application retrieves similar embeddings to the prompt and generates the response. Given a prompt, the language model calculates the probability of the next sequence of words in the context of human language. These methods are indeed encouraging, as they can give customised assistive communication, which helps to lift up the horizon of the impaired. The models could provide a contextually relevant sentence. The results verify the capability of such technology to be supportive to the hearing impaired, making their activities more manageable (Gadiraju et al., 2023). With the advance of foundation model, like FLA-LLM, it is expected to have nearly a 15% improvement in time management under various sectors. These models have huge scope in making assistive technologies for the hearing impaired.

Multi-modal large language models can use databases to retrieve specific information, enhancing the accuracy. The users experience more inductive responses. It can make sense of context better by linking information from different types of data. These models can perform real-time interpretation of sign language for those with hearing challenges (Hutchinson et al., 2020). With the numerous benefits, there are potential risks with LLM-based applications. There needs to be a mechanism to take care of the security and privacy of the data. Currently, the bias in the LLM-based chatbots is severe, as it needs more diverse data to comprehend generalised responses. It is quite integral to add hearing-impaired people in the design and development of such architectures (Lever & Geurts, 2016).

12.4.7 Communication

Sign language is one of the popular communication methods used by the deaf and hard of hearing for communication. As this language is not popular among people with hearing, the deaf people find it difficult to interact with society. Sign language translation is the popular solution recommended to solve this difficulty. In many day-to-day life situations, like in hospitals, railway stations, and airports, the communication barrier makes their lives difficult. Sign language is a complete language with its own grammar and structure. The body language, hand movements, hand positions, and facial expressions are part of its grammar. To effectively translate sign language to text and text to sign language, all these grammar parts need to be considered. Also, the regional variations of sign language are large, which makes its vocabulary huge. American Sign Language is a standardised language that is popular in many countries. But different countries use different sign languages, like British Sign Language, Indian Sign Language, Spanish Sign Language, Irish Sign Language, etc. Some of these sign languages use both hands to sign, while some others use only one hand. Even within the countries, the sign language varies (Singh et al., 2022). Though many studies are in progress, an efficient sign language to text converter is yet to be developed. The multi-modal GenAI, which accepts various modalities of input, can be promising in developing an efficient sign language translator that can capture the hand shapes, gestures, positions, facial expressions, and body language.

Converting text to sound, describing images, converting sound to text or image, or a combination of any of these can enable PwD to communicate effectively, considering their individual preferences. When a multi-modal input is given to the model and the output also has text in it, the readability can be improved to suit the persons with dyslexia. ChatGPT is found to be an effective companion for many PwD who suffer cognitive difficulties in organising thoughts and retrieving vital details by simplifying the text, highlighting key points, and providing audio descriptions. Many of the visually changed persons depend on Braille-to-text and text-to-Braille converters. This tool, powered with AI, helps them to access all the online content and effectively communicate with others through tools like WhatsApp. Amyotrophic lateral sclerosis (ALS) is a neurological condition that affects nerves in the brain and spine. The GenAI-powered devices can make the communication tasks of these people by predicting words or phrases combined with audio or video.

The GenAI-powered tools are available to make the unclear speech patterns to aid persons who have less speech clarity. The bedridden patients and visually challenged can use to operate various devices integrated with IoT, which will work on voice commands. The deaf usually find it difficult to participate in group discussions as many speak simultaneously from various directions during the discussion. The real-time transcription in simplified language along with images may be helpful to overcome this challenge. The advantages of PwD with the usage of GenAI tools are summarised in Fig. 12.8.

12.4.8 Emotional Support and Mental Health

"Transforming mental health for all" is the theme of the World Mental Health Report 2022 published by WHO. This indicates the importance of mental health in improving the quality of one's life. The awareness programmes are initiated by various institutions and need to be accessible as the information is important and useful

Fig. 12.8 Advantages of using multi-modal GenAI tools in improving communication

to all. For this, the information should be available in formats that are suitable to all individuals irrespective of the physical challenges they face. Multi-modal GenAI can be used to provide information in the format, which the user needs it. The techniques like text-to-speech synthesis, visual assistance, and alternative communication techniques can ensure that the people can access therapeutic materials, self-help books, and crisis support services in ways that best fit their preferred communication styles and sensory capacities.

One of the key elements in autism care is behaviour analysis. The GenAI tools can identify the behaviour patterns effectively and continuously. These patterns may not be easy for humans to perceive. These tools can be used to identify the triggers that might end in problematic behaviour or anxiety issues, which is very common in people with autism (Singh et al., 2020). The intervention strategies can also be planned by analysing the data that is collected by these devices.

The autistic people without cognitive impairment generally face mental depression and anxiety (Lever & Geurts, 2016). Multi-modal GenAI models can be used to identify and interpret emotional cues from people with disabilities. The speech analysis, facial expression recognition, and other sensory inputs can be used for this detection. By offering feedback on interpersonal interactions and emotional states, it assists people in gaining understanding of their own emotions and communication patterns. People are able to develop stronger relationships and handle social situations more skilfully as a result. Applications that make use of multi-modal GenAI can assist individuals with impairments in keeping an eye on their mood swings and self-care routines. Multi-modal GenAI systems can recognise signs of crisis or distress in individuals with impairments and provide timely support and intervention. By analysing communication patterns, sentiment trends, and risk indicators, they are able to identify individuals who are at danger of suicide or self-harm and link them to appropriate resources, such as crisis hotlines, mental health specialists, or emergency services. Through further innovation and collaboration, these

Table 12.5 Advantages of using GenAI tools for emotional and mental support

Method	Advantages
Behaviour Analysis	• Identification of behaviour pattern • Identification of triggers that cause anxiety issues • Customised intervention strategies
Emotional Analysis	• Identification and interpretation of emotional cues • Feedback on interpersonal interactions and states • Keep track of mood swings and self-care routines • Timely support at the time of crisis
Mental Health Support	• Access to therapy resources, self-help books in preferred format, and sensory capabilities • Early identification of persons at risk of self-harm or suicide • Timely connection to helplines or therapists or emergency services • Access to therapist resources

technologies have the potential to completely transform the way mental health treatment and support systems are provided, ensuring that individuals with disabilities have access to the resources and support they need to live fulfilling lives. The advantages are tabulated in Table 12.5.

12.4.9 Independent Living

With the revolutionary advancement of technology, more applications that support independent living of people with disabilities are developed. The assistive devices are developed with an aim to make the lives of the needy capable, independent, and self-sufficient (Manjari et al., 2020). With multi-modal GenAI, the people with disabilities like blindness, deafness, autism, and physical difficulties can have customised solutions provided to meet particular requirements. The unique feature of multi-modal GenAI to input and output data of different modalities enables it for object identification, assisted navigation, and the conversion of visual content into accessible formats such as audio descriptions or Braille for blind people. Real-time captioning along with sign language interpretation and visual alerts for critical messages are helpful for deaf people.

Training in social skills, routine management, and emotion recognition are ways that autistic people are supported. GenAI is used by people with physical disabilities for assistive robotics, virtual rehabilitation, and smart home automation. The GenAI-powered navigation tools are helpful for the visually challenged and the mobility challenged. Independent living makes one more self-esteemed and self-determined. As they are the ones who can assess their needs better than any other person, living independently without the support from anyone else will make their social life more successful. Considering all this, multi-modal GenAI enables people with disabilities to live more independent lives by offering individualised support, information access, and assistance for a variety of everyday tasks.

12.5 Challenges and Considerations

The advantages of multi-modal GenAI-based applications are very promising, but the challenges associated with these applications also need to be considered. As mentioned earlier, 16% of the world population is persons with significant disabilities. Though this is a considerable size, the datasets needed for training the multimodal data can be limited. For an efficient AI model, the availability of data and the quality of data are very important. If the model is trained with inadequate data or with no quality, it will generate undesirable results. This is true with multi-modal GenAI models too. As the data is obtained in different modalities, the scarcity of data in one modality may result in poor results. As in any other AI model, data collection and data size are crucial in the performance of the model. Effective methods need to be taken to ensure adequate size of data for the accurate prediction mechanism. Training data with inherent biases and a lack of data in particular modalities might produce skewed and incorrect results. Multi-modal generative AI can reinforce and magnify pre-existing biases if it is trained on biased input, producing discriminating results. Careful usage of bias mitigation strategies needs to be used to avoid biases while training. For applications to be fair and ethical, careful data selection and bias reduction techniques are essential (Singh & Singh, 2023).

The chances of generating false information, which is popularly known as hallucinating data, are another challenge in using multi-modal and other GenAI models. This usually happens when the training data is insufficient or has large gaps that make the model inefficient. This may result in unrealistic outputs. So, measures need to be taken to ensure the quality of data. The data that need to be collected can be highly personal and sensitive. This may raise privacy concerns. So, measures to secure privacy need to be given more importance.

Since cross-modal interpretations are required, the computation time and cost can be greater. New, efficient, resource-friendly training methods need to be developed to handle this limitation. The complex structure of the model often gives opaque decisions. This will make it difficult to understand the predictions and potential biases. Making the model transparent using interpretable and explainable techniques will be challenging. The ability to produce realistic visuals, audio, and video can be misused to disseminate false information and produce deep fakes, which may have an impact on social and political environments. The metrics for evaluating multi-modal GenAI models are not yet standardised.

12.6 Future Directions and Conclusion

To improve the living conditions of PwD, accessibility options need to be improved. By including multiple modalities like speech, gestures, and images, the accessibility can be improved. This can be achieved by developing customisable AI solutions so that individual and specific needs can be considered. Advanced natural language processing can be used for identifying complex and nuanced language input of PwD. The interaction between PwD and AI systems can be made simple using the multi-modality inputs and multiple modes of communication like speech recognition, sign language recognition, sentiment analysis, and Braille recognition. Better understanding of facial expressions, gesture recognition and emotions, and advanced AI-powered assistive devices with advanced sensory perception capabilities can be developed. The training of the model can be improved from the continuous feedback of the user, which in turn helps to adapt the device to the user behaviour pattern. To provide interactive experiences through virtual assistants and stimulators, the possibilities of integrating augmented reality/virtual reality with these devices need to be explored.

Though much research is happening in this area, some promising developments need to be mentioned. The latest AI model, Gemini by Google Research, is one among them. It outperforms human capabilities in Massive Multitask Language Understanding (MMLU) applications. To test the problem-solving and general knowledge skills, it uses more than 57 subjects, which include mathematics, science, and reasoning. Its multi-modal capabilities make it exceptional in complex logical and reasoning abilities. Instead of using the common approach of training the various modality data separately and then fusing them together, Gemini tries to pre-train the model with data of different modalities from the start (Saeidnia, 2023; Perera & Lankathilake, 2023; Gemini Team et al., 2024). To improve the effectiveness, it is further trained with additional multi-modal data. This helps to understand and reason all kinds of data from the ground up. Google claims Gemini to be a scalable, reliable, safe, and efficient model.

Microsoft Research also developed GenAI-powered applications like Microsoft Translator, which provide translation of speech, text, and images. Microsoft Soundscape and Microsoft Seeing AI are two applications that give more accessibility to visually challenged persons. Microsoft Soundscape uses 3D cues to give spatial awareness and sound-based route finder for navigation assistance. Microsoft Seeing AI provides auditory feedback to users recognising objects and obstacles in the surroundings. Microsoft Office 365 also has many accessibility features like Immersive Reader, Dictation, and Accessibility Checker.

Smart glasses that make the navigation of the visually challenged easier by providing audio descriptions are another innovation (Ananth et al., 2023; Pydala et al., 2023). This is a topic of interest for many studies, which includes image recognition, face recognition, and lip movement recognition. AI-powered smart glasses can convert text to audio, which is useful for the visually challenged, and convert speech to sign language or visual representations for the deaf. These glasses provide easy navigation with real-time transcription of the environment. AI-powered gadgets like Rabbit RI are getting popular with their ability to control all the apps. The model is trained on how to use other apps to get work done. This single-interface device can operate on voice commands or text commands. The PwD can add other modalities of operations accruing to their needs.

Though the research in multi-modal GenAI applications is advancing, the primary focus on creating benchmark datasets and standardising them can be a future scope in this domain. Developing an ethical framework to ensure data privacy while deploying GenAI applications is another key area of focus. From the industry's point of view, many workforces may get replaced by these applications. The employees can be sensitised, and giving training on the new technologies is an area to focus on. The potential of using multi-modal GenAI-based applications to improve the lives of PwD needs to be recognised, and the participation of end users in the development needs to be ensured. The needs of PwD vary from individual to individual. A more universal solution can be designed with the potential of multi-modal GenAI technique.

References

Adão, T., Oliveira, J., Shahrabadi, S., Jesus, H., Fernandes, M., Costa, Â., Ferreira, V., Gonçalves, M. F., Lopéz, M. A. G., Peres, E., & Magalhães, L. G. (2023). Empowering deaf-hearing communication: Exploring synergies between predictive and generative AI-based strategies towards (Portuguese) sign language interpretation. *Journal of Imaging, 9*, 11. https://doi.org/10.3390/jimaging9110235

Allen, M. L., Hartley, C., & Cain, K. (2016). iPads and the use of "apps" by children with autism spectrum disorder: Do they promote learning? *Frontiers in Psychology, 7*(AUG), 1–7. https://doi.org/10.3389/fpsyg.2016.01305

Ananth, S., Balaji, N. G., Prasad, P., Bhargavi, L. N., & Iyyanar, D. (2023). *Design and implementation of smart guided glass for visually impaired people. International Journal of Electrical and Computer Engineering, 5*(11), 1691–1704.

Aydin, O., & Diken, I. H. (2020). Studies comparing augmentative and alternative communication systems (AAC) applications for individuals with autism spectrum disorder: A systematic review and meta-analysis. *Education and Training in Autism and Developmental Disabilities, 55*(2), 119–141.

Clinton, S. (2020). Challenges and opportunities for teaching students with disabilities during the COVID-19 pandemic. *International Journal of Multidisciplinary Perspectives in Higher Education, 5*(1), 167–173.

Gadiraju, V., Kane, S., Dev, S., Taylor, A., Wang, D., Denton, E., & Brewer, R. (2023). "I wouldn't say offensive but···": Disability-centered perspectives on large language models. *ACM International Conference Proceeding Series*, 205–216. https://doi.org/10.1145/3593013.3593989

Ganle, J. K., Otupiri, E., Obeng, B., Edusie, A. K., Ankomah, A., & Adanu, R. (2016). Challenges women with disability face in accessing and using maternal healthcare services in Ghana: A qualitative study. *PLoS One, 11*(6), 1–13. https://doi.org/10.1371/journal.pone.0158361

García-Peñalvo, F. J., & Vázquez-Ingelmo, A. (2023). What do we mean by GenAI? A systematic mapping of the evolution, trends, and techniques involved in generative AI. *International Journal of Interactive Multimedia and Artificial Intelligence, 8*(4), 7–16. https://doi.org/10.9781/ijimai.2023.07.006

Gong, J., Foo, L. G., He, Y., Rahmani, H. & Liu, J. (2024) "*LLMs are Good Sign Language Translators*," IEEE/CVF Conference on Computer Vision and Pattern Recognition (CVPR), Seattle, WA, USA, 2024, pp. 18362-18372, https://doi.org/10.1109/CVPR52733.2024.01738

Goodfellow, I., Pouget-Abadie, J., Mirza, M., Xu, B., Warde-Farley, D., Ozair, S., Courville, A., & Bengio, Y. (2014). Generative adversarial networks. *Communications of the ACM, 63*(11), 139–144. https://doi.org/10.1145/3422622

Hendricks, D. (2010). Employment and adults with autism spectrum disorders: Challenges and strategies for success. *Journal of Vocational Rehabilitation, 32*(2), 125–134. https://doi.org/10.3233/JVR-2010-0502

Hutchinson, B., Prabhakaran, V., Denton, E., Webster, K., Zhong, Y., & Denuyl, S. (2020). Unintended machine learning biases as social barriers for persons with disabilitiess. *ACM SIGACCESS Accessibility and Computing, 125*, 1–1. https://doi.org/10.1145/3386296.3386305

Khairnar, D. P., Karad, R. B., Kapse, A., Kale, G., & Jadhav, P. (2020). PARTHA: A visually impaired assistance system. *2020 3rd International Conference on Communication Systems, Computing and IT Applications, CSCITA 2020—Proceedings*, *April*, 32–37. https://doi.org/10.1109/CSCITA47329.2020.9137791.

Khan, S., Nazir, S., & Khan, H. U. (2021). Analysis of navigation assistants for blind and visually impaired people: A systematic review. *IEEE Access, 9*, 26712–26734. https://doi.org/10.1109/ACCESS.2021.3052415

Kosanic, A., Petzold, J., Martín-López, B., & Razanajatovo, M. (2022). An inclusive future: Disabled populations in the context of climate and environmental change. *Current Opinion in Environmental Sustainability, 55*, 1–11. https://doi.org/10.1016/j.cosust.2022.101159

Lederberg, A. R., Schick, B., & Spencer, P. E. (2013). Language and literacy development of deaf and hard-of-hearing children: Successes and challenges. *Developmental Psychology, 49*(1), 15–30. https://doi.org/10.1037/a0029558

Lever, A. G., & Geurts, H. M. (2016). Psychiatric co-occurring symptoms and disorders in young, middle-aged, and older adults with autism Spectrum disorder. *Journal of Autism and Developmental Disorders, 46*(6), 1916–1930. https://doi.org/10.1007/s10803-016-2722-8

Lintangsari, A. P., & Emaliana, I. (2020). Inclusive education services for the blind: Values, roles, and challenges of university EFL teachers. *International Journal of Evaluation and Research in Education, 9*(2), 439–447. https://doi.org/10.11591/ijere.v9i2.20436

Manjari, K., Verma, M., & Singal, G. (2020). A survey on assistive technology for visually impaired. *Internet of Things (Netherlands), 11*. https://doi.org/10.1016/j.iot.2020.100188

Mariani, M., & Dwivedi, Y. K. (2024). Generative artificial intelligence in innovation management: A preview of future research developments. *Journal of Business Research, 175*(January), 114542. https://doi.org/10.1016/j.jbusres.2024.114542

Moon, N, W., Todd, R. L., Morton, D. L., & Ivey, E. (2012). Accommodating students with disabilities in science, technology, engineering, and mathematics (STEM). In *SciTrain: Science and Maths for All*. https://hourofcode.com/files/accommodating-students-with-disabilities.pdf

Nair, V., Olmschenk, G., Seiple, W. H., & Zhu, Z. (2022). ASSIST: Evaluating the usability and performance of an indoor navigation assistant for blind and visually impaired people. *Assistive Technology, 34*(3), 289–299. https://doi.org/10.1080/10400435.2020.1809553

Narayanan, S. (2018). A study on challenges faced by disabled people at workplace in Malaysia. *International Journal for Studies on Children, Women, Elderly And Disabled, 5*(July), 85–82.

Oladipo, M. B., Owei, O. B., & Drobinna, V. C. (2008). The challenges encountered by mobility-challenged persons in accessing facilities in public buildings. (a study of selected public universities in Rivers state). *International Journal of Advances in Engineering and Management (IJAEM), 2*(10), 257. https://doi.org/10.35629/5252-0210257264

Pereira, J. A., Pereira, J. A., & Fidalgo, R. D. N. (2021). Caregivers acceptance of using semantic communication boards for teaching children with complex communication needs. *Congresso Brasileiro de Informática Na Educação, Cbie*, 642–654. https://doi.org/10.5753/sbie.2021.218141.

Perera, P., & Lankathilake, M. (2023). Preparing to Revolutionize Education with the Multi-Model GenAI Tool Google Gemini? A Journey towards Effective Policy Making. Journal of Advances in Education and Philosophy, 7(08), 246–253. https://doi.org/10.36348/jaep.2023.v07i08.001

Pydala, B., Kumar, T. P., Baseer, K. K., & Author, C. (2023). *SMART_EYE: A navigation and obstacle detection for*. *Journal of Applied Engineering and Technological Science, 4*(2), 992–1011. https://doi.org/10.37385/jaets.v4i2.2013

Rights of Persons with Disabilities Act, 2016. (2016). *Act of Parliament, India, 49*.

Saeidnia, H. R. (2023). Welcome to the Gemini era: Google DeepMind and the information industry. *Library Hi Tech News, December.* https://doi.org/10.1108/LHTN-12-2023-0214

Salminen, A. L., & Karhula, M. E. (2014). Young persons with visual impairment: Challenges of participation. *Scandinavian Journal of Occupational Therapy, 21*(4), 267–276. https://doi.org/10.3109/11038128.2014.899622

Scott, M., & Sedgewick, F. (2021). 'I have more control over my life': A qualitative exploration of challenges, opportunities, and support needs among autistic university students. *Autism and Developmental Language Impairments, 6.* https://doi.org/10.1177/23969415211010419

Singh, A., & Saravanan, V. (2024). XAI decision MODELS: Programming models for decentralized BlockXAI. In *Convergence of blockchain and explainable artificial intelligence* (pp. 15–22). River Publishers.

Singh, A., and Singh, K. K. (2023). YORES: An ensemble YOLO and Resnet network for vehicle detection and classification. In ProfIT AI, pp. 26–37.

Singh, A., Sharma, A., Singh, K. K., & Dhull, A. (2020). Sentiment analysis of social networking data using categorized dictionary. *Journal of Information Technology Management, 12*(4), 105–120.

Singh, A., Li, P., Singh, K. K., & Saravana, V. (2021). Real-time intelligent image processing for security applications. *Journal of Real-Time Image Processing, 18*, 1787–1788.

Singh, A., Singh, K. K., Greguš, M., & Izonin, I. (2022). CNGOD-an improved convolution neural network with grasshopper optimization for detection of COVID-19. *Mathematical Biosciences and Engineering, 9*, 12518–12531.

Tapu, R., Mocanu, B., & Zaharia, T. (2020). Wearable assistive devices for visually impaired: A state of the art survey. *Pattern Recognition Letters, 137*, 37–52. https://doi.org/10.1016/j.patrec.2018.10.031

Team, G. (2024). *Gemini: A family of highly capable multimodal models*. https://doi.org/10.48550/arXiv.2312.11805.

Wei, J., Kim, S., Jung, H., & Kim, Y.-H. (2023). *Leveraging large language models to power chatbots for collecting user self-reported data*. http://arxiv.org/abs/2301.05843

Yang, J., & Zhang, H. (2024). Development and challenges of generative artificial intelligence in education and art, *Highlights in Science, Engineering and Technology, 85*, 1334–1347. https://doi.org/10.54097/vaeav407

Chapter 13
Single Modality to Multi-modality: The Evolutionary Trajectory of Artificial Intelligence in Integrating Diverse Data Streams for Enhanced Cognitive Capabilities

Hardeep Kaur, C. Kishor Kumar Reddy, D. Manoj Kumar Reddy, and Marlia Mohad Hanafiah

Abstract The evolution of artificial intelligence (AI) towards multi-modality represents a significant advance in the field by integrating multiple sensory inputs to enhance machine understanding and affected communication. This research provides a broad overview of these processes, tracing the history of the development of artificial intelligence from its early stages to the rise of machine learning. It delves into the foundations of multi-modal AI, covering its concepts, values, and early stages. Core enabling technologies, including neural networks, transformers, tracking mechanisms, multi-modal fusion technology, and cross-modal learning, are examined. Multi-modal AI systems, such as vision-language models (e.g., CLIP and DALL-E), audio-visual models, text and image synthesis models, and diversity of reference standards, are explored along with their applications in natural language processing, computer vision, speech and visualisation, robotics, and healthcare. Developments in multi-disciplinary interaction are discussed with a focus on user interaction, human-computer interaction, and augmented and virtual reality. The impact of multi-modal intelligence on creative industries such as art, design, film, animation, music, and games is also examined. Ethical and social aspects such

H. Kaur (✉) · C. Kishor Kumar Reddy
Department of Computer Science and Engineering, Stanley College of Engineering and Technology for Women, Hyderabad, India

D. Manoj Kumar Reddy
Department of Computer Science and Engineering, Stanley College of Engineering and Technology for Women, Hyderabad, India

M. M. Hanafiah
Department of Electrical and Electronics Engineering, Vardhaman College of Engineering, Hyderabad, India

Centre for Tropical Climate Change System, Institute of Climate Change, Faculty of Science and Technology, Universiti Kebangsaan Malaysia, Bangi, Malaysia

A. Singh, K. K. Singh (eds.), *Multimodal Generative AI*,
https://doi.org/10.1007/978-981-96-2355-6_13

as impartiality, fairness, privacy, transparency, and efficiency are examined. Future directions and innovations in multi-modal AI are explored, focusing on advances in learning algorithms, integration with new technologies, achievements over time, and personal development and adaptation. The study aims to provide a better understanding of change and the problems that need to be solved.

Keywords Artificial intelligence · Ethical aspects · Human–computer interaction · Neural networks · Transformers · Visual-language models

13.1 Introduction

The development of artificial intelligence (AI) has gone through milestones and successes, evolving from simple computational models to complex systems that can make decisions and learn. The roots of artificial intelligence can be traced to the mid-twentieth century, with the emergence of computer science and the development of the first digital computers. Early AI techniques focused on symbolic thinking and reasoning, exemplified by the work of pioneers such as Alan Turing and John McCarthy. When Turing's concept of "universal machines" laid the theoretical foundation of artificial intelligence, McCarthy introduced the term "artificial intelligence" and held the Dartmouth Conference in 1956, which is considered the beginning of AI as a field. Limitations of early AI systems, especially in dealing with complexity and the variability of real-world data, led to the transition to machine learning (ML). In the 1980s and 1990s, AI research increasingly focused on developing algorithms that could learn from data and improve over time. During this time, many types of education emerged, including observed learning, unsupervised learning, and extended learning. Major advances include the introduction of neural networks and backpropagation algorithms, which enable computations to learn and make predictions from large data sets. The rise of machine learning is a great example of exhibiting the major shift from rule-based processing to data-driven processing, enhancing approach, ability, and use of intelligence (Das et al., 2020).

The dawn of twenty-first century brought a deep learning revolution; a giant leap in intelligence is driven by advancements in computing power, big data, and new algorithms. Deep learning is a category of machine learning that uses multiple neural layers in a network to process complex patterns in data, as shown in Fig. 13.1. This revolution brought success in image recognition processes using convolution neural networks (CNNs) by analysing the performance of the AlexNet model in the ImageNet 2012 competition. Due to this, there is rapid progress in natural language processing, development, and games, which helped in achieving better performance than before. Revolutionary learning not only pushes the limits of what AI can achieve but also allows for the development of multi-modal intelligence and

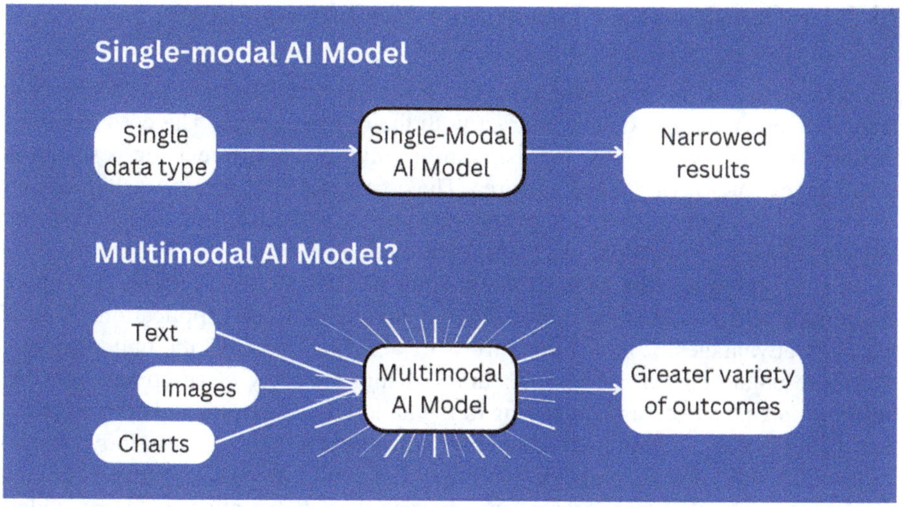

Fig. 13.1 Comparison of single-modal and multi-modal AI models

insights to create strong and diverse mental abilities. The transformation of artificial intelligence into multi-modality is important in the field of artificial intelligence (Das et al., 2020). Multi-modal AI systems can achieve better understanding and context by combining different information to achieve efficiency in different ways. As research and development in this area continues, we can see a more powerful revolution in artificial intelligence that will push the limits of what is possible, finally bringing us closer to the version of a truly intelligent machine (Sharma et al., 2022).

13.2 Foundations of Multi-modal AI

The core of multi-modal AI is the study of combining data into different formats, such as text, images, video, and audio, to improve understanding, decision-making, and interaction in AI. This integration includes areas such as computer vision, natural language processing, speech recognition, and voice processing to create AI models that can identify and interpret a variety of data. The development of algorithms and technologies in multi-modal intelligence can integrate data in different steps, which include pre-processing raw data, extracting relevant features, and creating a design that can combine features for better understanding. The main challenges are adapting data in different formats, handling differences and changes between different devices, and ensuring stability and generalisation across multiple platforms (Trescak et al., 2018).

13.2.1 Definition and Importance of Multi-modality

Multi-modality is an important concept in artificial intelligence to change how machines see, understand, and interact with their environment. The core of multi-modality involves combining and using information from different sources such as text, images, audio, video, and gestures. The combination of these variables allows AI systems to capture the richness and complexity of human communication and understanding and to reflect the natural interaction of humans with their environment. By combining data from different sources, AI systems can make data more meaningful and thus improve performance in various tasks and applications. One of the main advantages of multi-modality is its ability to improve the understanding and representation of intelligent machines. Using a variety of additional data, AI models can gain a deep understanding of the underlying structure and content of the data (Singh et al., 2024a). For example, consider the case of descriptive visual images in a task such as image description. This combination allows the AI engine to generate not only clear instructions but also details that allow a deeper understanding of the image.

In addition, multi-modal AI improves the robustness and generalisation of machine learning models. By combining information from different sources, AI systems can reduce the limitations of a single source and increase their ability to adapt to different locations and situations. This efficiency allows AI systems to perform well across a wide range of tasks and controllers, contributing to reliability and efficiency in real-world applications. Multi-modal AI also supports interaction between humans and machines (Trescak et al., 2018). By supporting various forms of communication such as speech, text, and gestures, AI systems can better understand and respond to user input, enabling more interactive experiences and better usability. These natural interactions improve the user experience, ensure that AI systems are more responsive to the needs and preferences of human users, and encourage the adoption of AI technologies. Multi-modality in general plays an important role in advancing artificial intelligence, allowing machines to perceive, understand, and interact with the world in a way that is more consistent with human knowledge and communication. With the development of research in this field, we can expect significant progress in AI technology and its application in many fields, ultimately shaping the future of human-machine interaction and collaboration (Munawar et al., 2018).

13.2.2 Early Multi-modal Approaches

Early multi-modal approaches to AI focused on integrating information from different perspectives to improve the performance and capabilities of AI. This startup process is often primitive compared to today's models but forms the basis of many different models. One of the early pioneers of multi-modal research is audiovisual speech recognition. Here, machines combine the audio signal of speech with visual cues of lip movement to improve speech recognition, especially at loud sounds. This combination increases efficiency and performance by allowing the system to benefit from the increased power of both. Another important part of the initial application was the graphics and text. The system begins to combine the annotation with the visual image to improve functions such as image storage and annotation. For example, it may be more accurate to combine image features with text and search for images based on their content, resulting in increasingly better understanding of visual information. Early approaches are also exploring the integration of different types of sensory information into robots. Robots are equipped with many sensors, such as cameras, microphones, and tactile sensors, that allow them to better interact with their environment. By processing and combining data from these different sources, robots can perform more complex tasks, such as navigating through environments, identifying objects, and better interacting with humans (Munawar et al., 2018).

These early multi-modal systems rely solely on manual features and heuristics to integrate information from different models, using fusion techniques (combining features extracted from different models) and decision-level fusion (combining single-modal classifiers to make decisions) and other technologies. However, this process faces many challenges: synchronising and organising data from different models is often difficult, especially in the case of dynamic or asynchronous inputs, making it difficult to create an integrated system. Despite these challenges, early multi-modal research paved the way for the development of more advanced technologies. With the advent of deep learning and increased budgeting, modern multi-modal methods have made significant progress. Modern models can use big data and neural network architectures, such as convolutional neural networks (CNNs) and recurrent neural networks (RNNs) for image processing, or transformers for data text arrays. Modern multi-modal systems use technologies such as multi-modal tracking systems to effectively combine and process information in different formats, enabling applications such as visual query answering (VQA), multi-dimensional analysis, and human-computer interaction. These machines can process large amounts of data, learn to solve problems at advanced levels, and excel at a variety of tasks. These basic efforts have evolved into advanced, powerful, and versatile multi-intelligence systems that are increasingly used in many applications today.

13.3 Core Technologies Enabling Multi-modality

The core technologies that support many aspects of AI are important for integrating and processing disparate data, thus supporting increasingly powerful AI systems. These technologies include advances in data fusion, deep learning, natural language processing, and sensor technology. The emergence of multi-modal AI represents the greatest advance in the field of AI. It ensures the operation of the process and combines information from different sources such as paper reading, photographs, audio, and video. Using the added power of different models, these systems can provide a more comprehensive and intuitive understanding of complex data, leading to more powerful and versatile applications. The key technologies driving this progress include advanced data processing techniques, deep learning models, advanced language customising, and new technologies. These advances have transformed the capabilities of artificial intelligence, making it possible to create machines that can interact with the world in ways that approximate the thoughts and experiences of human (Daubert et al., 2021).

13.3.1 Neural Networks and Deep Learning

Neural networks and deep learning are technologies of modern intelligence and have been successful in many applications. Neural networks are communication patterns that arise from the structure and function of the human brain, consisting of a series of connections or neurons that process information. These networks can be trained to learn complex patterns and representations from data, allowing them to perform tasks such as classification, propagation, and pattern recognition.

13.3.1.1 Neural Networks

The basis of a neural network is the neuron, a simple computational unit that takes input, processes it with weighted sums, uses an activation function, and produces it. Neural networks usually have an input layer, one or more hidden layers, and an output layer. Each layer changes the input data, allowing the network to learn the hierarchical representation. During training, optimisation techniques such as gradient descent are used to adjust the connection weight of neurons guided by the loss, which evaluates the performance of the network.

13.3.1.2 Deep Learning

Deep learning is a subfield of machine learning that focuses on multi-layered neural networks called deep neural networks (DNN). The depth of these networks allows them to model complex, abstract data, making them particularly powerful for high-dimensional data such as images, audio, and text. Deep learning has led to advances in many fields such as computer vision, language processing, and speech recognition.

Training a deep neural network requires feeding a large amount of labelled data, adjusting the weights via backpropagation, and repeating the regression. Use techniques such as dropout, batch normalisation, and data augmentation to improve generalisation and prevent overfitting. The advent of powerful GPUs and specialised hardware accelerators has made it possible to train large deep learning models on large datasets.

Neural networks and deep learning have transformed many fields, from working to reaching human levels or better performance on certain tasks. In the field of computer vision, deep learning models of electronic images and video recognition are used in driving, security, and healthcare. They enable complex language interpretation, logic analysis, and communication in natural language processing. They are making advances in speech recognitions, virtual assistants, and instant transcription services. Neural networks and deep learning collectively form the backbone of modern artificial intelligence and provide the tools and techniques necessary to create intelligent machines that can understand and interact with the world in complex and meaningful ways. As research continues, this technology is expected to reveal more functionality and applications in the future (Crompton & Burke, 2023).

13.3.2 Transformers and Attention Mechanisms

Transformers and attention mechanisms are cutting-edge technologies in the field of artificial intelligence that improve the capability of natural language processing (NLP) and other fields that involve information. This technique addresses the fundamental limitations of previous models such as recurrent neural networks (RNNs) and long short-term memory (LSTM) networks by enabling a long and careful look at their long-range dependencies to get better.

13.3.2.1 Transformers

Transformer is a deep learning program introduced by Vaswani et al. They have revolutionised natural language processing (NLP) by overcoming the limitations of previous models such as recurrent neural network (RNN) and long short-term memory (LSTM) networks. Unlike sequential models, the transformer processes all components simultaneously through a process called self-monitoring, which allows for more efficient remote control and ensures uniformity (Munawar et al., 2018). This innovation has led to significant advances in areas such as machine translation, text generation, and query answering, making transformers a cornerstone of modern AI research and use.

The transformer consists of an encoder–decoder structure:

Encoder: The encoder performs the input processing and creates a group representation for each source. It has multiple layers; each layer has a mechanism for self-monitoring and a feedforward neural network.

Decoder: The decoder produces an output sequence based on the encoded representation. It also has many, including self-tracking, encoder-decoder tracking mechanism, and feed-forward neural network.

13.3.2.2 Attention Mechanisms

The attention mechanism is at the core of the functionality of transformer and many other neural network models. They allow the model to focus on different parts of the input sequence while generating each part of the output. The main idea is to calculate the weighted sum of the input representation, where the weight is determined by the learning function of the input data.

Types of Attention

Self-Attention: It is also called intra-attention; this mechanism allows the model to determine the relationship between different tasks in a sequence. Each element in a sequence is projected onto each other, allowing the model to capture progress regardless of their distance in the sequence.

Cross-Attention: It is used in the transformer's decoder; cross-attention allows the model to focus on the encoder's output while producing the output. This process helps generate ideas and products periodically.

One of the innovations in the Transformer architecture is multi-head listening. Instead of using a single headset, multiple headphones are used together. Each head works on different parts of the representation, allowing the model to capture different objects. The results of these heads are combined and linearly transformed to create the final representation. Since transformers process entire arrays in parallel, they need a way to combine the actions of the arrays. This is done with positional coding, where the position is fixed or learned to add information to the input. This encoding preserves order information, allowing the model to distinguish between different functions in the system. Transformers and attention mechanisms have revolutionised natural language processing and are used in many tasks, including machine translation, text summarisation, text generation, and question answering (Daubert et al., 2021). The effectiveness and power of transformers and tracking systems have extended beyond NLP to fields such as computer vision, where models such as vision transformers (ViT) have been developed using similar models such as painting. Together, transformers and maintenance mechanisms represent a major advance in intelligence, becoming more powerful, effective, and flexible models capable of performing many tasks involving difficult, sequential materials. These technologies drive innovation and set new standards in multiple domains.

13.3.3 Multi-modal Fusion Techniques

Multi-modal fusion technology is essential for integrating and processing information from various sources such as text, images, audio, and video. This technology enables artificial intelligence (AI) processes to use additional data from different sources for better understanding and decision-making. There are many key approaches to multi-modal fusion, each with its unique advantages and uses.

13.3.3.1 Early Fusion

Early fusion, also known as feature-level fusion, involves combining raw or extracted features from different models at an early stage. This approach integrates information before important actions or decisions are made, allowing the model to learn how to bring information together. Early fusion is very effective in eliminating unrelated patterns because it allows the model to learn complex patterns of different types of objects, which leads to further fusion. However, this approach must be done carefully to ensure that products from different sources are compatible and consistent with the standard. Bias or differences in measurement and classification may result in decreased performance (Crompton & Burke, 2023). Additionally, early fusion can create an extremely challenging work environment, which can be computationally intensive and require large amounts of training data.

13.3.3.2 Late Fusion

Late fusion, also known as decision-level fusion, combines the outputs of separate structures that undergo each mutation independently. In this way, all changes are carried out by different models or connections, and their predictions or decisions are combined in the next stage, usually by methods such as averaging, voting, or multiple pooling methods. Post-production fusion allows the use of the most appropriate technology and design to handle any changes, making it flexible and easy to use. It is less dependent on competition and modelling of raw data and can combine different types of predictions or decisions. However, late fusion may miss some of the deeper parts of the structure that early fusion can capture. Because each variable is processed independently, the model will not learn the joint representations needed to understand the interaction.

13.3.3.3 Hybrid Fusion

Hybrid fusion technology combines elements of early fusion and late fusion, integrating various stages of pipeline processing. This approach is to balance the benefits of capturing correlations with the ease of processing the model itself. Hybrid fusion can involve multiple layers of integration, where fusion at the initialised level is followed by the operator, followed by fusion at the decision level. This framework captures the interaction of change while remaining modular and flexible. However, hybrid convergence can be difficult to implement as it requires creating and maintaining multiple levels of integration. Ensuring that each stage contributes to overall performance requires careful testing and optimisation.

Multi-modal fusion technology is gaining importance in terms of integrating and processing information from different sources such as text, image, audio, and video, increasing the potential of artificial intelligence in many areas. Monitoring systems play an important role in multiple integrations by focusing on the most important aspects of each change, improving the ability to capture relevant information, and providing greater predictability and context awareness (Rai et al., 2020). But creating effective attention mechanisms requires careful consideration of how weights are calculated and applied. Canonical Correlation Analysis (CCA) is one of the best ways to examine covariates to capture the relationship between variables, but it may not capture features as well as modern deep learning. Deep learning techniques such as multi-modal transformers, multi-modal autoencoders, and graph-based models lead to many effects. This method is good at using self-correcting, encoding and decoding processes, and graphical representation to capture the relationship and hence is good at handling difficult data. Applications for this fusion technology span various fields, including healthcare, where medical images, patient information, and genealogical information are integrated, making self-diagnosis more accurate and improving user connectivity by providing data from multiple sensors for human-computer safety (Guan et al., 2020). In all, multi-modal fusion technology is essential for the creation of intelligent tools that can efficiently process and combine different information to achieve a more comprehensive and detailed understanding, thereby driving the advancement of many applications and industries.

13.4 Key Multi-modal AI Systems

Essentially, multi-modal AI systems combine various data transformations, such as image, language, and audio to create powerful and versatile applications. Visual language models such as CLIP and DALL-E are designed to understand and create information that combines visual information and textual data to perform tasks such as image classification, search, and generation definition based on textual description. Audiovisual modelling improves understanding and content creation by combining audio and visual information, which is particularly useful for applications such as video analysis and speech recognition in noisy environments. Text and image linking models can create links from images and vice versa, simplifying automatic content creation and user interaction. In addition, multi-modal embeddings create a unified representation that combines information from different modalities into a unified space, thus improving tasks such as visualising questions answered and cross-modal retrieval. Together, these systems represent advances in artificial intelligence that use the integration of multiple types of data to deliver greater situational awareness and technological capabilities (Horakova et al., 2017).

13.4.1 Vision-Language Models

Visual language models are designed to understand and create visual and textual information. CLIP (comparative language imagery pretraining) and DALL-E are prime examples of this. It works like zero-shot image distribution and image search, leveraging the ability to associate images with their descriptions without requiring special training explanation. It demonstrates the potential of artificial intelligence in combining visual patterns and language by leveraging transformer-based architecture to create new and coherent images based on text.

13.4.2 Audio-Visual Models

Audio-visual models combine audio and visual information to enhance understanding and create rich content. This model is particularly useful in applications such as video analysis, multimedia content creation, and speech recognition in noisy environments. By combining lip movements with speech sounds, this model increases the accuracy of speech recognition, especially in difficult situations such as noisy backgrounds. This model also analyses facial expressions and tone of voice to accurately recognise and understand human behaviour to implement its abilities in applications from customer service to wellness applications.

13.4.3 Text and Image Synthesis Models

Text and image synthesis models can create composite text from images and vice versa. They facilitate a wide range of applications, from automatic content creation to effective user communication. Models such as neural image captioning generator create captions based on image content. This model uses convolutional neural networks (CNNs) for image feature extraction and recurrent neural networks (RNNs) or transformers for text.

Multi-modal embedding is a unified representation that integrates information from multiple variables into a shared latent space, thus facilitating the integration and processing of multi-modal information. Models such as VisualBERT and LXMERT create a unified environment where both visual and data are organised into a common representation, allowing models to conceptualise visuals and data simultaneously and improve the performance of functions such as visual response (VQA) and image recovery. Using multi-modal embeddings, AI systems can perform cross-iterative tasks such as finding images that match descriptions or storing text based on input images (Horakova et al., 2017). This functionality is important for applications such as multimedia search engines and digital asset management systems, as shown in Table 13.1. This important multi-modal AI framework is

Table 13.1 Overview of multi-modal models

Model types	Description	Examples
Vision-Language Models	Understand and create visual and textual information	CLIP, DAL-E, image search, image generation
Audio-Visual Models	Combines audio and visual data for better understanding of situations	Video analysis, speech recognition in noise
Text and Image Synthesis	Creates text from images and vice versa	Image captioning, automatic content creation

making progress in integrating and leveraging multiple data transformations, resulting in more powerful, versatile, and context-aware AI applications.

13.5 Applications of Multi-modal AI

The integration of multi-modal intelligence is revolutionising various fields, facilitating the combination of diverse data types like text, images, audio, and sensor information. This integration enhances AI systems' ability to comprehend and carry out intricate tasks, leading to notable advancements in the effectiveness and interpretation of numerous applications. Multi-modal AI harnesses the potential of multi-modality to enhance situational awareness, precision, and user engagement, propelling advancements in natural language processing, computer vision technology, speech and voice recognition, robotics, autonomous systems, and healthcare. This cooperation not only enhances current technologies but also unlocks new possibilities for creating more intelligent, adaptable, and user-friendly artificial intelligence (Anisha et al., 2022).

13.5.1 Natural Language Processing and Understanding

Natural language processing can be improved by combining data with other formats such as images and audio. The combination allows for more abstract language patterns and concepts that are more understandable. In visual question answering models, text and images are processed to understand and generate clear answers. Understanding visual content is important to answering questions correctly in educational tools. The accuracy of the translation can be increased by including visual elements. If there are images of animals or baseball bats, you can influence the definition of a sentence. This approach ensures that the translation is appropriate and accurate. A deeper understanding of the emotional content in communication can be provided with multi-modal emotion analysis (Singh & Singh, 2023b). This assessment is important for customer service applications, where integration of the user's voice and facial expressions during the call can help customer service representatives evaluate customer satisfaction and respond appropriately. Additionally, multi-modal NLP models can be used to create virtual assistants and interactive agents. By using forward vision technology, these systems can better understand and adapt to their environment. A virtual assistant can be more intuitive if it can see the user's environment or the objects they interact with. This holistic analysis provides a deep understanding of public opinion, helping brands and organisations better understand what their target audiences are thinking and feeling. Moreover, multi-modal NLP can increase accessibility for people with disabilities. Communication training that converts speech into text will enable deaf and hard-of-hearing people to communicate more effectively. In general, multi-modal intelligence and understanding in natural language processing expand the capabilities of language models, making them more familiar, accurate, and useful, thereby improving applications in many places (Joshi et al., 2021).

13.5.2 Computer Vision

In computer vision, multi-modal AI systems enhance image and video understanding by combining visual information with other formats such as text, audio, and sometimes even tactile output ideas. This multi-modal integration leads to better and more accurate evaluation of the visual image, improving the performance of a variety of tasks. AI produces coherent and contextual narratives based on insight. For example, models such as Show and Tell and Show, and Attend and Show use convolutional neural networks (CNNs) to extract features from images and recurrent neural networks (RNNs) or transformers to generate descriptive text and the creation of detailed reports as shown in Fig. 13.2. These features are useful for applications such as automatic content creation, improving accessibility for visually impaired users by providing descriptions or text, and setting up discovery and uncovering the bigger picture from questions.

In object detection and situational awareness, combining audio and visual information can improve accuracy and situational awareness. For example, identifying a ringing phone, a barking dog, or sound in a movie can provide additional details to help determine a match. Audiovisual integration is particularly important in

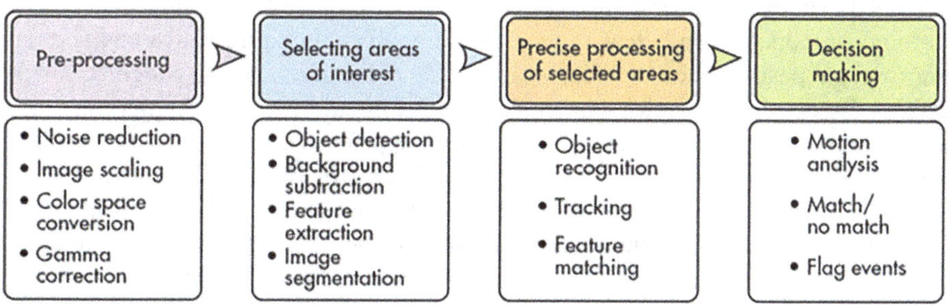

Fig. 13.2 Analogy of computer vision

environmental monitoring, where visual information alone may be insufficient or ambiguous, and automatic understanding of audio content can help determine the correct situation. This multi-modal approach is also useful for augmented reality (AR) and virtual reality (VR). Multi-modal AI systems create more interactive experiences by combining visual, auditory, and sometimes even haptic feedback. For example, overlaying visual cues with audio cues in AR applications can guide users more effectively, improving navigation and performance in real-time environments. Incorporating these models into VR can create realistic experiments that enhance education, gaming experience, and clinical applications by providing users with an experience similar to interactions in the world (Mondal et al., 2020). Additionally, many computer vision applications are used to combine data from cameras, lidar, radar, and microphones to enhance environmental perception and decision-making in autonomous vehicles. This integrated information improves safety and reliability by improving the vehicle's ability to detect and react to road dynamics. In healthcare, multi-modal AI systems integrate medical information with patient data and other sensor data to provide accurate diagnoses and treatment plans. For example, combining an MRI or CT scan with a patient's medical history and genetic information allows for more detailed diagnosis and treatment of disease, thereby improving patient outcomes. In general, the integration of various types of information into computer vision can improve the ability and performance of intelligence to be accurate, context-sensitive, and applicable to various situations (Reddy et al., 2024).

13.5.3 Speech and Audio Processing

Multi-modal AI supports speech and language by combining audio information with visual and textual input to improve accuracy and understanding of content. A key part of this combination is in speech recognition models that combine lip movements and speech signals to improve speech recognition, especially in noisy locations. The technology, called speech reading, allows the system to use the sense of lip movement to eliminate sounds that may be missed due to background noise. These features are important for applications such as instant transcription services that need to convert speech to text, as well as for affordable hearing aids that benefit from noise reduction and clear speech. Additionally, combining voice and text can improve the performance of voice assistants and communicators (Singh & Singh, 2023a). By combining audio with contextual information, these systems can better understand users' intentions and produce more appropriate responses. For example, a voice assistant equipped with camera feedback can better interpret user commands. This multi-modal approach enhances the user experience through greater interactivity and environmental responsibility. This multi-modal approach enhances the user experience through greater interactivity and environmental responsibility. Moreover, multi-modal AI is revolutionising automatic speech recognition (ASR) by integrating visual data. Traditional ASR systems rely entirely on audio data. This can be difficult in an environment with noise or conflicts. By using data such as lip movements and facial expressions, these systems can achieve better and more powerful results. These advancements are useful in situations such as broadcasts with background noise or public announcements (Ali et al., 2020).

Educational environments also use multi-modal AI that supports language and literacy learning by combining audio, visual, and text. For instance, interactive courses may use feedback to help students to improve their speaking and comprehension skills. These various lessons can make learning easy, meaningful, and interesting. In security and surveillance, combining audio and video data can improve the investigation and interpretation of incidents. This integration allows for better monitoring and response to potential security threats. Additionally, in healthcare, multi-modal intelligence can help diagnose and monitor speech- and hearing-related conditions. For example, combining a patient's speech with facial expressions can help detect diseases as early as possible. Overall, combining different types of information in speech and audio can lead to more accurate, meaningful content with greater intelligence. These advances have increased the use of talent and skills in many areas, from technology to education, security to healthcare, and have provided time for better problem-solving and improved user experiences.

13.5.4 Robotics and Autonomous Systems

In the field of robotics and autonomous systems, multi-modal AI plays an important role in improving understanding, decision-making, and interaction capabilities. Multi-modal AI enables robots to better understand and navigate their environment by integrating multiple senses such as vision, hearing, and feedback. Robots are equipped with many sensors, such as camera, microphone, and lidar, which can understand the environment better and thus detect objects, avoid obstacles, and direct them correctly. An important application of multi-modal intelligence in robotics is autonomous vehicles, where combining data from cameras, lidar, and GPS can improve situational awareness and decision-making. By combining data from these different sources, autonomous vehicles can see their surroundings, detect potential threats, and act safely by making informed decisions (Jaiswal & Arun, 2021). This not only improves the safety of passengers and pedestrians, but also increases the efficiency and reliability of their transportation system. Additionally, multi-modal AI is helping human engineering collaborate more intuitively and efficiently in industries ranging from manufacturing to healthcare. By combining vision, hearing, and feedback, robots can interact more closely with humans. For example, in a manufacturing environment, robots equipped with multi-modal sensors can complete tasks in parallel with humans, understanding, pointing, and responding to instructions. The integration provides productivity, flexibility, and security for the business environment. In healthcare, multi-modal AI-powered robotic systems can help doctors perform a variety of tasks, from patient care to surgery. These systems include visual, auditory, and tactile inputs to navigate the hospital environment, assist with patient care, and even perform minor surgeries with precision (Raghunath et al., 2022). Seamless integration of multi-modal inputs enhances the capabilities of robotics assistants, allowing them to deliver timely and personalised care while reducing the risk of errors. Overall, multi-modal AI is revolutionising robotics and machine control, allowing robots to perceive, understand, and interact with their environments in a variety of ways. From self-driving cars to collaborative robots in manufacturing and healthcare, the integration of multiple needs improves safety, efficiency, and effectiveness, increasing trust and paving the way for a future where robots work seamlessly with humans.

13.5.5 Healthcare and Medical Diagnosis

Multi-modal AI has the potential to improve treatment and diagnosis by combining dissimilar data such as medical images, patient data, and genetic information. This integration leads to more effective, more accurate diagnoses and treatments, providing better health results for patients and personalised care. An important application of multi-modal AI in healthcare is the analysis of medical images. Artificial intelligence systems can provide more accurate and detailed diagnoses by combining various measurements, such as MRU scans, X-rays, and CT scans, with patient history and information treatment. Medical tests can detect minor abnormalities or early symptoms of disease that are invisible to the human eye (Renz & Hilbig, 2020). By comparing these predictions with the patient's symptoms and medical history, the AI system can generate recommendations for further evaluation or planning of treatment. An important part of the use of multiple intelligences in medicine is the analysis of medical images. Artificial intelligence machines can provide accurate and detailed diagnoses by combining various tests, such as MRI scans, X-rays, and CT scans, with the patient's medical history and medical records. For example, AI algorithms can analyse medical images to detect small abnormalities or early signs of disease that are invisible to the human eye. By comparing predictions with the patient's symptoms and medical history, intelligent machines can produce recommendations for further evaluation or treatment planning.

Additionally, multi-modal intelligence improves mobile and telehealth services by combining data from wearable devices with patient-reported outcomes. By using sensors to monitor vital signs, activity levels, and other health metrics, AI systems can obtain immediate information about a patient's health and notify or replace the physician. This regular monitoring can detect health problems early and facilitate timely intervention, even in remote or underserved areas where care is not available. Moreover, multi-modal AI supports the integration of telemedicine platforms and AI-based diagnostic tools to support consultation and diagnosis. By analysing patient data from multiple sources, including diagnoses, test results, and patient reports, AI systems can help doctors get more information and treatment recommendations regardless of distance. Overall, multi-modal AI has the potential to revolutionise healthcare and diagnostics, making them more accurate, personalised, and effective. By seamlessly integrating different data models and leveraging advanced AI algorithms, multi-modal AI systems enable healthcare providers to deliver better outcomes and improve the patient's experience, ultimately leading to healthier people and healthier consumption.

13.6 Advancements in Multi-modal Interaction

Advances in interactive communication include the development of artificial intelligence that can integrate and interpret different types of input, such as voice, text, images, and gestures, providing more user information and better experiences. These advances enable human-computer relationships, allowing users to communicate with technology in the same way that it enables human-to-human relationships. For example, modern virtual assistants can understand and respond to commands while also interpreting messages from the user's environment. Additionally, advances in the field are helping create augmented reality (AR) and virtual reality (VR) experiences that combine the ideas of sight, sound, and motion to enhance user engagement. These innovations not only increase accessibility for people with disabilities by providing a variety of equipment and products but also support the development of areas where good communication and work are important, such as the use of human resource services, education, and entertainment.

Multi-modal user interfaces (MUI) represent a major breakthrough in human–computer interaction by integrating multiple senses such as touch, sound, pointing, and visual signals. Such interfaces lead to more intuitive and natural interactions that suit different users' preferences and abilities. MUI offers different models to enable the richness of interaction and make technology more efficient and effective. For example, users can use voice commands to interact with smart devices, and they can also use gestures or visual suggestions to adjust their devices (Zawacki-Richter et al., 2019). This integration enables better and more efficient communication, increases customer satisfaction, and expands the app's usability across multiple locations and devices. Human–computer interaction (HCI) has evolved with the emergence of multi-modal intelligence and is changing the way users interact with technology. Traditional HCI was truly based on text or images, but today's development includes many changes towards creating stronger and more powerful information. Multi-modal human–computer interaction uses speech recognition, eye tracking, facial analysis, and haptic feedback to create a more immersive experience for users. This flexibility allows for personalised and flexible interactions where the system can instantly understand and respond to user needs. Applications range from virtual assistants that can translate complex commands to interactive learning tools that adapt to different learning styles. Multi-modal AI supports augmented reality (AR) and virtual reality (VR) to provide more interactive experiences. Multi-sensor integration in AR allows digital information to be embedded in the physical world and generates feedback from a variety of sensors such as cameras, microphones, and measurement devices. This creates a connection between virtual devices and the real world, improving applications in areas such as education, healthcare, and entertainment. VR leverages all types of interaction by providing an optimal environment that responds to the user's body movements, gestures, and voice commands. This creates a more collaborative and realistic environment that supports applications from virtual meetings and remote collaboration to simulation and advanced training. The integration of multi-modal AI with AR and VR not only increases user engagement but also expands the potential for new cross-industry applications.

The integration of different types of intelligence with user interface, human–computer interaction, and augmented and virtual reality represents a revolution in technology, as shown in Table 13.2. This integration leads to more communication and better information to meet customer needs and preferences. This development democratises technology, making it easier and more accessible to a wider audience. Advances in human–computer interaction further enhance the user experience by combining voice, gesture, and visual cues to enable real-time social and personal interaction. This change is particularly evident in learning tools and virtual assistants that can tailor responses to individual users, making these systems efficient and engaging. Multi-modal AI transforms user engagement by creating unique experiences. AR applications use the integration of digital and physical worlds to easily access and understand complex information in areas such as education and healthcare. VR, on the other hand, uses a variety of devices to create experiences that faithfully replicate real situations. It is proving invaluable in areas such as training, simulation, and distance integration. Overall, no progress has yet been made on multi-tasking. By being interactive, intuitive, and flexible, multi-modal artificial intelligence is changing our relationship with technology, pushing the boundaries of what is possible, and laying the foundations for a future where human-machine collaboration is seamless and ubiquitous. These advances highlight the potential for multi-modal intelligence and herald a future in which technology will be more responsive, inclusive, and integrated into our daily lives (Colchester et al., 2017).

Table 13.2 Multi-modal AI applications across interaction domains

S.no	Types	Description	Examples/applications
1	Multi-modal-user interfaces (MUIs)	MUI combines various senses such as touch, sound, gesture, and visual symbols to provide more intuitive and natural human-computer interactions.	Users interact with smart devices using voice commands and gestures, increasing the richness of interactions and extending usability across multiple locations and devices.
2	Human–computer interaction (HCI)	Multi-modal AI has transformed HCI by integrating speech recognition, eye tracking, and haptic feedback to create dynamic response systems.	Virtual assistants that interpret complex commands, interactive tools that adapt to different learning styles, and systems that instantly respond to user needs.
3	Augmented and virtual reality (AR/VR)	Multi-modal AI enhances AR and VR by combining input from multiple sensors to create an interactive experience.	Main applications involve virtual meetings, remote collaboration, advanced stimulations, and training programs.

13.7 Multi-modal AI in Creative Industries

Multi-modal AI is transforming the creative industry by delivering new tools and technologies that enhance human creativity, performance, and push the boundaries of art. In the world of art and design, platforms like DALL-E and DeepArt allow artists to explore new ideas by combining images into beautiful things such as descriptions or performances. In movies and TV shows, advanced technology and AI-powered content analysis tools increase visibility and improve marketing strategies, resulting in a more immersive movie experience. In music and sound production, AI-supported composition algorithms and sound synthesis models can create original and realistic sounds, while AI-supported tools can increase productivity. Multi-modal AI has revolutionised the creative industry, ushering in a new era of innovation and artistry by providing creators with new tools to create creative products and collaboration (Malik et al., 2017).

13.7.1 Art and Design

With the integration of different types of skills, art and design have undergone significant changes, providing designers with new tools and technologies that redefine traditional art standards. Through platforms like DALL-E and DeepArt, artists have unprecedented freedom to create complex images from text or combine different elements of films, expanding their creativity and encouraging them to try new ideas. Multi-modal AI algorithms not only enable artists to collaborate with intelligent machines but also enable human creativity to create works of art through the integration of intelligent technology skills, supporting the connection between human thought and computer interaction (Sharma et al., 2022). Additionally, the emergence of smart tools for image editing, image processing, and content creation has revolutionised the creative process and helped artists accelerate and streamline their ideas. Artists can explore a variety of artistic styles, from creating complex works using a variety of skills to adding depth and meaning in art. The AI-powered system helps discover new videos by allowing creators to push traditional boundaries and experiment with unconventional mediums, styles, and ideas. In addition, multi-modal AI provides free access to advanced tools, enabling artists of all skill levels to master their art. Whether with the help of machine content or with the help of artists, multi-modal AI is helping them express themselves more effectively than before (Muzaffar et al., 2021). Fundamentally, the combination of multi-modal intelligence with art and design not only enhances the creative process but also opens new avenues for discovery and teaching. Multi-modal AI is transforming the art by bridging the gap between human creativity and machine intelligence, enabling artists to push the boundaries of storytelling and usher in a new era of innovation and development in art and design.

13.7.2 Film and Animation

Multi-modal intelligence in film and video development offers tools and techniques that enhance the creative and efficient process of visual storytelling. In film production, technology supports all aspects of film production, from pre-production to post-production. AI aids in pre-production script analysis, storytelling, and film planning, allowing filmmakers to better visualise and create narrative shows. AI tools also help improve the casting process by analysing tapes and predicting the right actors for the desired role. In theatre, multi-modal intelligence enhances the creation of static characters and realistic environments. AI- powered motion capture and animation tools help animators create natural movements and details that make characters more relatable and immersive. Advanced rendering technology powered by AI reduces the time and resources required to produce high-quality videos. AI algorithms handle complex tasks like lighting, shading, and texture mapping, allowing animators to focus on the creative side of their work. Additionally, AI-enhanced visual effects (VFX) have revolutionised the film industry by allowing the creation of cost-effective special effects that were previously impossible or limited. AI technology can take visual identification to a new level of realism by simulating realistic physical conditions such as explosions, weather effects, and water quality. Additionally, AI-powered colour grading and video editing tools ensure that the final product meets the best visual standard. Multi-modal AI also plays an essential role in automated post-production, video editing, audio production, and subtitling. AI-powered editing software can analyse multiple images to identify the best shots, match shots based on continuity, and even suggest adjustments that will enhance the narrative spectacle. In audio production, smart devices can synchronise audio and video, create sound, and improve speech clarity. AI-powered subtitles and translation tools enable movies and TV shows to reach global audiences, breaking down language barriers and expanding access to creative content (Malik et al., 2017).

13.7.3 Music and Audio Production

Multi-modal intelligence is revolutionising music and audio production by providing advanced tools and techniques that enhance creativity, enhance performance, and improve overall sound quality. Artificial intelligence music creation algorithms used by platforms such as AIVA and Amper Music help composers create original works in different genres and formats. These algorithms analyse lots of music to create melodies, harmonies, and tunes that support or form the basis of new songs. This ability allows musicians to try different musical ideas quickly and effectively, encouraging a better exploration of creativity. Artificial intelligence tools in audio production make many tasks easier, from mixing to identifying words or sounds (Singh et al., 2024b). An AI-powered digital audio workstation (DAW) automatically mixes, optimises, analyses, and equalises tracks, using appropriate settings for equalisation and compression. This automation not only saves time but also ensures uniformity and high efficiency in the final product. Additionally, advanced technology for audio reproduction can improve the quality of audio content by creating realistic voices and ambient sounds. Multi-modal AI also plays an important role in improving the quality and creation of audio content. For example, artificial intelligence-supported sound synthesis models developed by companies such as OpenAI and Google can create realistic music, instruments, and soundtracks. These models allow producers to create unique sounds that are difficult or impossible to create using traditional techniques. Additionally, AI has created tools to tune and improve the sound, removing bad stuff and ensuring accuracy and integrity of the sound. Multi-modal AI can also increase collaboration and efficiency in music and audio production. The AI-powered collaboration allows musicians and producers to collaborate remotely, instantly share ideas, and edit their work. These platforms often include features such as instant translation and integration tools to eliminate geographic and language barriers. Additionally, professional music education and training tools provide personal training and guidance to help musicians improve their skills. Multi-modal AI in media and interactive media opens up new possibilities for audience creativity and engagement. AI technology can instantly identify audio components and display real-time or interactive content that responds to music, creating dynamic and powerful performance (Subbarayudu et al., 2017). These innovations not only improved living conditions but also provided new ways of presenting art and engaging with audiences. Overall, multi-modal AI is revolutionising music and sound design by providing new tools that enhance creativity, increase efficiency, and improve audio content quality. By integrating technology into every stage of the production process, musicians and producers can push the boundaries of their craft and create meaningful experiences that connect with audiences around the world (Mondal et al., 2020).

13.7.4 Gaming and Interactive Media

Multi-modal AI is changing games and media, introducing new solutions that improve gameplay, create beautiful environments, and deliver personalised experiences. Technology supported by artificial intelligence in game development is changing the way game worlds are created. Technology automatically created a vast, complex, and diverse environment, making each game a unique experience. By analysing patterns and using complex algorithms, intelligence can reduce the time and resources required in the design process by creating a variety of harmonious environments, events, and situations. The AI-powered system also improves the behaviour and skills of non-characters (NPCs), allowing them to be more responsive and adapt to players' behaviour. Artificial intelligence algorithms help NPCs display realistic behaviour, learn from player interactions, and adapt their thinking to create cooperation and play. This type of interaction keeps people interested for a long time, keeping the game unexplored and exciting. AI algorithms analyse player behaviour, preferences, and gaming patterns to match game content and challenges to each player (Singh et al., 2024a). This individual variation extends to level difficulty, description, and individual recommendations for in-game purchases or content. Artificial intelligence understands and responds to people's preferences, improving the game as a whole and making it more efficient and effective. Artificial intelligence also plays an important role in improving the quality and sound of games and interactive media (Singh & Singh, 2023b). Advanced processing technology powered by artificial intelligence for more effective voice use. Multi-modal AI has also supported the development of augmented reality (AR) and virtual reality (VR) experiences that have become popular in games and social media. Artificial intelligence enhances these experiences by providing realistic interactions, responsive environments, and intuitive user experiences. In reality, AI-powered machines can embed digital content into the physical world to create seamless interactions. In VR, AI can create intuitive virtual environments that instantly respond to user actions, providing unparalleled interactivity and immersion. Overall, multi-modal AI is changing the design, personalisation, and understanding of digital experiences in games and interactive media (Alshareef et al., 2021). By combining advances in AI technologies, product developers and interactive designers can push the boundaries of creativity and innovation to deliver better, more authentic, and more personal experiences that appeal to a global audience.

13.7.5 *Ethical and Societal Implications*

The rapid development of multi-modal AI raises ethical and social issues that need to be carefully considered and managed. As these technologies become increasingly integrated into all aspects of daily life, from healthcare and education to business and entertainment, they raise important questions about corruption and fairness, privacy, transparency, and future performance. Addressing these issues is important to ensure that smarter machines are designed and used ethically. By focusing on creating fair, transparent, and privacy-friendly AI systems and planning for their impact on the market, we can harness the benefits of AI while reducing the potential for harm and encouraging greater participation and future justice technology (Raghunath et al., 2022). Bias and fairness are critical issues in many AI systems because these technologies often exploit and encourage bias in information. This can lead to discrimination against particularly vulnerable groups. Ensuring integrity involves developing strategies to identify, reduce, and prevent bias in AI models. This includes the use of different sources and intermediaries, the use of bias detection methods, and regular monitoring of AI systems for inappropriate behavior. Addressing these issues is critical for creating equitable AI solutions that serve all users equally and prevent social inequality.

Privacy issues in many areas arise from the collection and integration of data from multiple sources, including voice, video, text, and data sensors. Collection of this information may result in the disclosure of an individual's sensitive information without the individual's consent. Protecting user privacy requires strong data management, including confidentiality, secure storage of data, and strict data governance. Additionally, data protection policies and user consent procedures are important to ensure that individuals understand how their data is used and have control over their personal information. Transparency and disclosure are key to building trust in many AI systems. Users and partners need to understand how AI models make decisions, especially in critical applications such as healthcare, finance, and criminal justice, as shown in Fig. 13.3. Achieving transparency requires creating processes that make AI processes and decisions open and explainable (Rai et al., 2020). Explaining AI techniques such as fuzzy modelling and visualisation helps demystify complex AI models and provides insight into their capabilities. This not only builds confidence but also helps identify and correct errors or biases in the model. Different types of artificial intelligence have a lot of impact on business, and opportunities and challenges coexist. Although artificial intelligence can increase efficiency and create new jobs, it also increases the risk of employee replacement, especially in jobs that can be done manually. Industrial jobs such as manufacturing, customer service, and transportation are particularly vulnerable to automation. To reduce negative impacts, it is important to invest in employee training to equip employees with skills relevant to job transitions (Trescak et al., 2018). Additionally, encouraging dialogue between business, government, and academia

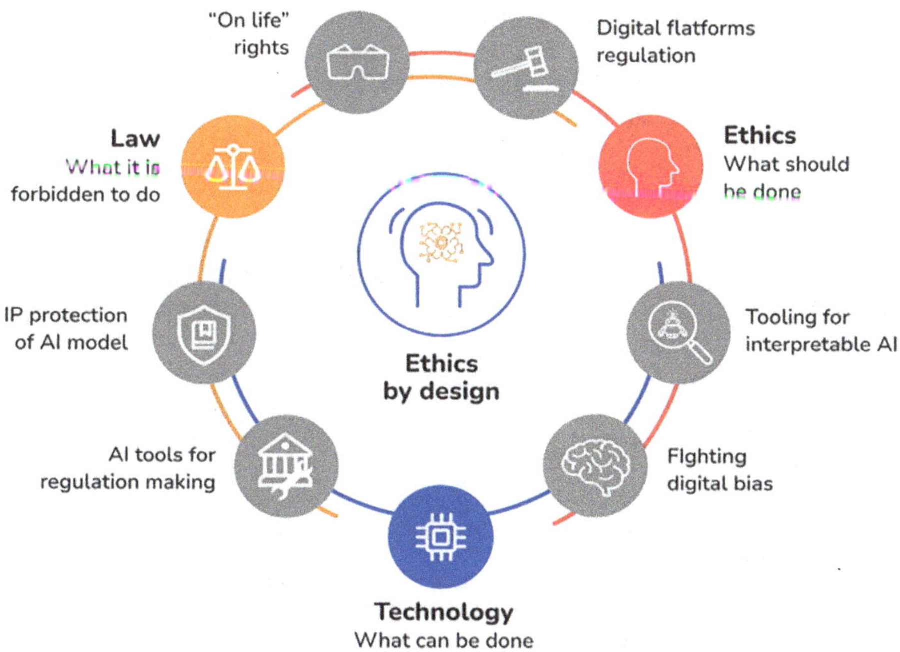

Fig. 13.3 Representation of ethical and societal implications on multi-modality of AI

can help develop strategies to spread the benefits of AI advances and empower people to work as they change.

13.8 Conclusion

The evolution of AI towards multi-modality is an important part of the development of AI and demonstrates the ability to collect and process different information with greater understanding. This research paper traces the historical development of artificial intelligence, from early tokens to the rise of machine learning and the evolution of deep learning, and shows how this progress forms the basis of multi-modal intelligence. Fundamental technologies such as neural networks, transformers, listening mechanisms, multi-modal fusion technology, and cross-modal learning make it possible to create complex systems that can understand and combine information from many sources. These capabilities have been demonstrated through fundamental processes such as facial models, vision models, and composite materials, which have found applications in fields such as natural language processing, computer vision, healthcare, and robotics. Artificial intelligence is big and transformative. In education, healthcare, creative industries, and user interaction, multi-modal AI increases efficiency and collaboration by enabling systems to better understand and meet people's needs. Additionally, advances in various interactions have improved the way users interact with technology, creating a positive and personal experience. But the integration of multi-modal AI also raises ethical and social issues. To ensure the benefits of multi-modal AI are realised and included at an exhibition, issues of data privacy, security, integrity, access, and AI bias need to be addressed. Continued innovations in multi-mode learning algorithms, flight performance capabilities, and integration with new technologies such as IoT and 5G will lead to the development and adaptation of more multi-mode systems. The development of artificial intelligence has expanded the scope of intelligence, also bringing new challenges and perspectives. By solving these problems and using the potential of multi-modal artificial intelligence, we can create smart machines that are stronger, more capable, and more responsive to human needs, and shape the future of technology and human relations.

References

Ali, M., Hussein, A., & Al-Chalabi, H. K. M. (2020). Pedagogical agents in an adaptive E-learning system. *SAR Journal-Science and Research.*

Alshareef, H. N., Majrashi, A., Helal, M., & Tahir, M. (2021). Knowledge extraction and data visualization: A proposed framework for secure decision-making using data mining. *International Journal of Advanced Computer Science and Applications (IJACSA).*

Anisha, P. R., Kishor Kumar Reddy, C., Nguyen, N. G., Bhushan, M., Kumar, A., & Hanafiah, M. M. (2022). *Intelligent systems and machine learning for industry: Advancements, challenges, and practices*. CRC Press.

Colchester, K., Hagras, H., Alghazzawi, D., & Aldabbagh, G. (2017). A survey of artificial intelligence techniques employed for adaptive educational systems within e-learning platforms. *Journal of Artificial Intelligence and Soft Computing Research*.

Crompton, H., & Burke, D. (2023). Artificial intelligence in higher education: The state of the field. *International Journal of Educational Technology in Higher Education*.

Das, S., Sharma, S., Dai, R., Bremond, F., & Thonnat, M. (2020). Vpn: Learning video-pose embedding for activities of daily living. In *European conference on computer vision*.

Daubert, M. A., Tailor, D., James, O. G., Shaw, L. J., Douglas, P. S., & Koweek, L. (2021). Multimodality cardiac imaging in the 21st century: Evolution, advances and future opportunities for innovation. *The British Journal of Radiology*.

Guan, C., Mou, J., & Jiang, Z. (2020). Artificial intelligence innovation in education: A twenty-year data-driven historical analysis. *International Journal of Innovation Studies*.

Horakova, T., Houska, M., & Domeova, L. (2017). Classification of the educational texts styles with the methods of artificial intelligence. *Journal of Baltic Science Education*.

Jaiswal, A., & Arun, C. J. (2021). Potential of artificial intelligence for transformation of the education system in India. *International Journal of Education and Development using Information and Communication Technology*.

Joshi, S., Rambola, R. K., & Churi, P. (2021). Evaluating artificial intelligence in education for next generation. *Journal of Physics: Conference Series*.

Malik, K. R., Mir, R. R., Farhan, M., Rafiq, T., & Aslam, M. (2017). Student query trend assessment with semantical annotation and artificial intelligent multi-agents. *Eurasia Journal of Mathematics, Science and Technology Education*.

Mondal, S. S., Mandal, N., Singh, A., & Singh, K. K. (2020). Blood vessel detection from retinal fundus images using GIFKCN classifier. *Procedia Computer Science, 167*, 2060–2069.

Munawar, S. K., Toor, M. A., & Hamid, M. (2018). Move to smart learning environment: exploratory research of challenges in computer laboratory and design intelligent virtual laboratory for eLearning technology. *Eurasia Journal of Mathematics, Science and Technology Education*. https://doi.org/10.29333/ejmste/85036

Muzaffar, A. W., Tahir, M., Anwar, M. W., Chaudry, Q., Mir, S. R., & Rasheed, Y. (2021). A systematic review of online exams solutions in E-learning: techniques, tools, and global adoption. *IEEE Access*.

Raghunath, K. K., Kumar, V. V., Venkatesan, M., Singh, K. K., Mahesh, T. R., & Singh, A. (2022). XGBoost regression classifier (XRC) model for cyber attack detection and classification using inception v4. *Journal of Web Engineering, 21*(4), 1295–1322.

Rai, A. K., et al. (2020). Landsat 8 OLI satellite image classification using convolutional neural network. *Procedia Computer Science, 167*, 987–993.

Reddy, C. K. K., Anisha, P. R., Hanafiah, M. M., Doss, S., & Lippert, K. (2024). *Intelligent systems and industrial internet of things for sustainable development*. CRC Press.

Renz, A., & Hilbig, R. (2020). Prerequisites for artificial intelligence in further education: Identification of drivers, barriers, and business models of educational technology companies. *International Journal of Educational Technology in Higher Education*.

Sharma, P., Singh, A., Singh, K. K., & Dhull, A. (2022). Vehicle identification using modified region based convolution network for intelligent transportation system. *Multimedia Tools and Applications, 81*(24), 34893–34917.

Singh, A., & Singh, K. K. (2023a). FedDDR: A federated improved DenseNet for classification of diabetic retinopathy. Proceedings. http://ceur-ws.org ISSN, 1613, 0073.

Singh, A., & Singh, K. K. (2023b). YORES: An ensemble YOLO and Resnet network for vehicle detection and classification.

Singh, A., Dhull, A., & Singh, K. K. (Eds.). (2024a). *Blockchain and deep learning for smart healthcare*. Wiley.

Singh, K. K., Rho, S., Singh, A., & Sergei, C. (2024b). Big data analytics and knowledge discovery for urban computing and intelligence. *Complex & Intelligent Systems, 10*(1), 1–2.

Subbarayudu, B., Lalitha Gayatri, L., Sai Nidhi, P., Ramesh, P., Gangadhar Reddy, R., & Reddy, C. K. (2017). Comparative analysis on sorting and searching algorithms. *International Journal of Civil Engineering and Technology*.

Trescak, R., Yang, B., Zio, E., & Chen, X. (2018). Artificial intelligence for fault diagnosis of rotating machinery: A review. *Mechanical Systems and Signal Processing*.

Zawacki-Richter, O., Marín, V. I., Bond, M., & Gouverneur, F. (2019). Systematic review of research on artificial intelligence applications in higher education–where are the educators? *International Journal of Educational Technology in Higher Education*.

Chapter 14
Interfacing Multi-modal AI with IoT: Unlocking New Frontiers

S. Delsi Robinsha and B. Amutha

Abstract The combination of artificial intelligence and the Internet of Things has recently altered our way of perceiving and interacting with the world around us. We examine the possible relationships between multi-modal AI and Internet of Things technologies in the chapter "Interfacing Multi-modal AI with IoT" from the book Unlocking the Potential. We aim to find the good changes that can come from integrating these two systems. Understanding and analysing complicated real-life occurrences is made easier by multi-modal approaches within the framework of artificial intelligence. These approaches integrate the competency of numerous data modalities, such as text, images, and sensor data. With the combination of multimodal AI and an abundance of data facilities powered by IoT devices, new possibilities for producing various forms of creativity and efficiency in various sectors emerge. An introduction to AI and the Internet of Things, outlining their fundamental concepts so that the reader may grasp their interconnection, is the first stop on this path. Our research also delves into the feasibility and efficacy of combining various forms of AI to address practical issues. Compatibility with the Internet of Things (IoT), real-time processing, data integration, and artificial intelligence (AI) are some of the problems that this book aims to solve. In addition, there is the investigation of how predictive analytics using automation that makes use of machine learning, AI, and data from the Internet of Things could revolutionise entire industries. All of this would be useful for learning about smart automation, predictive maintenance, and customer-specific services. This article presents reasons and examples to show how the health sector, manufacturing, transportation, and smart cities can all be transformed by integrating AI and the Internet of Things. When we think about the future of AI and the Internet of Things working together, some methodologies and trends are starting to emerge, such as quantum computing, federated learning, and edge computing. Furthermore, AS and personalised services backed by AI and IoT bring up ethical questions, as well as potential dangers and invasions

S. Delsi Robinsha (✉) · B. Amutha
Department of Computing Technologies, School of Computing, Faculty of Engineering and Technology, SRM Institute of Science and Technology, Chengalpattu, Tamil Nadu, India
e-mail: ds8912@srmist.edu.in; amuthab@srmist.edu.in

A. Singh, K. K. Singh (eds.), *Multimodal Generative AI*,
https://doi.org/10.1007/978-981-96-2355-6_14

of privacy. Combining multi-modal AI with the Internet of Things poses a disruptive danger, and this book provides a general view or outlook on that risk. In the coming age of the digital, this document could serve as a valuable resource for academics, technocrats, and legislators looking for a forum to debate and plan the further development of AI and the Internet of Things, as well as their potential applications as problem-solving tools and sources of economic benefit.

Keywords Artificial Intelligence · Internet of Things · AI-enabled automation · Edge computing · Federated learning · Text · Images · Data · Multi-model

14.1 Introduction

The Internet of Things (IoT) and multi-modal artificial intelligence (AI) have been the most innovative technologies that have been seen contributing across domains in the last few years. Multi-modal AI refers to computer systems that are capable of perceiving, interpreting, and making intelligent use of languages, images, sounds, and data collected from environments. While the SC is a small set of interconnected computers used for managing manufacturing processes and supply chains, as it receives and processes data, the IoT is a huge network of connected sensors, actuators, and devices that can send and receive data in real time. This new formation of multi-modal AI with IoT is paving newer forms of adaptiveness that we have never seen before in our interaction with technology. This opens up hitherto unimagined frontiers of possibilities as we seek to develop newborn, smarter, more efficient, and autonomous systems through integrating AI with IoT. Applications of artificial intelligence that are based on the ability of the system to analyse and learn from different types of data are called multi-modal AI applications. These systems are able to obtain a more thorough understanding of their context owing to the fact that they combine information from multiple domains, such as text and image analysis or data from the sensors with sound input. Its use numbers a host of advanced applications, including context-aware decision-making, predictive analysis, superior picture recognition, and natural language comprehension (Ghayvat et al., 2024).

The rise of IoT devices has led to the growth of interconnected and "smart" devices ranging from home appliances, smart clothes, wearable technology gadgets, iPhones and smart watches, city infrastructure, industrial equipment and many more. These devices are always on and capturing information about a lot from the places they occupy and the things they observe. With the help of analytical capabilities and implemented decision-making support made possible by IoT systems, companies are able to think operationally, achieve efficiency, as well as engage in operational optimisation and proactive maintenance intervention where necessary. It is crucial to state that the possibilities of the symbiosis of multi-modal AI and the Internet of Things are vast and could have an impact on quite a few various markets. Example: In smart homes, voice control can come in the ability of an artificial intelligence-based assistant to understand and execute voice commands and the

Internet of Things that allow the monitoring of power consumption, adjustments of the climate control systems, and management of home appliances for the provision of the best experience and comfort. In a production context, potential maintenance costs and lost time can be identified using the algorithms that will scrutinise data received from sensors to predict when a piece of equipment will fail. They agree that some issues are still encountered when both multi-modal AI and Internet of Things technologies are implemented (Morales-García et al., 2024). The requirements that need to be met in order for integrated systems to work optimally are as follows: Liabilities, incompatibility problems, data security, Privacy, and new, larger structures. The various opportunities that are inestimable in comparison with the evils include having extraordinary chances of increasing efficiency, improving standard of living, and meeting significant social and ecological issues. It is necessary to note that the prospective synthesis of multi-modal AI and the Internet of Things would require an understanding of the framework, the real-world considerations, and the new directions towards the creation of this fusion. It is the aim of this book to provide an overview of these advancements in the context of the interaction between multi-modal AI and IoT on the one hand and the possibilities on the other hand that these technologies offer in the field of intelligent systems and environment.

14.2 Fundamentals of Multi-modal AI

Multi-modal artificial intelligence (AI) involves several interfaces and a combination of text, pictures, speech, and sensors to make it possible to understand and process data in multiple ways. Concisely, multi-modal AI systems are an integration of diverse modes of operation encompassed by the theories and findings in this section. By applying complementing properties of several modalities introduced in AI systems, the analysis and interpretation of multi-modal data could be performed in parallel, offering a different perspective towards the input. More accurate, detailed, and contextualised exegeses are conceivable because the algorithms of such an AI system can apply this procedure to unite information from multiple sensory modalities (Gadey et al., 2024). Considering the multi-modal data processing, it should be mentioned that the use of the neural network architecture and deep learning models in this case seems to be rather effective. Transformer models and recurrent neural networks (RNNs) are well-suited for breaking down sequential information like text and sound, while convolutional neural networks (CNNs) excel on images. The principles of GNNs, or graph networks, and multi-modal transformers are some examples of successful attempts at developing architectures that can enhance the management of interactions between numerous modalities. The neutrality of data acquired from several senses and comparable ways of constructing the encoding must be consistent for multi-modal AI systems. It is common to use feature fusion and late fusion approaches here. Feature fusion aimed at merging features of several modalities, whereas late fusion combined different modalities, processing each of them separately and then merging the result.

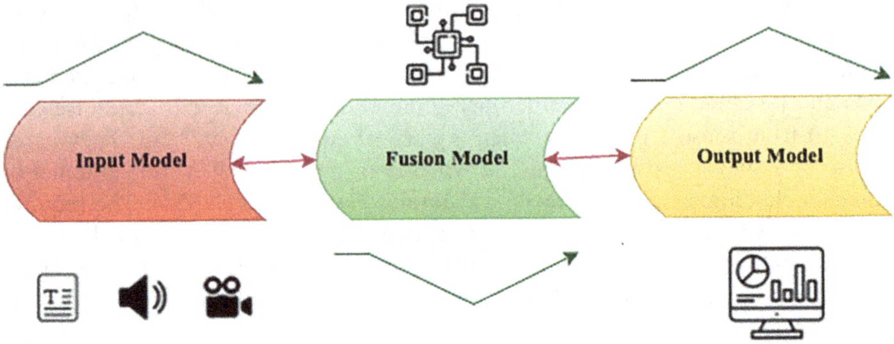

Fig. 14.1 Fundamentals of multi-modal AI

Notably, the deep learning models for the neural network-based models have remarkably interfacial-level performance to handle the multi-modal data. When concerned with sequential input, particularly text or speech, alternatives like CNNs and the transformer model are more useful; RNNs and CNNs are especially good at graphical input. To reunify the operation of separated modalities, architectures like multi-modal graph networks and multi-modal transformers have been proposed. Every multimodal AI model example including mobility training makes use of a multimodal dataset, or data gathered from several modalities. In this work, the model performs multiple (2+) modalities in parallel and enables it to learn anticipation from the different outcomes. The paper has also suggested that additional tuning, based on the particular domain, results in good performance after generalisation on a large, multi-modal raw dataset. Figure 14.1 shows the fundamentals of multi-modal AI.

Multi-modal AI is useful in many different fields, such as:

- Captioning Images: Making textual descriptions of pictures.
- To answer questions involving pictures, one uses visual question answering (VQA).
- Extracting emotional states from audio signals: speech emotion recognition.
- Using text, photos, and videos to draw conclusions about a topic is known as multi-modal sentiment analysis (Burns & Lambert, 2024).
- Connecting sensors, cameras, and global positioning systems to enable autonomous vehicles to navigate and perceive their surroundings.

14.3 Technologies Powering Multi-modal AI

Multi-modal AI can therefore be said to have been developed through the aggregation of the knowledge in various AI domains. The ability to store and process data of many types and modalities has been a trend that runs very favourably for AI researchers and practitioners in recent years.

14.3.1 Deep Learning

One of the branches of artificial intelligence, which is called deep learning, utilises an artificial neural network, which is an algorithmic set applied to complex problem-solving. Normally, the numbers in question are neural numbers, and in the present generative AI revolution, transformers—a neural architecture in detail—work. Integrating this module in the flexible framework of human-like AI is still a long shot, which must be encouraged to cover more ground in the future. There is the need to embark on new methods in combining methods and improving the transformers' existing methods.

14.3.2 Natural Language Processing (NLP)

NLP is incredibly important in AI because of the power it brings to the table, allowing computers to understand human language. This fields is as interdisciplinary as is can be, and is rapidly changing human interaction through the means of allowing computers to comprehend, translate, and, in some cases, generate human language. (Bimonte et al., 2024). Regarding generative AI models, which can be especially multi-modal, NLP is a must to ensure great performance during its application. This is because text is the main way of interacting with the machines. Text is used in communication with machines because it is the main input and output used in the process of interacting with the machines.

14.3.3 Computer Vision

Image analysis is among the computer visions, which is a set of techniques that enables a computer to "see" and understand a scene. In this area, technology has developed sophisticated methods that allow developing multi-modal artificial intelligence that can analyse video and picture data and provide results in such formats.

14.3.4 Audio Processing

With some of the current powerful generative AI models, it is possible to feed it with audio playbacks and fetch specific audio outputs. It is quite easy to accomplish voice mail processing, music production, and simultaneous translation to add to the list.

14.4 Applications of Multi-modal AI

By learning from a variety of sources, or "modalities," computers can improve their accuracy and interpretability. A flood of new applications are flooding in from all walks of life and all kinds of industries thanks to these powers, including:

14.4.1 Augmented Generative AI

Primarily able to receive user input in the form of text and respond with text, the majority of the initial crop of generative AI models were text-to-text. Some examples of multi-modal models are GPT-4 Turbo, Google Gemini, and DALL-E. These models open up new avenues for improving the input and output user experience. Whether it's responding to commands in different ways or creating material in different forms, the potential of multi-modal AI agents appears to have no bounds.

14.4.2 Autonomous Cars

Multiple modalities of artificial intelligence are crucial to self-driving vehicles. All sorts of sensors allow these vehicles to take in data from their environments and process it in different ways. Multiple vehicles can't make intelligent decisions in real time without multi-modal learning, which combines multiple data efficiently (Robinsha & Amutha, 2024).

14.4.3 Biomedicine

Biomedical data is becoming more accessible through biobanks, EHRs, clinical imaging, medical sensors, and genetic data, which is driving the development of multi-modal AI models for use in healthcare. In order to make informed clinical decisions and better understand human health and illness, these models may interpret data from a wide variety of sources, including numerous modalities.

14.4.4 Earth Science and Climate Change

Imaging tools such as ground sensors, drones, satellite data, and other measuring tools are rapidly growing, which on the same respects is improving our ability to make sense of the world. Therefore, to integrate this data appropriately and build new applications and tools from multi-modal AI that can also help us in several fields, such as in agriculture, tackling emissions of greenhouse gases, or in handling severe weathers, is crucial.

14.5 Fundamentals of IoT

Through the help of IoT technologies, users may automate, analyse, and integrate their systems to a greater extent. Both the range and precision of these areas are enhanced by them. Sensors, networks, and robotics are all part of the IoT, which makes use of both current and future technologies (Dayananda et al., 2024). The Internet of Things takes use of evolving mindsets towards technology, lowering hardware prices, and new software advancements. Its cutting-edge features cause enormous shifts in the distribution of commodities and services, as well as in the resulting social, economic, and political climate.

Principal attributes of the Internet of Things are artificial intelligence, connection, sensors, active involvement, and the usage of small devices, which are the most essential elements of the Internet of Things (IoT). The fundamentals of IoT are shown in Fig. 14.2. This section provides a concise overview of these features.

(i) *AI*: The Internet of Things (IoT) and artificial intelligence (AI) enable almost everything to be "smart," improving daily life in every way possible through the integration of data gathering, AI algorithms, and networks. This can be as easy as setting up an automated system in your kitchen to notify you when milk and your favourite cereal are running low, so you can automatically place an order with your chosen grocery store.
(ii) *Connectivity*: New networking enabling technologies, particularly networking for the Internet of Things, have made networks less reliant on big carriers. Networks can still operate at lower, less costly scales. Such local networks are formed by interconnected devices in an IoT system.
(iii) *Sensors*: The Internet of Things becomes notional without sensors. They are the defining instruments that make the Internet of Things (IoT) more than just a network of devices; they make it an active system that can be integrated with the actual world.
(iv) *Active Engagement*: Modern linked technology is largely used in a passive manner. Through the Internet of Things (IoT), a new paradigm for interactive content, product, or service consumption is revealed (Singhal, 2024).
(v) *Small Devices*: Technology has progressed as expected, with devices getting smaller, cheaper, and more powerful with time. Internet of Things (IoT) accuracy, scalability, and adaptability are brought to you via tiny, purpose-built devices.

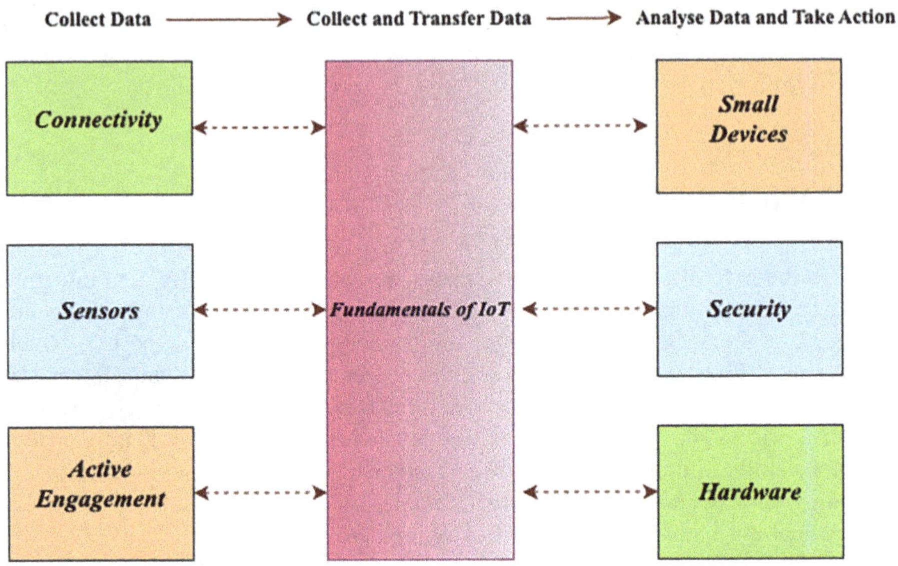

Fig. 14.2 Fundamentals of IoT

14.5.1 IoT Architecture

14.5.1.1 Sensing Layer

Figure 14.3 shows the layers involved in IoT architecture. The most fundamental component of this architecture consists of sensor-enabled smart objects. The sensors make it possible to gather and process data in real-time by connecting the digital and physical realms. Many different kinds of sensors are available, each with its own unique function. Among the many variables that these sensors can record are air quality, temperature, velocity, humidity, pressure, flow, motion, and electricity, among others. Sometimes they can even remember a few things, which means they can record a specific amount of measures. An instrument may interpret the signal that a sensor produces after measuring the physical property. Sensors are classified based on their specific function; for example, there are sensors for the environment, the human body, household appliances, telematics in vehicles, and so on.

It is necessary to link the majority of sensors to the sensor gateways. A LAN, like Ethernet or Wi-Fi, or a PAN, like ZigBee, Bluetooth, or Ultra-Wideband (UWB), can be the type of network this takes. Sensors that don't need to be connected to aggregators can still be connected to backend servers and apps using WANs like GSM, GPRS, and LTE. A basic example of a sensor network is a wireless sensor network (WSN), which consists of connected devices that require little power and communicate at low data rates. WSNs are becoming more popular since they can cover a lot of ground, have a long battery life, and support a high number of sensor nodes.

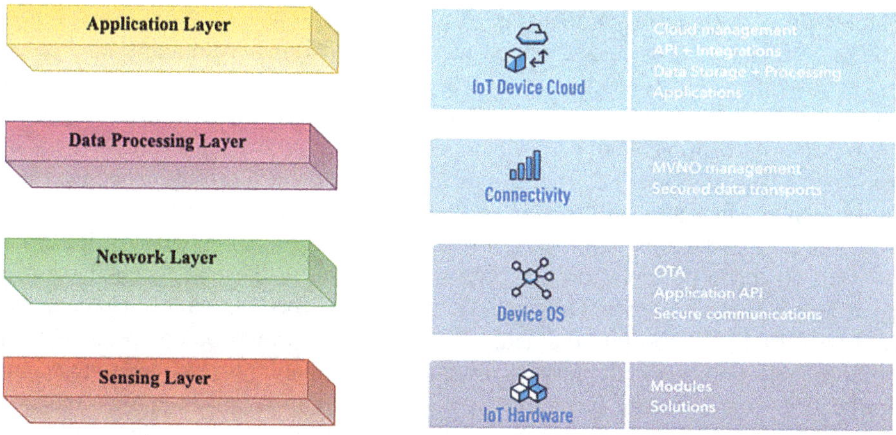

Fig. 14.3 Layers involved in IoT architecture

14.5.1.2 Network Layer

These tiny sensors will generate massive amounts of data, necessitating a fast and reliable network infrastructure, whether it's wired or wireless. The machine-to-machine (M2M) networks and the applications they enable have relied on existing networks, which are frequently bound together using various protocols. More and more Internet of Things (IoT) services and applications, including context-aware apps, high-speed transactional services, and others, necessitate heterogeneous configurations of networks using different technologies and access protocols to meet user demand. Constructed to accommodate latency, capacity, or security-related communication needs, these networks might take the shape of private, public, or hybrid models. Multiple gateways and gateway networks (Wi-Fi, GSM, GPRS, etc.) in a variety of physical forms

14.5.1.3 Data Processing Layer

Data processing is made feasible by the management service's analytics, security controls, process modelling, and device management. The engines for business and process rules are an integral part of the management service layer. The Internet of Things (IoT) allows for the interconnection and interaction of many items and systems, which in turn provides data in the form of events or contextual information, such as the present location, traffic conditions, and product temperatures. While some of these occurrences necessitate processing in a later stage, like collecting periodic sensory data, others demand urgent action, such as responding to medical emergencies. In order to make the Internet of Things (IoT) system more responsive, rule engines help with the creation of decision logics and set off automatic and interactive processes. Analytics makes use of a variety of analytics technologies to quickly process enormous amounts of raw data in order to extract important information. Data caching in random access memory (RAM) instead of physical discs is made possible by analytics like in-memory analytics. Data query time is reduced and decision-making speed is enhanced with in-memory analytics. Another type of analytics is streaming analytics, which require real-time analysis of data (also known as data-in-motion) to make choices in seconds. The ability to control the movement of data and information is known as data management (Robinsha & Amutha, 2023a; Mondal et al., 2020). Access, integration, and control of information are all made possible through data management in the management service layer. It is possible to prevent higher-level apps from processing unneeded data and to lessen the likelihood that the data source's privacy would be compromised. In order to make sure that only the most important information is provided to the right applications, data filtering techniques, including data anonymisation, data integration, and data synchronisation, are employed. To achieve more agility and cross-domain reuse, businesses can benefit from data abstraction, which involves extracting information to create a shared view of data. From the smart object layer all the way down to the application layer, security must be ensured across the complete IoT architecture. System security reduces risks by preventing unauthorised persons from hacking or compromising the system.

14.5.1.4 Application Layer

The application layer is where the user interacts with the system and where the services tailored to the application are provided. One possible example is a smart home app that lets users control appliances like coffee makers with the touch of a button. Another example is a dashboard that displays the current state of all the components in a system. Smart cities, smart homes, and smart health are just a few examples of the numerous possible deployments of the Internet of Things (Robinsha & Amutha, 2023b; Singh & Singh, 2016).

14.5.2 Communication Protocols

An IoT communication protocol is a collection of rules and standards that governs the manner in which data is processed and communicated between devices that are connected to the Internet of Things (IoT). IoT devices are able to communicate with one another thanks to this definition.

14.5.3 Message Queue Telemetry Transport (MQTT)

MQTT, which stands for Message Queue Telemetry Transport, operates according to the publish-subscribe concept. Within the framework of this approach, the gadgets, which are referred to as publishers, do not transmit information directly to the readers. Certain customers, who are referred to as subscribers, are the only ones who can access the information. It is certain that the delivery will take place because the MQTT communication channels are based on TCP. Devices with limited bandwidth and networks with limited bandwidth can benefit from using MQTT.

14.5.4 Constrained Application Protocol

The client–server architecture that supports the RESTful principles is employed by the constrained application protocol, often known as CoAP. GET, POST, PUT, and DELETE, are the types of HTTP requests that can be made to the server by any client Internet of Things device. The resource can be accessed using a URL. Due to the fact that the communication channels established by CoAP are based on UDP, the delivery is not assured. CoAP is able to support network configurations that are extremely crowded as well as Internet of Things devices that have limited resources, such as low-power sensors or embedded systems.

14.5.5 HyperText Transfer Protocol (HTTP)

In the world of the Internet of Things (IoT), this protocol is being employed because it was the first method of data communication for the World Wide Web (WWW). In spite of this, it is not optimised for it due to the following problems: In order to obtain information, it is time-consuming and energy-consuming to link multiple sensors because the hypertext transfer protocol (HTTP) was designed for two systems to communicate with each other at the same time. HTTP is a unidirectional protocol, which means that it is designed for one system (the client) to transmit a message to another system (the server). As a result, it is rather challenging to elevate an Internet of Things solution. Battery-powered apps are not a good fit for the hypertext transfer protocol (HTTP), which is dependent on transmission control protocol (TCP), which requires a significant amount of computer resources (Zhao et al., 2023; Singh et al., 2013).

14.5.6 ZigBee

In a manner comparable to that of Bluetooth, ZigBee is primarily utilised in commercial and industrial environments. While it is well positioned to take advantage of wireless control and sensor networks in Internet of Things applications, it possesses a number of key advantages in complex systems, including low-power operation, high security, robustness, and high Scalability. Essentially, the different ZigBee wireless standards have been consolidated into a single standard, which is the most recent version of ZigBee, which was just recently released as version 3.0.

14.5.7 Z-Wave

Z-Wave is an Internet of Things (IoT) technology that is primarily designed for home automation, with items such as lamp controllers and sensors, among many other devices, being made possible by its low-power radio frequency (RF) connections. Sigma Designs is the only manufacturer of chips for Z-Wave, in contrast to the various suppliers that are available for other wireless technologies such as ZigBee and others. Z-Wave employs a simpler protocol than some others, which can enable faster and simpler developer development.

14.6 Possibilities and Obstacles of Combining Multi-modal AI with the Internet of Things

Multi-modal artificial intelligence (AI) and Internet of Things (IoT) are on the edge of providing huge improvements in several sectors. This is because it fosters the idea of smart systems that are intelligent enough to decipher complex situations based on a combination of many data forms. Still, this integration leads to some issues that arise as follows, which should be effectively addressed: At the same time, it creates numerous opportunities for the appearance of brand new apps and further evolution of the existing functions (Iyer et al., 2023). An enlarged look at the opportunities and problems that present themselves when combining these cutting-edge technologies is as follows:

14.6.1 Data Complexity and Volume

Despite this, IoT gives rise to a wide range of data kinds that includes everything from simple temperature readings to complex video streams; furthermore, there are several extra layers of challenge in the process of data processing and analysis. The magnitude of this strain is even more complicated by the fact that new data is being generated at an unprecedented rate by an endless array of devices. The coordination of this complexity and the assurance that multi-modal AI systems can operate in a beneficial way require solutions like enhanced spatial statistics algorithms, edge computing, and data slimming.

14.7 Challenges

14.7.1 Interoperability and Standardisation

This is a big compatibility issue that occurs when two products do not use compatible standards of data exchange. This means that artificial intelligence systems do not seamlessly perform well when it comes to combining and analysing information from many sources. To address these problems, it is useful to define and employ standardised communication patterns and message structures. This might be achieved by regulation of the industry or global legal standards.

14.7.2 Latency and Processing Power

Real-time data processing poses challenges because of the brief time within which answers have to be provided and since many of the Internet of Things devices have constrained processing capacities and since there might be delays in delivering data. The incorporation of more potent edge computing solutions and the adoption of 5G networks can help reduce latency levels and ramp up the processing within or in close proximity to the data source.

14.7.3 Privacy and Security

The integration of multi-modal artificial intelligence leads to the introduction of various levels of data, some of which may contain personal data; this increases the possibility of data compromise and leakage. The measures of trustworthy authentication techniques, safe data governing and disposal, and secure end-to-end encryptions are Most Essential. Another essential requirement is to maintain GDPR compliance, as well as other privacy standards, in order to protect users' information.

14.7.4 Scalability

This means that as the IoT systems become more complex, these systems need to handle more devices and data without the added advantage of having more time on their side. That is, the use of cloud computing, virtualisation, and large and scalable models of artificial intelligence would allow one to address the issues connected with a growing number of data and devices, thus making it possible to increase or decrease the size of the system as such.

14.8 Opportunities

14.8.1 Enhanced Decision-Making

If data from several sensors is collected and analysed, the decision-making becomes more complete and therefore usually more accurate and contextual. Subsequently, it results in greater efficiency and effectiveness of tasks and outputs. Images and information gathered from surveillance sites or from the climate could be processed by the traffic control systems to enhance traffic circulation and reduce delays in real time from the automobile telematics.

14.8.2 Predictive Maintenance and Operations

The use of multi-modal data can provide early indications of equipment breakdown, which enables preventative maintenance and reduces the amount of time that equipment is offline.Through the analysis of sound, vibration, and operational data, artificial intelligence can be used in manufacturing to forecast the breakdown of machines, allowing maintenance to be scheduled only when it is required.

14.8.3 Personalised User Experience

IoT devices that are powered by multi-modal artificial intelligence have the ability to adapt to individual behaviours and environmental settings, hence giving features that are tailored to the user. Based on the resident's mood, which may be determined by facial expressions and voice tones, the lighting, temperature, and music in smart houses can be adjusted accordingly.

14.8.4 Innovative Services and Products

Integrated multi-modal artificial intelligence and Internet of Things systems have the potential to lead to the development of new and innovative goods that can completely transform existing markets. Technologies that improve virtual reality by adjusting experiences based on real-time physiological and environmental data in order to achieve a higher level of immersion.

14.8.5 Operational Efficiency

In terms of resource utilisation, systems have the potential to become more efficient, hence lowering waste and enhancing service delivery. Intelligent grids that make use of Internet of Things devices have the ability to dynamically balance load and generation, which results in a significant improvement in energy efficiency.

14.9 Interfacing Techniques and Protocols

The general guidelines and policies regarding interactions are instrumental in the harmonisation of multiple AI industries with IoT. Such techniques and protocols help in achieving high levels of efficient, safe, and successful communication between the devices and systems. To stress the need for reliable interface solutions, let us note that IoT devices can deliver all sorts of data, ranging from basic temperature sensors to highly complex video security cameras, and the latter are just as valid an example of what IoT can offer as the former (Golec et al., 2023; Singha & Singhb, 2023). Since different parts of a software system may need to work together to accomplish tasks, APIs offer a standardised way for multiple parts of a system to transfer data to one another and collaborate effectively within this design. This is especially beneficial due to their capability of managing infinite data formats and communication termination commands usual in Internet of Things systems.

This allows devices to work in unison within the system and ecosystem that is assumed to be interconnected. APIs have a friend in middleware solutions because they are not only a layer of consideration between various gadgets and the multi-modal AI systems but also a way to control the interactions between those devices. In this process, raw data collected from the wireless sensor network is transformed and pre-processed into models that artificial intelligence algorithms can easily work upon. This makes it possible to arrive at decisions in real time employing full data analyses carried out concurrently. In large-scale operations, middleware can also undertake competencies like device management, data caching, and load balancing, which can be very crucial in ensuring that the large-scale system keeps on running efficiently.

Communication protocols, on the other hand, refer to the set of rules or structures that were put in place to direct the procedures of how data flows over the network. This guarantees that devices of all classes can interact harmoniously since some of the gadgets may have a different ability to access resources. Two specific examples of such standards are MQTT and CoAP, both of which were created with the IoT constraints in mind. These protocols aim at what could be referred to as "constrained bandwidth usage" and "intermittent connectivity," which are often characteristics of IoT systems that are portable or are located at great distances. The integration of inputs that are initiated in parallel with audio, video and sensor data in an Internet of Things system is achieved here with help of protocols like WebSockets that uphold live-stream data transmission. Such protocols are essential for applications like the real time and the interactive ones. In terms of the integration of multi-modal AI with the Internet of Things, or more broadly IoT, there is one critical field that stays as a priority—security. Other examples include Transmission Layer Security (TLS/SSL) and Internet Protocol Security (IPsec), which provide solid encryption and authentication mechanisms to safeguard data integrity and confidentiality across respective pathways. This is a must not only for the protection of personal as well as commercial data but also for keeping the IoT ecosystem from intrusion as well as attacks that may potentially harm the whole structure of the IoT.

The interface techniques and protocols used in integrating multi-modal artificial intelligence with the IoT provide the IoT with those ingredients that are requisite in making the IoT system safe, seamless, and efficient. In addition, they open the number of opportunities for the numerous circulation and exchange of the data between the IoT tools; they also provide the prospects for multi-modal artificial intelligence in the utilisation of its application to the IoT solutions in the more effective, precise, and intricate ways—all of which combined results in the attempts to create the more efficient, superior, and adaptive Internet of Things (Wang et al., 2023; Singh & Singh, 2023). Planners, designers, and implementers of the Internet of Things, as well as individuals and companies that use its products, get to increase the value and scope of their implementations in one way or another by making use of these products. It will create a future for new high-IQ apps and services to begin. Thus, the following opportunities can be realised.

14.10 Applications of Multi-modal AI and IoT Integration

In enhancing the server operations, the methods of real-time decision-making and offering specific and special consumer solutions, the integration of multi-modal artificial intelligence and the Internet of Things is changing the face of several business fields dramatically. In the subsequent four paragraphs, I will explore some of these applications in a more detailed manner and also show that the clustering of these technologies is leading to remarkable advancements.

14.10.1 Healthcare: Improving Patient Tracking and Medical Attention

In the healthcare industry, the combined application of multi-modal AI and IoT has brought about drastic change, especially in such aspects as distant diagnostics as well as attending to old and frail people. The flow of health data is ongoing, received from people's homes or wearable devices through the network of things, which are equipped with numerous sensors. Such parameters as the pulse, blood pressure, activity rhythm, and even the quality of slumber constitute this data. And therefore, performing an analysis of this information is a multi-modal artificial intelligence system. Such systems are capable of processing and interpreting multitudes of data, which can range from current sensors to the patient record. For instance, independent learning frameworks can be employed for RPM with a view of predicting adverse health trends. These algorithms are capable of filtering small changes that an observer might find hard to notice from the data fed back by the sensors. This capability allows timely interventions, so one can prevent hospital readmissions or emergency scenarios when the illness acts up again. Smart home sensors can also alert the system if there is any abnormal movement or lack of movement in a room, indicating the occupant of a room has fallen or is not well. This can prompt quick notifications to be sent to carers or medical professionals. Smart home sensors are also used in the care of older people.

In Table 14.1, we can see the results of comparing different metrics in several applications, including healthcare, smart cities, industries, and agriculture. In the control group, we didn't integrate multi-modal AI with IoT. In the experimental group, we did. The experimental group's percentage rise or decrease relative to the control group is shown in the improvement percentage column.

Table 14.1 Potential research table for analysing the results of combining multi-modal AI with the Internet of Things in different applications

Metric	Control group (no integration)	Experimental group (integration)	Improvement (%)
Energy efficiency	24.5 kWh/day	18.2 kWh/day	25.71%
Response time	3.5 s	1.2 s	65.71%
User satisfaction (1–10 scale)	6.2	8.4	35.48%
Anomaly detection accuracy	78.5%	92.3%	17.55%

14.10.2 "Smart Cities": Intelligent Management of Urban Infrastructure

The integration of multi-modal artificial intelligence with the Internet of Things improves the management and optimisation of urban infrastructure in smart cities, particularly in the areas of traffic management and waste management systems. Traffic lights and sensors that are equipped with intelligence gather information on the flow of vehicles, the number of pedestrians, and even the weather conditions. In order to alleviate congestion and enhance traffic flow, multi-modal artificial intelligence examines this data and makes adjustments to traffic lights in real time. This can greatly reduce the amount of time spent commuting and the amount of pollution that is produced. Another domain in which this connection is advantageous is waste management in smart cities. Bins that are equipped with Internet of Things capabilities are able to monitor the amount of rubbish and relay this information to central management systems. To determine the most efficient routes to follow when transporting the garbage-filled bins, multi-modal artificial intelligence analyses this information in combination with the time of previous pickups, traffic information, and bin locations. Because fewer unnecessary pickups and vehicle emissions are generated through this process, this enhances the efficiency of waste collection while at the same time IRA negatively impacted because operational costs are reduced.

14.10.3 Industrial Robotics: Keeping Things Running Smoothly and Efficiently

Multi-modal artificial intelligence simplifies the assessment in predictive maintenance and quality control, whereas Internet of Things sensors are crucial in the industrial sector for monitoring both the equipment and the conditions around them. With artificial intelligence, one is able to detect or diagnose that there are signs of impending equipment breakdown since data are gathered from sensors that are fitted into the machine. For instance, vibrations or temperatures that are not from normal ranges can mean that there is a problem with the equipment, wear, or any imminent failure that the maintenance crew can attend to before a disaster that is costly happens. Artificial intelligence and IoT can also be used in manufacturing process, for example, through the use of cameras and sensors to enhance quality control. This makes it possible that weaknesses on the products are easily seen at various stages of the production process such as the assembly line. Today artificial intelligence sub-systems are used to make real-time photo analysis and sensor data with distinguishing such mistakes as it is impossible to see from the view of people. What is more, the overall effectiveness of the entire process of creating products is enhanced because of this, as well as the quality of the final product that is produced is made better.

14.10.4 Farming: Accurate Soil Mapping and Animal Tracking

Being an industry that involves both precision farming and livestock monitoring, agriculture has much to gain from a synthesis of multi-modal artificial intelligence and the Internet of Things. Satellite images provide an overall view of crops along with their health in vast regions, while IoT sensors provide information regarding various elements and types of soils installed in farms across a large area. Enhancing the productivity of crops and efficiency of resources can be achieved by using the multi-modal artificial intelligence for anticipating different data sources in order to plan the help. They include knowing when to plant crops, predicting the most suitable time in the year to implement an irrigation programme and offering information on the right approach in crop management to avoid depriving one crop as a result of prioritising another. The IoT sensors in this case are used to track the livestock health indices and the overall behaviour of the animals. To make an early negative health outcome indicator prediction, artificial intelligence does this data analysis. This means that animals can easily be checked and treated before arising to more serious issues, which not only improves their health but will also boost the productivity of animal businesses. Artificial intelligence and the Internet of Things are two general concepts that can be integrated into a single system, and although these specific examples present a considerable number of prospects throughout many fields, it is evident that the idea is vast, and there can be numerous other means of implementation. Thus, systems that are smarter and more responsive are of considerable advantage to and for persons, agencies, and authorities. When more technology is available and more data is given, it increases the opportunity for innovation and thus results in innovation systems.

14.11 Success Stories and Real-Life Examples

The combination of multi-modal artificial intelligence with the Internet of Things has given birth to countless new innovations in today's modern industry. In an effort to indicate how some of these integrated approaches affect practical real-world settings, we will now look at a number of examples and success stories that will articulate areas of practical application and benefits or breakthroughs that have been realised.

14.11.1 Medical Care: System for Tracking Patients from a Distance

14.11.1.1 Analysis of VitalConnect's VitalPatch as a Case Study

Multi-modal artificial intelligence coupled with IoT in the healthcare industry is evident through the VitalConnect's VitalPatch. As the device is tracking eight different physiological parameters, including heart rate, respiration rate, and body temperature, through its eight sensory spots, it can give the correct information. To achieve this, the large amounts of data collected are processed through artificial intelligence algorithms with a view of giving timely information as well as alarms. For instance, the system was integrated into a large hospital where it helped decrease the number of patients who fell by alerting the handlers about any shifting in the condition of the patient that could imply that the patient is prone to falling.

14.11.1.2 Success Story

As demonstrated by the case of VitalPatch in several healthcare organisation implementations, there are impressive enhancements in both outcomes for the patients as well as improvements in the overall functioning of the healthcare organisation. In this regard, it can be stated that the need for the physical monitoring of the systems has become redundant, and hence the healthcare staff has been able to divert their attention to other tasks. Also, the number of patients who subsequently need to be readmitted to the hospital has significantly been brought down through offering effective and constant attendance of the patients.

14.11.2 Smart Cities: A System for Managing Traffic

14.11.2.1 Case Study of IntelliStreets

IntelliStreets is a system that combines IoT smart lighting and multiple intelligent modalities in order to enhance the control of urbanisation. It is equipped with sensors and cameras that can collect data on speed, trajectory of moving cars, and state of the environment together with the motion of pedestrians. AI constantly processes streetlight statistics at intervals to determine the most effective way to light or darken a certain area to avoid spending so much energy but, at the same time, ensure the safety of the public.

14.11.2.2 Success Story

IntelliStreets has gained acceptance and was implemented in cities like Las Vegas, where it serves as a security enhancement tool by illuminating streets during the night and also assists in managing traffic flow more efficiently at peak times and during special events, hence reducing congestion and emissions levels. Apart from this, it has also been instrumental in providing cities with quantitative information on some aspects of their society, which they could use for planning and even for disaster management.

14.11.3 Industrial Robotics: Predictive UpKeep

14.11.3.1 Wind Turbines Manufactured by Siemens Gamesa: A Case Study

Siemens Gamesa utilises IoT sensors together with machine learning algorithms to predict when their wind turbines are likely to need servicing. In this case, the information that is obtained through the sensors regarding operational parameters like vibration, temperature, and torque is examined by artificial intelligence with a view of identifying potential future faults before they occur. The outlined technique for the predictive maintenance enables repair works to be done at required intervals, thereby increasing the life span of the turbines, not mentioning the minimal time they will take to be down.

14.11.3.2 Success Story

The application of this strategy has led to the decrease of the total number of hours on unscheduled maintenance by up to 30%, a corresponding decrease in the repair costs, and an overall increase in the effectiveness of energy output. The information provided by artificial intelligence has also improved the aspects of the design and construction of future turbines, which have led to the production of renewable energy sources that are more reliable and productive.

14.11.4 Agrarian Practices: Accurate Cropping

14.11.4.1 John Deere's Farm Sight: A Case Study or Example

IoT devices and analytics powered by AI systems are embedded in John Deere's Farm Sight, John Deere. This data is compiled from the sensors mounted on different machines and implemented in the fields that include the state of the soil, the health of the crop, and the performance of the machinery. It is then utilised to analyse this data to provide farmers with advice on planting, watering, and even the right time to harvest.

14.11.4.2 Success Story

Farm Sight has shown how the usage of water and fertilisers for cultivation, for instance, can be minimised, yet the productivity can be boosted. This has been done through perfecting farming activities for efficient farming for different crop areas. This not only has an impact on the profits and balance sheet of a farm but also allocates resources in a more sustainable manner that can harm the ecosystem to such a degree.

14.12 Looking Ahead: Current and Next Steps

In the future, when others, such as multi-modal artificial intelligence, are incorporated into the Internet of Things, certain topics emerge that are expected to define the shape of technology and innovation in the future. First of all, one can talk about distributed computational platforms, where one of the most significant trends is edge computing. Due to a continuous increase in the unique signature device commonly referred to as the Internet of Things, it becomes more and more crucial to deal with the data at the edge level. In edge computing, computation is brought nearer to where the data is being generated, thus reducing the number of hops required to transmit the data from one point to another, reducing the bandwidth needed, and also protecting the privacy of data by processing the data at the same place where it is gathered. This could lead to more use of artificial intelligence algorithms that are applied on top of the Internet of Things devices. This will enable the possibility of analysing the data and even making a decision with the help of the networked devices without having to depend on centralised servers always. Standardisation and the integration of multiple models of several artificial intelligence are also expected to exert a significant impact on the integration of multimodal artificial intelligence and the Internet of Things in the future. The IoT comprises numerous devices and platforms; it is for this reason that standards and protocols that are global will be of help in communication and incorporation. The positive effects of these allowances will include the enhancement of IoT settings as being more resilient and scalable amid settings that foster innovation and collaboration. More specifically, they are about to bring out the new possibilities for creating multi-modal AI systems that utilise trends in deep learning and natural language processing. These enhancements will bring better and more reactive applications that will allow for a deeper understanding of the data and the integration of multiple modalities into the application. It will be advanced, respond to user voice commands, and be aware of its environment, which will make smart homes, offices, and public areas much better to use.

However, with the advancement of such technologies, issues to do with security, privacy, and ethics will grow more dominant. To fully ensure that AI-powered systems used are fair, transparent, and accountable additional strict or enhanced legal restrictions or norms will be necessary. In this context, it will be possible to solve problems with pre-determined biases in AI algorithms, ensure the protection of personal data of users, and introduce clear guidelines for creating and implementing ethical AI technologies. There are paradoxes within the two: multi-modal artificial intelligence and the Internet of Things provide virtually endless opportunities to transform sectors and increase sustainability while they come with their set of constraints. There is a concern that the future of artificial intelligence, specifically passive multi-modal and the Internet of Things, may lead to a society where things are smarter and more connected, but it is not necessarily bad. This could be achieved through using digital energy in smart cities, creating a new model of healthcare, and introducing new industrial applications. Evaluating the role of artificial intelligence and the Internet of Things, it is multi-modal artificial intelligence and the Internet of Things that will therefore hold the key to the future and thus determine the level of technological advancement that will impact on improving the lives of people globally.

14.13 Conclusion

Multimodal artificial intelligence and the Internet of Things are very disruptive when they are combined together in industries and value generators, and they are even helping to make people's lives better. In the process of looking into the future, some features that, more or less, can be considered as tendencies are emerging. Some of these trends are the progress in the edge computing concept, emphasis made on interfacing and standardisation as well as the advancements made in the AI methods. These tendencies can result in the freeing of abilities, earlier out of reach by human skill, and introducing improvements in a broad range of other industries. There is, however, what has to do with security, privacy, and ethics that emerge as these technologies enhance their capacity in performing such tasks. This can be done by establishing clear guidelines and norms that can let us ensure that AI is being created and implemented in a proper way and with justice and openness on its basis. It opens a great opportunity for the industries bringing in new and progressive changes, enhancing sustainability aspects, and making society smarter and connected through new technologies of multimodal artificial intelligence and the Internet of Things. Therefore, to embrace new opportunities and to walk the road of positive change for tomorrow's generations, one has to learn to combine the possibilities offered by these technologies and then face the challenges.

References

Bimonte, S., Coulibaly, F. A., & Rizzi, S. (2024). An approach to on-demand extension of multidimensional cubes in multi-model settings: Application to IoT-based agro-ecology. *Data & Knowledge Engineering, 150*, 102267.

Burns, D., & Lambert, A. (2024). Enhancing cybersecurity through multi-model AI tracking systems. *Innovative Computer Sciences Journal, 10*(1), 1–9.

Dayananda, C., Hemashree, P., Impu, D., Janavi, S., & Gururaja, H. S. (2024). A multi-model ensemble approach for proactive student mental health assessment. In *2024 IEEE International Conference for Women in Innovation, Technology & Entrepreneurship (ICWITE)* (pp. 494–500). IEEE.

Gadey, N., Pande, S. D., & Khamparia, A. (2024). Enhancing 5G and IoT network security: A multi-model deep learning approach for attack classification. In *Networks attack detection on 5G networks using data mining techniques* (pp. 1–23). CRC Press.

Ghayvat, H., Geddam, R., Awais, M., Tiwari, P., Milrad, M., & Kumar, N. (2024). *Aicaregaitrehabilitationbyfusion: Multi-model and multi-modalities sensor data fusion for AI-IoT enabled realtime electrical stimulation device for pre-fog and post-fog to person with Parkinson's*. Available at SSRN 4690089.

Golec, M., Gill, S. S., Golec, M., Xu, M., Ghosh, S. K., Kanhere, S. S., et al. (2023). BlockFaaS: Blockchain-enabled serverless computing framework for AI-driven IoT healthcare applications. *Journal of Grid Computing, 21*(4), 63.

Iyer, V., Lee, S., Lee, S., Kim, J. J., Kim, H., & Shin, Y. (2023). Automated backend allocation for multi-model, on-device AI inference. *Proceedings of the ACM on Measurement and Analysis of Computing Systems, 7*(3), 1–33.

Mondal, S. S., Mandal, N., Singh, A., & Singh, K. K. (2020). Blood vessel detection from retinal fundas images using GIFKCN classifier. *Procedia Computer Science, 167*, 2060–2069.

Morales-García, J., Terroso-Sáenz, F., & Cecilia, J. M. (2024). A multi-model deep learning approach to address prediction imbalances in smart greenhouses. *Computers and Electronics in Agriculture, 216*, 108537.

Robinsha, S. D., & Amutha, B. (2023a). IoT architecture for energy management in smart cities. *International Journal of Services Operations and Informatics, 12*(4), 325–343.

Robinsha, S. D., & Amutha, B. (2023b, November). IoT revolutionizing healthcare: A survey of smart healthcare system architectures. In *2023 international conference on Research Methodologies in Knowledge Management, Artificial Intelligence and Telecommunication Engineering (RMKMATE)* (pp. 1–5). IEEE.

Robinsha, S. D., & Amutha, B. (2024). Velocious: A resilient IoT architecture for 6G based intelligent transportation system with expeditious movement mechanism. *Wireless Personal Communications*, 1–22.

Singh, K. K., & Singh, A. (2016). Detection of 2011 Sikkim earthquake-induced landslides using neuro-fuzzy classifier and digital elevation model. *Natural Hazards, 83*, 1027–1044.

Singh, A., & Singh, K. K. (2023). YORES: An ensemble YOLO and Resnet network for vehicle detection and classification.

Singh, K. K., Mehrotra, A., Nigam, M. J., & Pal, K. (2013). Unsupervised change detection from remote sensing images using hybrid genetic FCM. In *2013 Students Conference on Engineering and Systems (SCES)* (pp. 1–5). IEEE.

Singha, A., & Singhb, K. K. (2023). FedDDR: A federated improved DenseNet for classification of diabetic retinopathy. *Proceedings*. http://ceur-ws.org ISSN, 1613, 0073.

Singhal, S. (2024). Real time detection, and tracking using multiple AI models and techniques in cybersecurity. *Transactions on Latest Trends in Health Sector, 16*(16).

Wang, Y. H., Li, J. J., & Su, W. H. (2023). An integrated multi-model fusion system for automatically diagnosing the severity of wheat fusarium head blight. *Agriculture, 13*(7), 1381.

Zhao, Z., Zhang, H., Wang, L., & Huang, H. (2023). A multi-model edge computing offloading framework for deep learning application based on Bayesian optimization. *IEEE Internet of Things Journal.*

Chapter 15
Enhancing Safety and Reliability in Vanets for Autonomous Vehicles by M-XAI (Multi-modal Explainable-AI)

Umesh Gupta, Ayushman Pranav, Ankit Dubey, Rajesh Kumar Modi, and Akansha Singh

Abstract Multi-modal explainable artificial intelligence (M-XAI) refers to the field of study that focuses on making sure that people understand the reasoning behind decisions made by machines. In simpler terms, M-XAI should produce explanations for AI findings that are plain and straightforward enough for human beings to comprehend with their limited knowledge of algorithms. This means that it is necessary to open up some parts of an opaque system if we want others to trust us, expose everything so that everybody can be held responsible, and let individuals cooperate among themselves as well as between them and artificial intelligence systems. In order to increase awareness both within individuals as well as organisations about responsible use and adoption while fostering local/global development on the same note, considering its potential influence across many industries like transportation, where intelligent transport systems have recently been introduced, leading to a new era characterised by high levels of efficiency combined with minimal congestion rates along major highways. This study scrutinises the utilisation of this (M-XAI) explainable artificial intelligence concept of self-driving independent vehicles. For instance, it analyses (M-XAI) approaches designed for self-driving cars and evaluates their strengths and weaknesses across different tasks. In addition to highlighting problems related to explainability here, we also suggest research areas that need more exploration.

Keyword Explainable AI · Multi-modal explainable artificial intelligence · Autonomous vehicles · VANET · Smart transportation

---

U. Gupta · A. Pranav · A. Dubey · A. Singh (✉)
SCSET, Bennett University, Greater Noida, India
e-mail: akansha1.singh@bennett.edu.in

R. K. Modi
Stuvalley Technologies Pvt. Ltd, Gurugram, India

A. Singh, K. K. Singh (eds.), *Multimodal Generative AI*,
https://doi.org/10.1007/978-981-96-2355-6_15

15.1 Introduction

Explainable artificial intelligence (M-XAI) is a very new emerging domain of research in artificial intelligence, which has now started growing at a very rapid pace; it then strives for a better understanding of (AI)-systems. It emphasises the development of tools and strategies for a more lucid explanation of the inner workings and reasoning behind the choices made by complex AI models, allowing for more transparency and trust in AI systems (Möhlenhof et al., 2021; Gohel et al., 2021).

M-XAI provides an explanation of how AI reaches a particular solution along with other "wh" questions about the black box. This type of intelligible AI is different from traditional AI. Explainability and comprehension are important in domains like defence, law and order, healthcare, self-driving vehicles, etc., where their usage can help gain trust and transparency. Several M-XAI techniques have been proposed to increase understanding of AI technologies in such domains.

In the context of autonomous vehicles, M-XAI can be used in the following ways:

- Drivers and passengers can understand how the autonomous vehicle is making decisions.
- It can also identify and address potential biases in the autonomous vehicle's decision-making process.
- This helps in improving the safety and dependability of autonomous cars by reducing the number of possible accidents.

15.1.1 Background and Significance of M-XAI in Autonomous Vehicles

Autonomous vehicles are increasingly becoming a new trend in today's world, which brings in the need for M-XAI. Once on the roads, these vehicles will have decision-making power, which can affect lives; hence, in order for this new technology to be fully accepted by individuals, there must be transparency and an understanding of how these systems operate and make decisions. Having a thorough knowledge of the decision-making procedures is one of the ways in which an individual will confide in them and ultimately adapt to this new technology (Mankodiya et al., 2021).

M-XAI can use this kind of reasoning to justify the vehicle's decisions. Also, for autonomous vehicles, M-XAI is a requirement. They detect and correct any biases that may occur while self-driving cars operate; in addition, they help prevent probable accidents. It allows for prompt and effective problem-solving, which leads to better system performance and operation efficiency (Sanneman & Shah, 2022; Madhav & Tyagi, 2022).

15.1.2 Overview of VANETs (Vehicular Ad Hoc Networks) for Autonomous Vehicles

Wireless networks that connect vehicles with each other as well as with roadside infrastructure are referred to as vehicular ad hoc networks (VANETs). Autonomous vehicles use this information to choose among different navigation strategies while on the road. These devices increase the situational awareness of the vehicle, thus ensuring safety during travel.

Another function of VANETs is providing M-XAI for autonomous vehicles. For instance, VANET data, such as the speed and position of neighbouring cars, current weather conditions, and the state of roads, can be used in explaining to drivers and passengers why the car made certain decisions. Traffic updates, road condition reports, and emergency alerts are among the essential items shared by VANETs with vehicles (Nwakanma et al., 2023; Adadi & Berrada, 2018).

15.1.3 Objective and Structure of the Chapter

This chapter of the book is going to talk about M-XAI in self-driving cars. It begins with an overview of M-XAI and its importance in self-driving cars. Following that, the chapter will discuss VANETs for M-XAI in self-driving cars. At last, this section ends by talking about the challenges and opportunities of using M-XAI in autonomous vehicles (Joshi et al., 2021).

15.2 M-XAI Techniques for Explainability in Autonomous Vehicles

15.2.1 Rule-Based Explanations

Rule-based explanations are one type of explanation given by M-XAI, which uses simply defined rules to explain AI decisions.

15.2.1.1 Definition and Formulation of Rule-Based Explanations

However, even though it is described as a sequential collection of reasons, the underlying artificial intelligence conclusion contains many different instructions reflecting the determination made by an artificial cognitive system. These rules can be expressed in natural language or more formal languages like mathematics and logic (Gerlings et al., 2020; Ali et al., 2023a).

15.2.1.2 Examples of Rule-Based Explanations in Autonomous Vehicles

There have been several examples of rule-based explanations used within autonomous vehicle systems:

For instance, when an autonomous vehicle stops at a red traffic light, it may provide a rule-based explanation for its decision-making process. The governing rule states that "If the light is red, the vehicle must stop moving." Such an approach increases transparency and helps users, as well as onlookers, understand what actions should be expected from the car under different circumstances.

An independent vehicle could also have explained why it moved lanes with a rule-based explanation. Here is an example of such a rule: "When a slower moving vehicle is detected in the next lane, it becomes necessary for the autonomous vehicle to quickly initiate a lane change maneuver."

15.2.1.3 Mathematical Representation of Rule-Based Explanations

Rule-based explanations can also be expressed mathematically. One commonly used approach is through decision trees, which are widely applied tools in practice. A decision tree is a very graphical representation of rules where each node represents one rule and branches show different outcomes that result from applying these rules. Such a framework allows for a systematic and exhaustive analysis of the decision-making process (Das & Rad, 2020; Byrne, 2019) (Fig. 15.1).

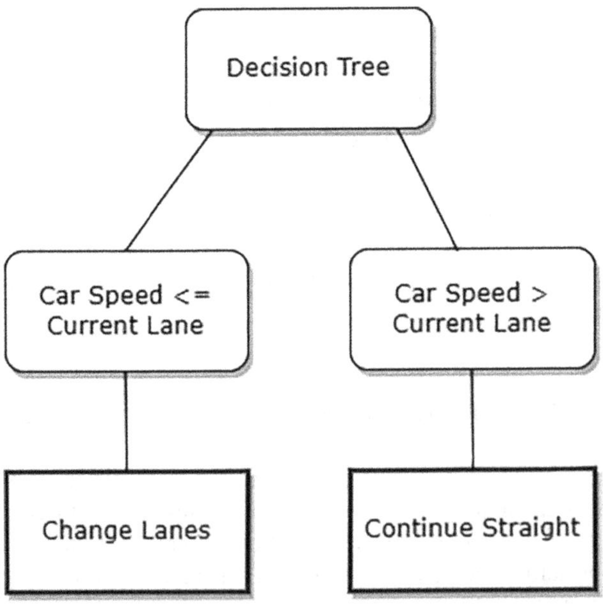

Fig. 15.1 Flowchart representing a decision tree for an autonomous vehicle's lane change decision

In this diagram, the initial condition is whether the car speed in the adjacent lane is slower than the current lane. If this type of condition holds true (represented by the left branch), then the vehicle changes the lanes; otherwise, if this type of condition is false (right branch), the vehicle continues straight (Google, 2024).

Mathematically, we can represent this rule as an inequality:

```
If car_speed_next_lane <= car_speed_current_lane, then
change_lanes.
```

This is the rule that is represented by the very decision of the tree visually, thus helping to gain an understanding of how the decisions are made. It totally represents the different conditions and actions of the vehicle.

15.2.2 Model-Based Explanations

15.2.2.1 Definition and Formulation of Model-Based Explanations

Model-based explanations are a type of explainable AI(M-XAI) system that uses the model of the AI system to give reasons for its decisions. This rising technology in artificial intelligence can be represented as an abstract mathematical structure, a computer simulation that imitates its mental processes, or a visual, diagrammatic representation that shows how it works as a system. Model-based explanations also show what makes these AIs tick and why they did that (Comma.ai, 2024; Holzinger et al., 2021).

15.2.2.2 Examples of Model-Based Explanations in Autonomous Vehicles

There are many model-based explanations for very different ways to use them for self-driving cars. For example, one can train a model of the AI system on a natural language corpus to generate natural language explanations. On this note, visual explanations could be employed to make autonomous vehicle decisions more comprehensible. In this case, an AI system model that has been trained intensively on large numbers of images and videos with diverse content is used to generate such an explanation. The mathematical representation of model-based explanations exhibits variation contingent upon the specific model employed (Yang et al., 2022). In order to explain something mathematically using a formal framework of equations and formulas, it involves employing highly intricate equations and formulas accordingly, which summarise principles or functioning rules underlying phenomena being investigated by means of applications created by developers or engineers based on these mathematical formulae developed in the subject domains like physics. For instance, this comprehensive approach ensures a very thorough elucidation of the model's operations and the behaviour. These types of equations capture the very relationship between the various variables and parameters that influence the decisions made by the AI system. For example, consider a simple mathematical model representing an autonomous vehicle's decision to change lanes:

```
if (distance_next_lane<threshold_distance) and (speed_next_
lane<threshold_speed) and (safety_score>threshold_safety):
     change_lanes
Else:
     Continue_straight
```

When it comes to lane changing, there are two things that must be considered: thresholds for using safety indicators and logical expressions manifested as mathematical inequalities.

Simulation: One way of understanding artificial intelligence's thinking is by making it go through various tests while creating models that can imitate the thought process. In this kind, we can use many modelled trajectories and simulated scenarios illustrating how dynamic these systems are and thereby exposing their operational complexities. Diagrammatic illustration: This type presents explanations in diagrams consisting of nodes and edges where nodes represent different components or entities, whereas edges show relationships between them. Graphs allow complex structures and AI system information flow to be visually portrayed. With pictures like this, we may see what happens behind them more clearly and how everything works dynamically together. Consequently, if we adopt graph representations, we can analyse and optimise such systems' performance very effectively. Because of this approach, the mechanism can become fully realised, which guarantees the best results, as is shown in the example below (Jin et al., 2022; Joshi et al., 2021).

Graphical Representation: This representation uses nodes and edges to display information. Nodes represent different parts or things, while edges denote their relations or connections between them. Graphs are great illustrations because they reveal how complex systems work by displaying their structures in a manner that helps us understand them better. These diagrams also show what is likely to happen next in the system under examination since they demonstrate how its various components fit together (Malandri et al., 2023). Graphical representations can be used to optimise AI systems more effectively than other approaches, as all necessary data is on a single page, enabling us to analyse each part's role within the broader context of the whole system. Such figures should be employed to expose hidden mechanisms and show dynamic relationships among entities at a particular time during some process. This will help identify prerequisites for obtaining certain outcomes, thus significantly improving optimization. Additionally, this trick paves the way for full use of any possibility inherent in it, which invariably leads to success.

For instance (Figs. 15.2 and 15.3),

Nodes represent lanes and vehicles, while edges represent connections between them in this graph; a decision to change lanes is represented by an outcome node.

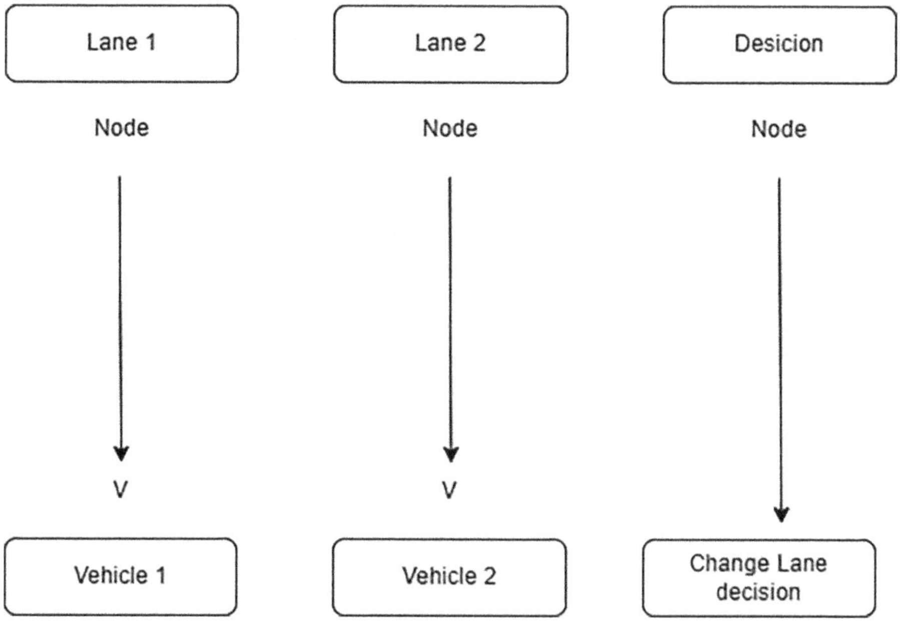

Fig. 15.2 Graphical representation of an autonomous vehicle's lane change decision

15.2.3 Example-Based Explanations

Explanations based on examples are a kind of explainable artificial intelligence (M-XAI) technique that, instead of revealing the complicated inner workings of the model, shows how AI reached its conclusion by using cases or examples. It can be done by showing what data influenced an AI system's decision.

Google Cloud, for example, has introduced example based explanations, which helps users improve the performance of their model by cleaning their data. It returns similar examples to new predictions or instances using approximate nearest neighbor service. This can help increase the interpretability of an AI model and, hence, cultivate trust in the AI system by the user (Sirapangi & Gopikrishnan, 2024).

15.2.3.1 Definition and Formulation of Example-Based Explanations

Example-based explanations are a technique in a set of examples used to explain the workings of an AI system. The examples can be presented either visually, such as by using images and videos, or through text.

One way to formulate example-based explanations is by using a counterfactual approach. In this approach, the system provides examples of data that would have led the AI model to reach a different decision. While waiting at a stoplight, an autonomous vehicle's system might note for the passenger the lack of oncoming

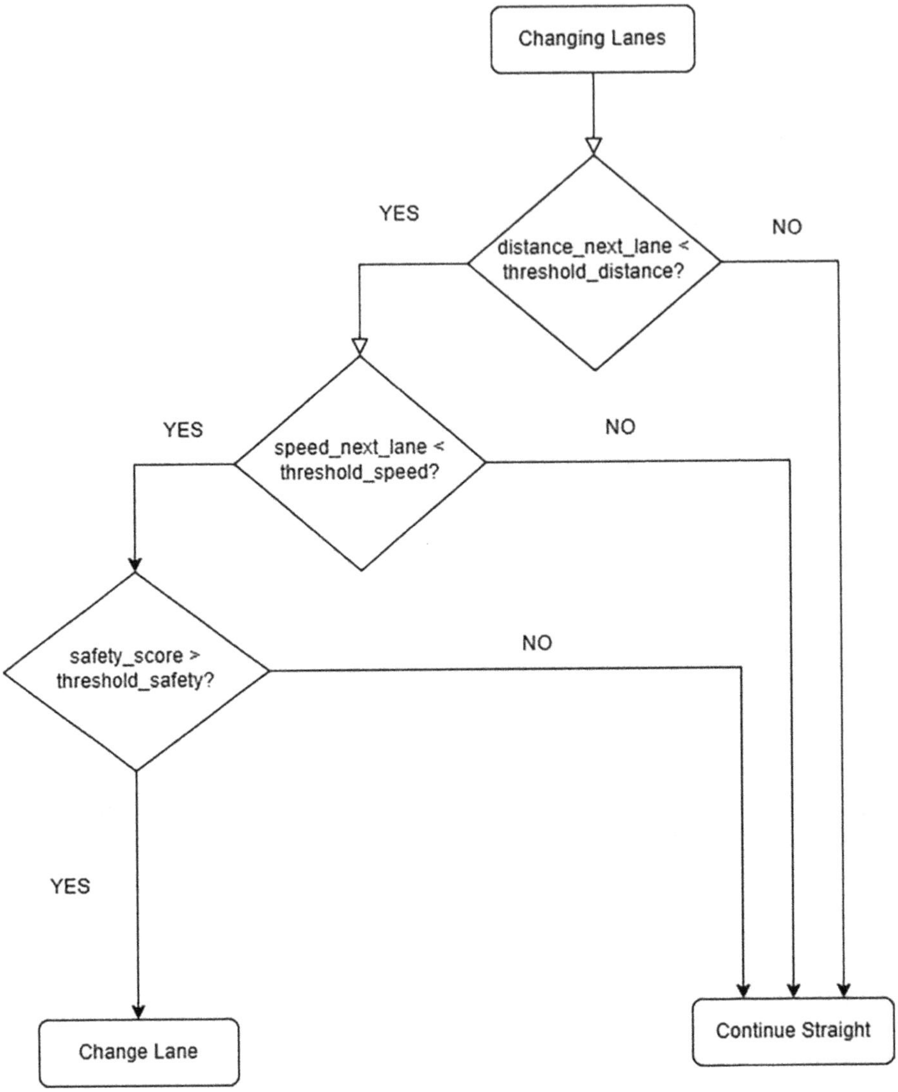

Fig. 15.3 Flowchart representing the model decision-making process for lane change

traffic or crossing pedestrians that, alternatively, would have prompted it to continue through the intersection unimpeded. This helps in understanding the conditions under which the decision of the model may vary. Hence, it creates transparency in its work (Kumar & Taylor, 2024).

Another way to formulate example-based explanations is by using a saliency map. Saliency Map: In this case, if a vehicle takes a turn, the saliency map can indicate which pixels in an image were most important in making such a decisive map point out the most influential or relevant features that contributed to a particular

Table 15.1 Matrix representing a saliency map of vehicles surroundings

0	0	0	0	0
0	0.2	0.6	0.2	0
0	0.6	1.0	0.6	0
0	0.2	0.6	0.2	0
0	0	0	0	0

decision made by an AI model. For example, if a car turns right at an intersection, we could use this method to determine which parts of the image were responsible for causing it to do so.

15.2.3.2 Examples of Example-Based Explanations in Autonomous Vehicles

Counterfactual Approach: These numbers could be represented mathematically through equations or formulae. Suppose there is an AV approaching a red light and it has stopped as one of its decision points.

Data point 1: [distance = 10 meters, speed = 30 km/h, light = red]
Data point 2: [distance = 5 meters, speed = 20 km/h, light = red]
Data point 3: [distance = 15 meters, speed = 40 km/h, light = red]

Each data point, in this case, refers to a different situation with different values of distance, velocity, and state of traffic light. These sets of data tell us why we stop at red lights. These values can be altered to see how different variables affect the decision-making process in self-driving cars.

Saliency Map Approach: Mathematically speaking, the saliency map approach represents example-based explanations as a matrix that indicates how much each pixel contributes to or reflects some important features within an image. This matrix assigns weights to pixels based on their significance for decision-making processes. Let's suppose there is a turn made by an autonomous vehicle (Table 15.1):

The matrix here is a simplified version of what a saliency map may look like for an image of the view around a car, with each dot numbered by how much or little it mattered when deciding whether to turn—bigger numbers indicate higher significance while smaller ones mean lower or no importance at all (Gupta et al., 2024; Ali et al., 2023b).

15.3 Performance Evaluation of M-XAI Techniques in Autonomous Vehicles

15.3.1 Metrics for Evaluating Explainability in Autonomous Vehicles

Different metrics can be used to evaluate the performance of M-XAI techniques in autonomous vehicles. These measures are divided into two main groups: accuracy measures, which evaluate the performance and efficiency of a model, and interpretability metrics, which aim at understanding and making transparent the decision-making process of a model.

- Accuracy metrics measure how clearly the M-XAI technique is able to explain the decision-making of autonomous vehicles. Some standard accuracy metrics include:
  - Accuracy: It is the percentage of times the M-XAI technique correctly explains the decision-making process of the autonomous vehicle.
  - Precision: The percentage of times the M-XAI technique correctly explains the decision-making process of the autonomous vehicle, given that it made the correct decision.
  - Recall: This is the percentage of times that an M-XAI technique correctly explains why an AI system made a wrong decision in an autonomous vehicle.
- Interpretability metrics measure how easily explanations are given by M-XAI and understood by humans. Some standard interpretability metrics include:
  - Transparency: It refers to how much information about the internal workings of an AI system is given by the M-XAI technique and can be understood by people.
  - Fidelity: This shows how explanations provided by the M-XAI method are accurate and complete.
  - Relevance: It shows to what extent explanations given by the M-XAI method relate to the decision-making process in self-driving cars.

The metrics to be used depend on the nature of the application being considered. For example, safety is a crucial consideration for autonomous vehicles; hence, in this case, value and relevance are greater for accuracy measures than interpretability ones. If the aim is to foster trust between people and self-driving cars, then it becomes necessary to pay more attention towards interpretability measures as opposed to accuracy measurements (White et al., 2023).

15.3.2 Accuracy and Interpretability Trade-off

Within such a given context, there is a balance between optimal interpretability and accuracy trade-off. More accurate M-XAI techniques are less interpretable, and vice versa. This means that M-XAI techniques, which are more accurate, tend to use complex models that humans find difficult to comprehend.

Finding a balance between accuracy and interpretability can be a challenging task, but achieving it is important in the context of autonomous vehicles. They need to be both safe and trustworthy. Safe autonomous vehicles mean they must be accurate and make correct decisions, while reliable autonomous vehicles must provide explanations humans can understand.

15.3.3 Quantitative Metrics for Assessing Explanations

In addition to the qualitative metrics described above, there are several quantitative metrics that are used to examine the quality of explanations yielded by M-XAI techniques. These metrics can measure explanations' accuracy, completeness, and relevance.

Some standard quantitative metrics for assessing explanations include:

Accuracy: The percentage of times the explanation correctly identifies the factors that influenced the decision-making process of the autonomous vehicle.

Completeness: This measures the degree to which the explanation covers the factors that influenced the decision-making process of the autonomous vehicle.

Relevance: It refers to how well the explanation corresponds with and adds to the autonomous vehicle's decision-making process.

15.4 Mathematical Formulation of Evaluation Metrics

The accuracy, completeness, and relevance of explanations can be mathematically formulated as follows:

- Accuracy = Total number of explanations/Number of correct explanations.
- Completeness = Total number of factors/Number of factors included in all explanations.
- Relevance = Total number of explanations/Number of relevant explanations.

These metrics are helpful in comparing the performance of different M-XAI techniques. The M-XAI technique with the highest accuracy, completeness, and relevance is considered the best.

15.5 Experimental Setup for Performance Evaluation

15.5.1 Dataset Collection and Preprocessing

The dataset employed in this research was obtained from Comma.ai: https://research.comma.ai/. It contains readings from the sensors on the vehicle, such as its velocity, position, and direction, alongside other metrics like decision-making algorithms for the driving behaviour of over 100 self-driving cars under different road conditions.

To ensure data consistency, preprocessing techniques were applied to the data before training. Then, a split was made between a training set and a test set. The work of the training set is to help the M-XAI approach learn and understand different patterns in the data, which leads to decision-making processes, hence making accurate predictions. On the other hand, the test set is used to assess how well M-XAI performs in terms of efficiency or effectiveness when dealing with new unseen data during its training (Kuznietsov et al., 2024; Dong et al., 2023).

15.5.2 Evaluation Framework for M-XAI Techniques

Various metrics, such as accuracy, explainability, and fairness, were used to evaluate how well M-XAI can predict and explain new or unseen information. Accuracy measures how good the M-XAI technique model is in predicting car decisions, while explainability measures how clearly this method is able to explain itself. Fairness tests whether it can make decisions fairly towards different groups of people.

The study evaluated performance across multiple decision-makings on the road, such as speed prediction for vehicles, steering angle, lane change, etc. This evaluation gave a thorough insight into the strengths and limits of the M-XAI approaches used. The M-XAI techniques were also evaluated on various road conditions, such as city, highway, and rural driving (Kasetti et al., 2024).

15.5.3 Mathematical Representation of Experimental Setup

Mathematically, the experimental setup for performance analysis is as shown below:

Where:

X is the dataset of sensor readings and car decisions.
Y is the set of labels for the car decisions.
f is the M-XAI technique.
e is the explanation of the M-XAI technique's decision.

a is the accuracy of the M-XAI technique.

The experimental setup can be used to evaluate different M-XAI techniques based on any dataset. The evaluation results of the different M-XAI techniques are used to compare the performance and pick the optimal one for a particular application (Fig. 15.4).

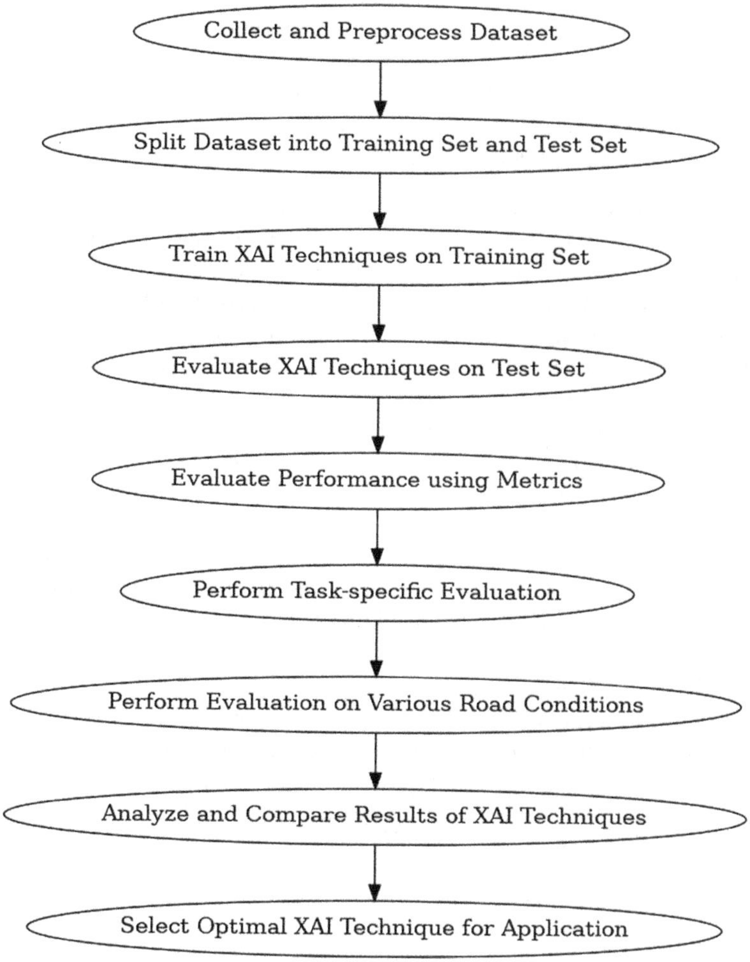

Fig. 15.4 Experimental setup for performance evaluation of M-XAI techniques using a dataset, including data collection, preprocessing, training, evaluation, and result analysis

15.6 Comparative Analysis of M-XAI Techniques

M-XAI techniques focus on providing reasoning behind AI systems' decisions, which is important in a number of ways, such as building user trust, debugging AI systems, and complying with regulations.

There have been several proposed explanation methods within M-XAI, but some include:

Rule-Based Explanations: Such explanations follow a predetermined list of rules when explaining why certain decisions were made. It may help in understanding decision-making logic, but generating comprehensive and accurate rules can be difficult.

Model-Based Explanations: These explanations give an insight into how an AI model makes decisions by revealing its internal workings. However, this approach may not be understandable without knowledge of machine learning.

Example-based Explanations: They provide examples of data or instances to demonstrate the behaviour of the model in different cases. This gives a more intuitive understanding of the model and how it is affected by this data.

Many studies have been done to evaluate M-XAI techniques. One of these was aimed at finding the best M-XAI for use in self-driving cars. The paper compared rule-based, model-based, and example-based explanations, among others, and found that rule-based explanations were the most effective M-XAI for autonomous vehicles. Model-based and example-based explanations were useful in other scenarios, such as when the user wants to understand why a certain decision was made or what types of data the model was trained on (Möhlenhof et al., 2021).

In another study, different explanation methods were tested against each other with regard to performance when used in intrusion detection systems (IDSs) for intelligent connected vehicles (ICVs). Just like autonomous vehicle systems, here also, rule-based explanations turned out to be better than any other method used as an explanation for decisions made by ICVs. However, it was also found that model-based and example-based explanations could be helpful in some cases, such as when users wanted to understand the specific features that were used to detect the intrusion or the types of data the model was trained on (Sanneman & Shah, 2022; Tekkesinoglu et al., 2024).

In total, some specific M-XAI techniques prove more useful for specific applications. Depending on the type of use case, different explanations may be better in certain situations, while others might work well in other contexts. Therefore, one should evaluate various M-XAI methods with respect to their application prior to settling on a particular method.

15.7 Challenges and Future Directions in M-XAI for Autonomous Vehicles

15.7.1 Complexity of Decision-Making in Autonomous Vehicles

The decision-making process of self-driving cars is multi-step and takes into account many factors like the environment around them, their sensors, and their overall goals, among others. This complexity of operation makes it hard to explain why an autonomous vehicle arrived at a given decision (Kaufman et al., 2024).

15.7.2 Mathematical Complexity of Decision-Making Algorithms

Autonomous vehicle decision-making algorithms are based on some pretty complicated math that can be hard for non-experts to follow. Experts in self-driving cars might struggle too, though, since many use deep learning methods, which are notoriously inscrutable even when they do work well enough.

15.7.3 Handling Uncertainty and Robustness in M-XAI

The world in which AVs find themselves is highly variable and uncertain. But explainable artificial intelligence systems also need to be robust under extreme conditions. This means that there should not only be robustness against the most challenging situations but also under mathematically complex decision-making algorithms.

15.7.4 Balancing Explainability with Accuracy

Designing M-XAI for self-driving cars always involves a trade-off between accuracy and interpretability. Explanations require models to be simpler than they otherwise would be, leading to decreased accuracy. Thus, more interpretable models usually have lower rates of success, while less so tend towards higher rates due to this explanation loss in accuracy.

15.7.5 Trade-Offs between Explainability and Performance

As one develops M-XAI for autonomous vehicles, there are several things that come into play in terms of trying to balance out accuracy and interpretability. One such factor might be model complexity, where more sophisticated models tend to be accurate but hard to explain, while simpler ones sacrifice a certain amount of accuracy while being explainable enough. Availability can also be another factor to consider, as well as the ease with which data can be obtained for thorough analysis leading to valid conclusion drawing depending on what we have around us regarding such stuffs or not so much.

15.7.6 Mathematical Optimisation Approaches for Balancing Explainability and Accuracy

There are different methods of optimising mathematics so that explanations and accuracy of M-XAI are balanced. For instance, adding a regularisation term into the objective function of the model. In this case, we penalise the model for being too complex, hence making it simple enough for us to know what it means. Another approach worth giving a shot is Bayesianism; here, probability is used as a tool for understanding and interpreting predictions made by models.

In this technique, a probabilistic distribution characterises how predictions are made by the model along with their associated uncertainties (Trivedi et al., 2024).

15.7.6.1 Regularisation Term

This involves representing the objective function mathematically by adding up regularisation terms into its goal functions, which makes them more interpretable because simplicity can be achieved through this approach if models were found to be very complicated, as:

$$\text{Objective Function} = \text{Minimise Loss}\left(\text{Data}\right) + \lambda^{*}\ \text{Complexity}\left(\text{Model}\right)$$

Error(Data) explains how wrong or how correct the model is according to the dataset, while the Complexity of (Model) tells us about how complicated the model is. The regularisation term multiplies a hyperparameter λ, which determines the balance between accuracy and complexity. One can adjust the λ value until they strike a balance between understandability and correctness (Nazat et al., 2024).

15.7.6.2 Bayesian Approach

The Bayesian framework can also be used. In this case, there is a probability distribution representing the model, and everything is expressed in terms of probabilities concerning parameters or predictions of models. This allows for much richer predictive understanding with more detail and nuance about what might happen as well as measuring uncertainty in judgements.

Mathematically, it involves expressing everything probabilistically by using Bayes' theorem, which states that:

$$P(\text{Model}|\text{Data}) = P(\text{Data}|\text{Model})^{*} P(\text{Model})$$

The probability of P(Model|Data) being true given some knowledge/data points is proportional to prior odds P(Data|Model) times P (Model) before gaining any information from looking at observations/dataset (prior), i.e., likelihood times prior equals posterior probability.

Probability distribution is taken into account by the Bayesian approach, which allows a more thorough comprehension of the model's predictions and uncertainty, therefore improving its explainability as well as accuracy.

In self-driving cars' M-XAI, explainability can be balanced with accuracy using mathematical optimisation methods, e.g., regularisation terms and Bayesians. The regularisation term enforces simpler models that are easier for humans to understand, while the Bayesian approach provides a framework for understanding the predictions and uncertainties in the form of probability.

15.8 Trust and Transparency in M-XAI for Autonomous Vehicles

15.8.1 Importance of Trust in Autonomous Vehicles

Trust building among people is critical if autonomous vehicles are to be widely adopted. Human drivers may still share roads with driverless cars, but passengers might not be willing to ride in them. There are several things that will help individuals embrace this new era of technology, such as:

- includes making sure that they are reliable and safe,
- have a transparent decision-making process, and,
- individuals can understand how these cars function.

15.8.2 Building Trust Through M-XAI in VANETs

Explainable artificial intelligence (M-XAI) can play a significant role in ensuring widespread acceptance and adoption of autonomous vehicles. M-XAI techniques combined with VANETs could give people an insight into how decisions were made by these machines, thereby promoting transparency and understanding, which ultimately leads to trust, thus increasing their adoption rates.

One way that VANETs can aid M-XAI is through sharing what it knows about the instant road conditions and other vehicle behaviour on the road. This data can then be used by drivers to help them make decisions based on a better understanding of their environment, thereby improving safety on our highways. Such information may include the speed of vehicles, directions being taken by cars, whether or not brakes have been applied, as well as intentions shown, among other things.

Another application of M-XAI involves giving people insight into why they made particular decisions while driving. For instance, if an autonomous car suddenly changes lanes without warning, its system could tell the driver all factors considered before shifting from one lane to another. This increases their comprehension about these machines, thus enabling them to come up with good choices about how best to use them in different driving situations (Cui et al., 2024).

15.8.3 Mathematical Models for Trust Assessment

Several mathematical models have been proposed for assessing autonomous vehicles' "trustworthiness." Trust and confidence are key factors that will contribute towards wider adoption and acceptance of self-driving cars. However, people might still be sceptical about this new technology.

Therefore, these models were designed with multiple criteria in order to evaluate trust levels in such systems, like simplicity of decision-making process, understandability by humans, reliability, etc. The models then use these factors to calculate a trust score for the vehicle.

Trust evaluation in mathematics could be illustrated by the Dempster-Shafer theory of evidence. In self-driving cars, this concept demonstrates doubt and vagueness. Many trust appraisal algorithms have been created based on this idea, which are useful in different situations.

The Bayesian network is another example of a mathematical model for trust assessment. They are graphical models used to represent the relationship between different variables but with some probabilistic aspect to them. Thus, they can handle complicated interdependencies that may exist among such factors as the reliability of a vehicle, transparency in making decisions, and human ability to understand or explain its behaviour (Dong, 2024; Raees et al., 2024).

15.9 M-XAI for Intrusion Detection and Mitigation in Intelligent Connected Vehicles (ICVs)

15.9.1 Overview of Intelligent Connected Vehicles (ICVs)

Intelligent connected vehicles (ICVs) have brought a revolution to the transport sector through advanced communication technologies. These vehicles are equipped with modern sensors, actuators, and communication devices that enable them to interact with their immediate environment as well as other nearby cars. This is important for enhancing safety on roads while at the same time improving efficiency, which is good for all parties involved.

15.9.2 Definition and Architecture of ICVs

These types of vehicles, intelligent connected vehicles (ICVs), are state-of-the-art cars designed to have an understanding of their environments using sophisticated sensors that help in detecting objects around them accurately, like knowing where another vehicle is, detecting pedestrians, and even giving an exact representation of what the road looks like.

Actuators control how fast they move, how they turn, and when they stop. These cars are fitted with very complex types of sensors that allow them to see everything around themselves and then make sense out of it all.

Typical ICV architectures contain:

Sensing elements: They are used to sense the immediate environment of an automobile.

Actuators: These assist in controlling a vehicle's speed, steering, and braking systems.

Communication techniques: The latter acts as the intermediaries between vehicles and other cars along the road or infrastructure.

Intrusion detection and fixing system (IDS): This identifies people or objects who break into a car system and restores them automatically.

15.10 Security Challenges in ICVs

ICVs face multiple security threats, such as:

Malware refers to software that is intended by professional developers for hacking into automotive control units, interfering with their normal functioning, causing unauthorised access, which could lead to operational anomalies, failures, or even accidents. Denial-of-service (DoS) attacks can be unleashed on networked software components of autonomous driving systems. Attacks like these aim at interfering with the smooth running of AVs by overwhelming them with requests until their communication becomes useless, thus introducing potential operational risks. An attacker might cause self-driving vehicles' systems to fail by breaking off these channels of communication (Zhang et al., 2024).

Eavesdropping basically involves malicious individuals overhearing and stealing private data from an automobile, such as where it is going to or its current position.

Spoofing, on the other hand, refers to leading another vehicle astray so that it gains unauthorised access to its systems or causes them to malfunction.

These security challenges are able to pose high risks to ICVs safety. Consequently, IDSs developments should be aimed at detecting and reducing these threats.

15.10.1 Explainable Artificial Intelligence (M-XAI) for Intrusion Detection and Mitigation in ICVs

M-XAI is information about how an AI system selects a decision process, which is crucial in ICVs because they improve the car's safety by making them more transparent and reliable.

M-XAI can work as IDS in ICVs. For instance, M-XAI can:

- Explain the decisions made by the IDS so that human operators can understand why the IDS have raised the alarm.
- Identify the features that the IDS used to make its decision so that engineers can improve the IDS's accuracy.
- Generate visualisations that can help human operators to understand the IDS's decision-making process.

Mitigation is where M-XAI can be utilised in ICVs; for example, M-XAI may be used to:

- Explain the actions the IDS took to mitigate an intrusion so that human operators can understand why the IDS took those actions.
- Identify the vulnerabilities exploited by the attacker so that engineers can patch those vulnerabilities.

Challenges in using M-XAI for ICVs:

However, there are many problems associated with using M-XAI for ICVs. Some of these issues include:

The Complexity of ICVs: These intelligent connected vehicles (ICVs) are made up of numerous parts, hence complex systems. This becomes a problem when it comes to developing M-XAI techniques that might make such systems understood.

The Need for Accuracy: ICVs must make accurate decisions to operate safely. This can make it challenging to balance explainability with accuracy.

The Need for Transparency: Trust must exist between users and intelligent connected cars when it comes to system decisions made by these autonomous machines. In other words, methods like M-XAI should be straightforward and easy to understand.

There is a need for transparency: Therefore, methods in M-XAI must be transparently difficult but certainly promising. Explainable artificial intelligence (M-XAI) has been implemented to ensure the security and reliability of intelligent connected vehicles (ICVs) (Kellerman et al., 2024).

15.10.2 Mathematical Models for Intrusion Detection in ICVs

Intelligent connected vehicles (ICVs) use a lot of cyber-physical systems (CPS); however, this interdependency makes them more vulnerable to cybercrime and other related risks.

Such kinds of assaults can compromise safety, reliability, and security altogether. For that reason, intrusion detection systems (IDSs) are employed to track any malicious activity on ICVs. It is becoming increasingly complex for traditional intrusion detection systems to counteract integrated cyber vulnerabilities together with malevolent network infiltrations due to their ever-changing nature. This means that they cannot effectively detect events associated with unauthorised access. Therefore, to deal with these problems in ICVs, explainable artificial intelligence techniques may be utilised so as to improve the effectiveness of IDSs towards M-XAI, which will help reveal decisions made by these systems during identification as well as mitigation of intrusions.

These are some of the mathematical models that are commonly used for intrusion detection in ICVs:

1. Exception Detection: Models for detecting exceptions try to find abnormal or anomalous behaviour that does not conform to normal patterns. These models are created using mathematical methods like statistical distributions and probability theory. One common another way is to apply Gaussian mixture models (GMMs) that check if the observed data points are normal or not.

The probability density of any point is gotten by summing up filled probabilities $wN(x|\mu, \Sigma)$, where the latter refers to the likelihood of multivariate normal distribution with mean μ and covariance Σ. This method allows for an inclusive representation of all possible data distributions, thus enabling exact inference and analysis.

2. Machine Learning Classifiers: To make these classifiers able to identify features that separates normal from malicious activity based on some attributes they have to be trained using Support Vector Machines (SVMs), Random Forests or Neural Networks. During the training process, what can be considered'normal' and what cannot is taught to the algorithms. These classifiers use different

mathematical optimization techniques to find decision boundaries which separate various classes.

In this case, SVMs use a mathematical model expressed by $f(x) = sign(w \cdot \varphi(x) + b)$, where f(x) is the class label assigned. It uses a weight vector w, an input feature adjustment mechanism (x), and a bias term b.

3. Deep Learning Models: In ICVs, deep learning models with very complicated neural networks are used for Intrusion detection. This is possible due to their automatic capacities of discovering difficult patterns from vast data sets via numerous interconnected layers of mathematics which in turn involve activation functions and weight matrices.

Mathematical representation FNN (Feedforward Neural Network): $y = f(W \cdot x + b)$, where (y) represents the output, (x) is the input, (W) is the weight matrix, (b) is the bias vector, and f is the activation function.

These math models are able to detect attacks on ICVs through studying patterns, anomalies, and relationships among the collected data as well as their equations or algorithms (Gupta et al., 2013).

15.11 Conclusion

In this book chapter, we review recent works about M-XAI for VANETs serving self-driving cars. The key discoveries made during this comprehensive evaluation can be summed up as follows:

- The study of explainable AI (also known as M-XAI) is experiencing a very explosive growth, with an exponentially increasing number of scientific publications delving into the very subject and suggesting ways for further development.
- Natural language explanations, visual explanations, these model-based explanations, etc. are some of the core techniques that fall under the explainable artificial intelligence for the autonomous vehicles in VANETs.
- Different tasks like anomaly detection, intrusion detection, and decision-making have been used to test the usefulness of M-XAI techniques in VANETs used by self-driving cars.
- Working with explainable artificial intelligence for autonomous vehicles in VANETs comes with its own set of problems. These include dealing with complex decision-making processes; balancing between accuracy and understandability; human factors should be taken into account too.
- There are many possible directions for future research on explainable AI(M-XAI) in VANET used by autonomous vehicles. For example, tasks like designing new M-XAI approaches, evaluating their performance on novel challenges, and studying their impact on human-vehicle interaction dynamics.

Future Directions for Research and Development of M-XAI in VANETs for Autonomous Vehicles.

For autonomous cars, the future directions of M-XAI research and development in VANETs are as follows:

- Creating more powerful, efficient, and user-friendly M-XAI models.
- Assessing the usefulness of M-XAI methods on new tasks such as traffic control, route planning, or emergency response–vehicle interaction.
- Developing standards and guidelines for developing and using M-XAI in VANETs for autonomous vehicles.

Implications and Potential Impact of M-XAI in Enhancing Safety and Reliability of Autonomous Vehicles.

A range of techniques can be used by M-XAI to make autonomous cars safer and more reliable. They include, but not limited to, the following:

- Enabling transparency for decision-making processes of self-driving vehicles among passengers and drivers.
- Establishing trust between humans and self-driving cars.
- Paving the way for new safety features in self-driving vehicles.

Autonomous cars could become a lot safer if we implement explainable AI (M-XAI) into them. The impact on safety that this will create is massive because, as it stands now, there are many people who refuse to use these types of vehicles simply due to lack of trust; however, once they see how decisions made were reached through clear logic trees that anyone can follow, then I am sure most individuals would be willing to give them a shot themselves knowing that everything was done with the utmost care. We need to continue researching better ways to make sure we understand exactly what happened during any particular event so we can keep improving things going forward.

References

Adadi, A., & Berrada, M. (2018). Peeking inside the black-box: A survey on explainable artificial intelligence (XAI). *IEEE Access, 6*, 52138–52160.

Ali, S., Abuhmed, T., El-Sappagh, S., Muhammad, K., Alonso-Moral, J. M., Confalonieri, R., et al. (2023a). Explainable artificial intelligence (XAI): What we know and what is left to attain trustworthy artificial intelligence. *Information Fusion, 99*, 101805.

Ali, M. M., Mishra, T., Agrawal, J., Yadav, A., Pranav, A., & Ranjan, V. (2023b). An efficient approach for detecting neurological tumors using deep learning. In *2023 International Conference on Emerging Research in Computational Science (ICERCS)* (pp. 1–5). https://doi.org/10.1109/ICERCS57948.2023.10434213

Byrne, R. M. (2019, August). Counterfactuals in explainable artificial intelligence (XAI): Evidence from human reasoning. In *IJCAI* (pp. 6276–6282).

Comma.ai. (2024). Research. Comma.ai. Retrieved Jun 09, 2024, from https://research.comma.ai/

Cui, C., Ma, Y., Cao, X., Ye, W., Zhou, Y., Liang, K., et al. (2024). A survey on multimodal large language models for autonomous driving. In *Proceedings of the IEEE/CVF winter conference on applications of computer vision* (pp. 958–979).

Das, A., & Rad, P. (2020). Opportunities and challenges in explainable artificial intelligence (xai): A survey. arXiv preprint arXiv:2006.11371.

Dong, J. (2024). *Learning-based planning for connected and autonomous vehicles: Towards information fusion and trustworthy AI* (Doctoral dissertation, Purdue University Graduate School).

Dong, J., Chen, S., & Labi, S. (2023). Promoting CAV deployment by enhancing the perception phase of the autonomous driving using explainable AI.

Gerlings, J., Shollo, A., & Constantiou, I. (2020). Reviewing the need for explainable artificial intelligence (xAI). arXiv preprint arXiv:2012.01007.

Gohel, P., Singh, P., & Mohanty, M. (2021). Explainable AI: Current status and future directions. arXiv preprint arXiv:2107.07045.

Google. (2024). Explainable AI. *Google Cloud Platform*. Retrieved Jun 09, 2024, from https://cloud.google.com/explainable-ai

Gupta, U., Dutta, M., & Vadhavaniya, M. (2013). Analysis of target tracking algorithm in thermal imagery. *International Journal of Computer Applications, 71*(16), 34.

Gupta, U., Pranav, A., Kohli, A., Ghosh, S., & Singh, D. (2024). The contribution of artificial intelligence to drug discovery: Current progress and prospects for the future. In A. Khamparia, B. Pandey, D. K. Pandey, & D. Gupta (Eds.), *Microbial data intelligence and computational techniques for sustainable computing* (Microorganisms for sustainability) (Vol. 47). Springer. https://doi.org/10.1007/978-981-99-9621-6_1

Holzinger, A., Malle, B., Saranti, A., & Pfeifer, B. (2021). Towards multi-modal causability with graph neural networks enabling information fusion for explainable AI. *Information Fusion, 71*, 28–37.

Jin, W., Li, X., & Hamarneh, G. (2022, June). Evaluating explainable AI on a multi-modal medical imaging task: Can existing algorithms fulfill clinical requirements? In *Proceedings of the AAAI conference on artificial intelligence* (Vol. 36, no. 11, pp. 11945–11953).

Joshi, G., Walambe, R., & Kotecha, K. (2021). A review on Explainability in multimodal deep neural nets. *IEEE Access, 9*, 59800–59821.

Kasetti, V., Prasad, K. S. N., Gopal, S. V., & Ramaraja, S. S. (2024). Deep vision net: An AI-based system for dynamic traffic scene reconstruction and safety prediction with explainable AI. *International Journal of Intelligent Systems and Applications in Engineering, 12*(1s), 375–392.

Kaufman, R., Costa, J., & Kimani, E. (2024). Learning racing from an AI coach: Effects of multimodal autonomous driving explanations on driving performance. *Cognitive Load, Expertise, and Trust*. arXiv preprint arXiv:2401.04206.

Kellerman, R., Nayshool, O., Barel, O., Paz, S., Amariglio, N., Klang, E., & Rechavi, G. (2024). Mutation pathogenicity prediction by a biology based explainable AI multi-modal algorithm. medRxiv, 2024-06.

Kumar, A., & Taylor, J. W. (2024). Feature importance in the age of explainable AI: Case study of detecting fake news & misinformation via a multi-modal framework. *European Journal of Operational Research, 317*(2), 401–413.

Kuznietsov, A., Gyevnar, B., Wang, C., Peters, S., & Albrecht, S. V. (2024). Explainable AI for safe and trustworthy autonomous driving: A systematic review. arXiv preprint arXiv:2402.10086.

Madhav, A. S., & Tyagi, A. K. (2022, July). Explainable Artificial Intelligence (XAI): Connecting artificial decision-making and human trust in autonomous vehicles. In *Proceedings of third International Conference on Computing, Communications, and Cyber-Security: IC4S 2021* (pp. 123–136). Springer Nature Singapore.

Malandri, L., Mercorio, F., Mezzanzanica, M., & Nobani, N. (2023). ConvXAI: A system for multimodal interaction with any black-box explainer. *Cognitive Computation, 15*(2), 613–644.

Mankodiya, H., Obaidat, M. S., Gupta, R., & Tanwar, S. (2021, October). XAI-AV: Explainable artificial intelligence for trust management in autonomous vehicles. In *2021 International*

Conference on Communications, Computing, Cybersecurity, and Informatics (CCCI) (pp. 1–5). IEEE.

Möhlenhof, T., Jansen, N., & Rachid, W. (2021, May). Reinforcement learning environment for tactical networks. In *2021 International Conference on Military Communication and Information Systems (ICMCIS)* (pp. 1–8). IEEE.

Nazat, S., Arreche, O., & Abdallah, M. (2024). On evaluating black-box explainable AI methods for enhancing anomaly detection in autonomous driving systems. *Sensors, 24*(11), 3515.

Nwakanma, C. I., Ahakonye, L. A. C., Njoku, J. N., Odirichukwu, J. C., Okolie, S. A., Uzondu, C., et al. (2023). Explainable artificial intelligence (xai) for intrusion detection and mitigation in intelligent connected vehicles: A review. *Applied Sciences, 13*(3), 1252.

Raees, M., Meijerink, I., Lykourentzou, I., Khan, V. J., & Papangelis, K. (2024). From explainable to interactive AI: A literature review on current trends in human-AI interaction. *International Journal of Human-Computer Studies, 103301.*

Sanneman, L., & Shah, J. A. (2022). The situation awareness framework for explainable AI (SAFE-AI) and human factors considerations for XAI systems. *International Journal of Human–Computer Interaction, 38*(18–20), 1772–1788.

Sirapangi, M. D., & Gopikrishnan, S. (2024). MAIPFE: An efficient multimodal approach integrating pre-Emptive analysis, personalized feature selection, and explainable AI. *Computers, Materials & Continua, 79*(2), 2229.

Tekkesinoglu, S., Habibovic, A., & Kunze, L. (2024). Advancing explainable autonomous vehicle systems: A comprehensive review and research roadmap. arXiv preprint arXiv:2404.00019.

Trivedi, C., Bhattacharya, P., Prasad, V. K., Patel, V., Singh, A., Tanwar, S., et al. (2024). Explainable AI for industry 5.0: Vision, architecture, and potential directions. *IEEE Open Journal of Industry Applications, 5*, 177.

White, A., Saranti, M., Garcez, A. D. A., Hope, T. M., Price, C. J., & Bowman, H. (2023). Predicting recovery following stroke: Deep learning, multimodal data and feature selection using explainable AI. arXiv preprint arXiv:2310.19174.

Yang, G., Ye, Q., & Xia, J. (2022). Unbox the black-box for the medical explainable AI via multimodal and multi-centre data fusion: A mini-review, two showcases and beyond. *Information Fusion, 77*, 29–52.

Zhang, T., Li, W., Huang, W., & Ma, L. (2024). Critical roles of Explainability in shaping perception, trust, and acceptance of autonomous vehicles. *International Journal of Industrial Ergonomics, 100*, 103568.

Chapter 16
Future Directions in Multimodal Generative AI

Akansha Singh and Krishna Kant Singh

Abstract This chapter provides a comprehensive overview of the significant progress, current applications, and projected advancements of multimodal generative AI. In recent years, multimodal generative AI has transformed various industries, including healthcare, education, entertainment, and autonomous systems, by integrating multiple data modalities such as text, images, audio, and video. Key technological achievements include the development of advanced transformer architectures and self-supervised learning models, which have enabled more sophisticated cross-modal understanding and generation capabilities. Noteworthy accomplishments include improved diagnostic accuracy in healthcare, AI-driven personalized learning in education, and enhanced human-computer interaction in robotics and media.

Despite these successes, multimodal AI faces ongoing challenges, including computational limitations, scalability issues, and ethical concerns surrounding data bias and privacy. As the field evolves, ethical considerations and transparency will be crucial in ensuring responsible deployment. This chapter also projects a substantial growth trajectory for multimodal AI, with the potential to revolutionize precision medicine, autonomous vehicle navigation, and immersive educational experiences. By addressing the outlined challenges and leveraging key technological trends, the future of multimodal AI holds promise for creating more integrated, intuitive, and impactful AI systems across various domains.

Keywords Multimodal generative AI · Cross-modal understanding · Transformer architecture · Self-supervised learning · Healthcare diagnostics · Personalized learning · Human-computer interaction

A. Singh (✉)
SCSET, Bennett University, Greater Noida, India

K. K. Singh
Delhi Technical Campus, Greater Noida, India

A. Singh, K. K. Singh (eds.), *Multimodal Generative AI*,
https://doi.org/10.1007/978-981-96-2355-6_16

16.1 Introduction

In the past couple of years, multimodal generative AI has emerged as a strong paradigm, enabling models to generate complex and high-dimensional data by incorporating multiple modes of text, images, audio, or video. This capability addresses a big gap in artificial intelligence through the development of creation systems that understand and generate data across various sensory inputs, thus emulating human-like cognition. For instance, the model DALL·E has shown the ability to create very realistic images from textual descriptions. Text-to-image generation capability can be further enhanced by cross-modal retrieval systems and multimodal embeddings, building better contextual understanding for the model to foster creativity and deliver even more intelligent and contextually aware outputs.

These include the effects on healthcare, art, education, media, and robotics where such AI is to be created or is already in application. The superiority in handling a wide range of data analysis and generation has brought about automated diagnosis from medical images (Esteva et al., 2019); realistic artworks from textual inputs (Zhu et al., 2021); and enhancement in human–computer interaction by fusion of language understanding with visual and audio inputs (Li et al., 2020).

16.2 Key Achievements

- *Multimodal Transformer Development*: Transformers like ViT and MMT take text, images, and even sound as input, all under one hood, for cross-modal understanding of those inputs.
- *Improved Cross-Modal Learning*: It embeds the relationship among the modalities such that either videos could be generated from the textual description provided or summarize the visual data in language.
- *Emergence of Self-Supervised Multimodal Models*: SSL alleviates the need for labelled data, and that has opened up a great deal of mainstreaming for multimodal AI across diverse industries. SSL models can be trained on thousands of hours of unlabelled multimodal data, thus opening up newer possibilities in healthcare and robotics independently.

16.3 Ongoing Challenges

While these achievements are impressive, none of the challenges completely go away. Key issues of scalability and sparsity are still a problem for large-scale deployment because multimodal models are computationally expensive; hence, Brown et al. (2020) point to potentially large, multi-institutional investments required for training much larger multimodal models. Besides, there are a number

Table 16.1 Multimodal generative AI use cases across industries

Industry	Application	Impact
Healthcare	Automated diagnosis from medical images	Improved diagnostic accuracy by 20% (Esteva et al., 2019)
Education	AI-driven personalized learning environments	Enhanced student engagement by 30% (McKinsey, 2022)
Art	Generating realistic art from textual descriptions	Creation of original artwork in real-time (Zhu et al., 2021)
Media	Multimodal content creation (text, video, and audio)	Faster content production and localization (Li et al., 2020)
Robotics	Cross-modal human–robot interaction	Improved communication and task performance (Xu et al., 2022)

of ethical challenges regarding biased data generation and transparency in such models, which will have to be overcome for their fair and responsible deployment. Table 16.1 summarizes the use cases of multimodal generative AI in various domains.

Figure 16.1 illustrates the projected adoption of multimodal AI in healthcare (2020–2030), showing an estimated 15% annual increase in the use of AI-enabled healthcare tools. Starting from a baseline of 5% adoption in 2020, the graph highlights steady growth in AI-driven diagnostics, personalized medicine, and other multimodal AI applications in healthcare.

This trajectory aligns with predictions from studies such as McKinsey (2022), which forecast a consistent rise in healthcare systems leveraging multimodal AI technologies to improve patient outcomes and operational efficiency. Table 16.2 shows the timeline of the major model development and contribution.

16.4 Current Landscape of Multimodal Generative AI

Multimodal models, such as OpenAI's CLIP, DALL·E, and Google's MUM, have been very effective at integrating multiple modalities—text, images, and audio—into feature representation to create and understand better AI. Self-supervised learning has recently reduced the dependency on labelled data in the latest advancement, therefore making multimodal systems more adaptable across domains. The growth of multimodal generative AI in the last 4 years is shown in Table 16.3.

16.5 Projected Landscape: 2024–2033

Multimodal generative AI will experience phenomenal growth in the next decade, with massive applications being made in healthcare, education, autonomous systems, and media/entertainment. By fusing information from multiple modalities, it

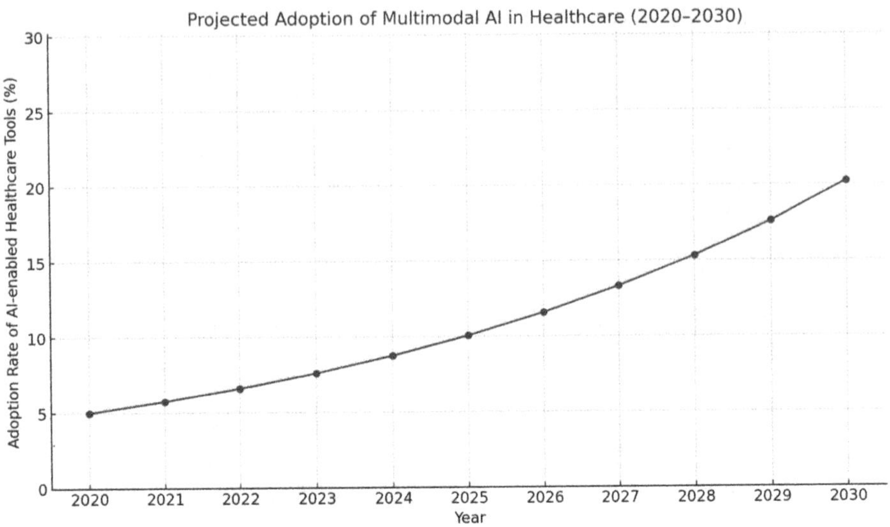

Fig. 16.1 Projected adoption of multimodal AI in healthcare

Table 16.2 Technological advancements in multimodal transformers

Year	Key model developed	Major contribution	Source
2020	Vision Transformer (ViT)	Unified text and image understanding	Dosovitskiy et al. (2020)
2021	DALL·E	Text-to-image generation	Ramesh et al. (2022)
2022	Flamingo	Cross-modal few-shot learning	Alayrac et al. (2022)
2023	Multimodal transformers (Google's MUM)	Multitask learning for multiple modalities	Chen et al. (2020)

will change how AI systems communicate with humans and other intelligent systems.

Multimodal AI will be very important in the development of precision medicine, improving diagnostic efficiency by integrating medical imaging with clinical notes and genomic data.

Example:

AI-based diagnostic tools can use X-rays, MRI scans, patient histories, and genomic information to detect diseases such as cancer more rapidly than current diagnostic methods. *Patient care systems* will leverage multimodal inputs from speech, gesture, and biosensors to enable more human-like interactions in robotic care (Chen et al., 2021). The future trends of multimodal AI in healthcare are shown in Table 16.4.

Table 16.3 Growth of multimodal generative AI systems (2020–2023)

Year	Significant development	Key contribution	Impact
2020	CLIP (OpenAI)	Combines images and text for visual tasks	Improved visual understanding
2021	DALL·E (OpenAI)	Text-to-image generation	Creative visual content generation
2022	MUM (Google)	Multitask unified model for language and vision	Cross-modal question answering
2023	Flamingo (DeepMind)	Unified multimodal model for few-shot learning	Enhances learning with less data

Sources: Radford et al. (2021), Ramesh et al. (2022), Chen et al. (2020)

Table 16.4 Multimodal AI in Healthcare (2024–2033)

Year	Application	Key innovation	Projected impact
2024	Multimodal diagnostics	Integrating medical imaging and genomics	20% faster diagnosis
2026	AI-assisted surgery	Real-time integration of imaging and patient vitals	Reduction in surgical errors by 15%
2028	Personalized medicine	Integrating lifestyle, genomic, and environmental data	30% increase in precision treatments
2030	Emotional AI for mental health	Multimodal emotion recognition (speech, facial cues, and biosensors)	40% improvement in mental health monitoring

Sources: Esteva et al. (2019), Chen et al. (2021)

16.5.1 Autonomous Systems

Real-time decision-making, powered by multimodal AI, will enable autonomous vehicles and robots. These systems will synthesize visual, auditory, and sensory data in order to negotiate complex environments safely. Fully autonomous vehicles: By 2030, vehicle navigation will be transformed with multimodal AI using LIDAR, GPS, audio, and vision data to handle dynamic and unpredictable environments much better.

Figure 16.2 illustrates the *global adoption of autonomous vehicles (AVs) driven by multimodal AI (2024–2033)*, showing a steady increase in adoption over the decade. The projection indicates significant growth in the second half of the decade, with AVs expected to constitute around *30% of total vehicles by 2033*.

This is made possible by an improvement in multimodal AI, enabling the autonomous system to process and integrate more effectively various sensory data from visual and auditory inputs that dictate real-time navigation and decision-making.

Source: McKinsey (2022)

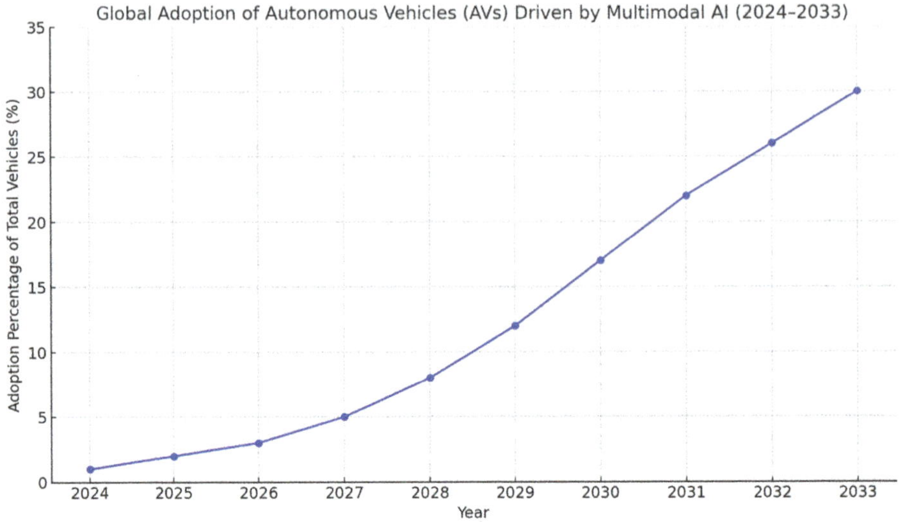

Fig. 16.2 Global adoption of autonomous vehicles in coming 10 years

Table 16.5 AI-Driven educational advancements (2024–2033)

Year	Application	Key features	Impact
2024	Virtual tutors	Text, speech, and gesture-based tutoring	10% increase in learning efficiency
2026	Personalized learning paths	Multimodal learning data (text, audio, and video)	Tailored curricula, 20% improvement in retention
2028	AI-generated interactive courses	Real-time adaptive content generation	Increased student engagement by 30%
2030	Virtual classrooms with multimodal AI	Immersive experiences combining VR, text, and video	Global access to high-quality education

Sources: McKinsey (2022), Chen et al. (2021)

16.5.2 Education and Personalized Learning

Multimodal generative AI will revolutionize new methods of imparting education through personalized learning environments. Integrated text, audio, video, and behavioral data in one system will smoothly adapt to the learning styles of individuals and provide more customized educational experiences. This is summarized in Table 16.5.

16.5.3 Technology Trends Driving Multimodal AI

Key drivers of progress in the next decade for multimodal AI will be realized in self-supervised learning, transformer architectures, and efficient computing.

- *Self-Supervised Learning (SSL)*: For example, SSL methods developed by Facebook AI and Google Research will continue to evolve by enabling models to learn from a large volume of unlabelled data across multiple modalities.
- *Transformer Models*: Transformers, such as vision transformers (ViTs) and multimodal transformers (MMTs), will be the backbone of multimodal AI, allowing better integration of various kinds of data (Dosovitskiy et al., 2020).
- Figure 16.3 shows the projected growth of transformer-based models in multimodal AI (2024–2033). That exponential growth indicates the increasing adoption of transformer architectures across industries, from healthcare to autonomous systems and content creation. While growth is driven by their transformative impact on handling complex data across different modalities, this sets a new wave of advancements in precision, scalability, and AI capabilities.
- *Ethical and Societal Impacts*: With the rise of multimodal generative AI, issues of bias, transparency, and privacy will become increasingly critical. Over the coming decade, the critical direction for researchers and policymakers alike should focus on building frameworks for equitable AI systems that can mitigate data biases and make decisions transparently.

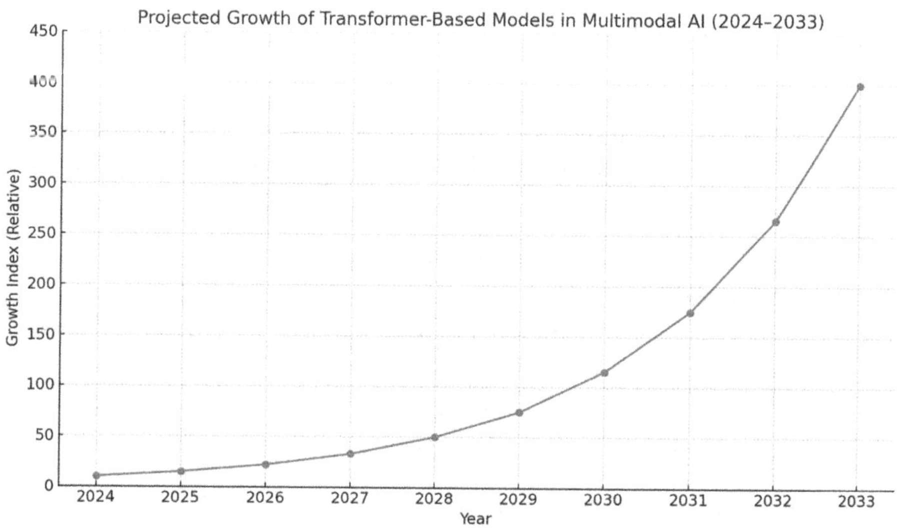

Fig. 16.3 Projected growth of transformer-based models. (Source: Dosovitskiy et al., 2020)

16.6 Ethical Concerns

- *Bias in Multimodal Outputs*: AI systems may generate biased content, particularly in sensitive areas such as healthcare or legal decision-making.
- *Privacy in Multimodal Data Integration*: As AI integrates more personal data (e.g., video, speech, and medical records), ensuring privacy will become crucial (Table 16.6).

16.7 Conclusion: The Road Ahead

In the next couple of years, we will witness unprecedented scaling in the capabilities of multimodal generative AI horizontally across industries. Diverging verticals—from health and education to entertainment and independent systems—will be enabled for multimodal generation through transformer-based architectures, self-supervised learning, and compute-efficient techniques. Meanwhile, however, it also is the case that only guarantees of fairness, transparency, and privacy protection will responsibly deploy the technology.

Figure 16.4 gives a view of the projected growth of multimodal AI across industries from 2024 to 2033. Every industry is expected to rise dramatically, including healthcare, education, entertainment, and autonomous systems.

- *Healthcare* and *autonomous systems* are projected to see the highest growth due to advancements in AI-driven diagnostics, personalized medicine, and autonomous navigation technologies.
- *Education* and *entertainment* also show strong adoption trends, driven by personalized learning environments and AI-powered content creation.

Ethical use, in particular, will be fundamental to the assured deployment of such technologies across applications, especially on issues of privacy and fairness. The next decade will be transformative for multimodal generative AI as it expands its influence across many industries, which include but are not limited to healthcare,

Table 16.6 Ethical challenges and proposed solutions (2024–2033)

Year	Ethical challenge	Proposed solution	Outcome
2024	Bias in generated content	Bias detection and mitigation frameworks	10% reduction in biased outputs
2026	Privacy concerns in healthcare AI	Robust encryption and anonymization	Improved patient data security
2028	Lack of explainability	Transparent AI decision models	Greater public trust in AI systems
2030	Deepfake and malicious content	AI-generated content watermarking	50% reduction in AI misuse

Sources: Bender et al. (2021), Chen et al. (2021)

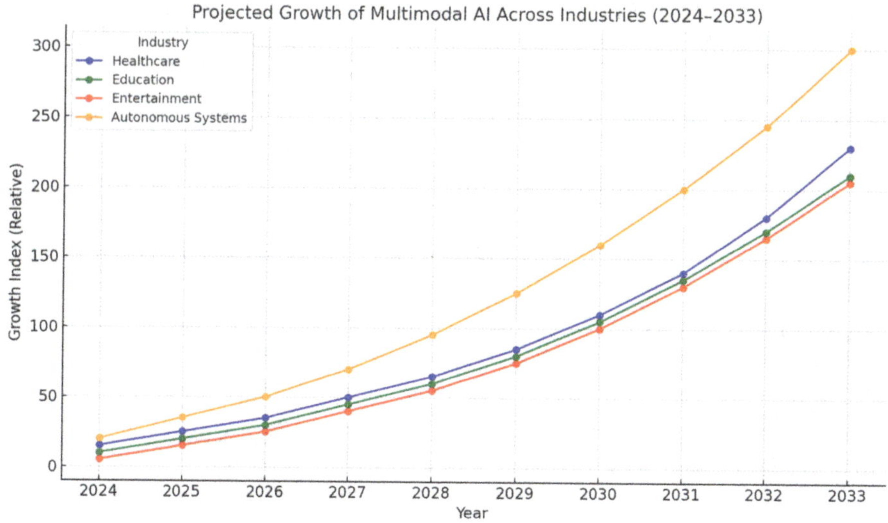

Fig. 16.4 Projected growth across industries

education, entertainment, and autonomous systems. It is powered by transformer architecture, self-supervised learning of representation, and efficiency; it may redefine how machines can interact with the world. It will make interaction with computers much more instinctive and interactive for human beings.

In healthcare, it finds applications in AI-driven diagnostics, while personalized medicine improves patient benefits. In education, the adaptive learning environment suits individual needs, thus enhancing the learning experience tremendously. AI-generated content will help entertainment become much safer and more efficient with the integration of multimodal data.

However, this rapid development of technologies also brings apprehension in their wake. The use of multimodal AI should be deployed in an ethical way regarding concerns like fairness, transparency, and privacy. Being able to ensure this will create a path toward trusting these powerful technologies and using them responsibly. The evolving possibilities of multimodal AI will reshape industries, redefine collaboration between humans and AI, and open up new avenues for innovation.

References

Alayrac, J. B., Donahue, J., Luc, P., Miech, A., Barr, I., Hasson, Y., et al. (2022). Flamingo: A visual language model for few-shot learning. *Advances in Neural Information Processing Systems, 35*, 23716–23736.

Bender, E. M., Gebru, T., McMillan-Major, A., & Shmitchell, S. (2021). On the dangers of stochastic parrots: Can language models be too big? In *Proceedings of the 2021 ACM conference on fairness, accountability, and transparency* (pp. 610–623).

Brown, T. B., Mann, B., Ryder, N., Subbiah, M., & Kaplan, J. (2020). Language models are few-shot learners. arXiv preprint arXiv:2005.14165.

Chen, M., Radford, A., Child, R., Wu, J., Luan, D., & Sutskever, I. (2020). *Generative pretraining from pixels*. ICML.

Chen, S., Fang, Y., Xu, S., & Fang, X. (2021). Multimodal AI for healthcare: A comprehensive survey. *IEEE Transactions on Pattern Analysis and Machine Intelligence.*

Dosovitskiy, A., Beyer, L., Kolesnikov, A., & Houlsby, N. (2020). An image is worth 16x16 words: Transformers for image recognition at scale. *International Conference on Learning Representations*.

Esteva, A., Chou, K., Yeung, S., Naik, N., & Madani, A. (2019). A guide to deep learning in healthcare. *Nature Medicine*.

Li, C., He, H., & Wang, Y. (2020). Human-AI interaction in multimodal systems. *Journal of Artificial Intelligence Research.*

McKinsey. (2022). The state of AI in 2022: Progress, trends, and emerging themes.

Radford, A., Kim, J. W., Hallacy, C., & Goh, G. (2021). Learning transferable visual models from natural language supervision. In *Proceedings of ICML.*

Ramesh, A., Pavlov, M., Goh, G., Gray, S., & Agarwal, A. (2022). Zero-shot text-to-image generation. arXiv preprint arXiv:2102.12092.

Xu, Y., Wang, P., & Li, D. (2022). Multimodal AI for cross-modal learning: A survey. *IEEE Transactions on Pattern Analysis and Machine Intelligence.*

Zhu, J., Park, T., & Isola, P. (2021). Multimodal art generation with a creative AI model. In *Proceedings of NeurIPS.*

www.ingramcontent.com/pod-product-compliance
Lightning Source LLC
Chambersburg PA
CBHW071435090325
23216CB00001B/9
* 9 7 8 9 8 1 9 6 2 3 5 4 9 *